Beauty in Abundance

Beauty in Abundance

Designs and Projects for Beautiful, Resilient Food Gardens, Farms, Home Landscapes, and Permaculture

Michael Hoag

With illustrations by Michael Hoag and Rebecca Stockert

Edited by Denise Miller

Transformative Adventures Cooperative
Change your life * Change the World
Fort Wayne, Indiana

Copyright 2021 by Michael Hoag
All rights reserved

No part of this book may be reproduced or transmitted in any form or by any means without written permission from the author.

Published in the United States in 2021; by
Transformative Adventures
1407 Spy Run
Fort Wayne, IN 46805

Inquiries:
269 350 3407
TransformativeMike@gmail.com

Photographs: Michael Hoag, unless otherwise attributed. Supplemental photographs via the Transformative Adventures Coop.
Illustrations, Michael Hoag and Rebecca Stockert

Table of Contents

Index of Useful Designs and Patterns   4

Part 1: A Beauty and Abundance Manifesto   8

Chapter 1: Introduction, Growing Abundance and Beauty   9
Adventure Step 1: Visioning, Top Priorities   25
Chapter 2: The Transformative Power of Beauty 27
Step 2: The Solid Base Self Assessment 44
Chapter 3: Models of Transformative Aesthetics 46

Part 2: Transformative Themes and Concepts   68
Chapter 4: Designing and Transforming the Landscape 69
Chapter 5: How a Sense of Play Builds a Better Life and a Better World 88
Chapter 6: The Permission to Get Weird 90
Chapter 7: The Nature of Beauty and Beauty of Nature 95
Chapter 8: Transcending the Conflict Between Intuition and Reason.   94
Chapter 9: Ghosts   101
Chapter 10: The Elements and Patterns of Transformative Aesthetics for Landscapes   106

Part 3: Transforming the Landscape, Beautifully   116
Chapter 11: The Power of Transformative Landscapes   117
Adventure Step 3: Water Wise Design   122
Chapter 12: Gardening Beautifully: Holistic Natural Management 127
Chapter 13: Holistic Intensive Gardening   151
Chapter 14: Nurse Log Patterns for Beauty, Ecology, and Productivity   166
Chapter 15: Transforming Soil? 174
Adventure Step 4: Plan your Fertility Systems   183
Chapter 16: Starting Gardens the Holistic Natural Way   188
Chapter 17: The Transformative Double Dig   192
Chapter 18: Sheet Mulching Revisited.   196
Chapter 19: Hugelkulturs that are Cool and Cultured   204
Chapter 20: Herb Spiraling out of Control   210
Chapter 21: Herb-Inspiraled Spin-offs   213
Chapter 22: Planting into Resident Vegetation   216

Part 4: Patterns and Projects for Abundant Beauty   218
Chapter 23: Overview Patterns for Beautiful Transformative Landscapes 218
Chapter 24: Thematic Patterns for Beautiful Gardens   221
Chapter 24: The Edible, Aesthetic, and Mighty Hedgerow, and/or Tapestry Hedge   247
Adventure Step 5: Begin listing Patterns you want in your landscape   256
Chapter 25: All-star Ornamental Edible Plant Profiles   257
Chapter 26: Selected Plant Profiles: 263
Adventure Step 6: Start Listing Major Perennial Crops and Plants 316
Chapter 27: Patterns for Sacredness   321
Chapter 28: Patterns for Garden Layout 337
Chapter 29: Basic Patterns for Garden Color   342
Chapter 30: Patterns for Attracting Natural Beauty to the Landscape   356
Chapter 31: Aesthetic patterns for ponds and water   364
Chapter 32: Patterns for Beautiful Paths 371
Chapter 33: Places to sit   388

Part 5: Aesthetic Guilds for the Beautiful, Abundant Garden   400
Chapter 34: Guild Design Basics   402
Chapter 35: Model Gardens for Inspiration   440
Growing in Beauty and Abundance Over Time   448

Index of Useful Designs and Patterns

1. Holistic Natural Management 127
2. Social Uses in the Garden: 26
3. Some Practical Ways to Build the Solid Base in the Landscape: 44
4. Solid Base Self Assessment 44
5. How to Observe a Site: 48
6. Patterns for Neo-Wabi sabi in the Garden: 54
7. Edible plants for a Coniferous forest garden aesthetic: 55
8. Plants for a Punk/Goth Edible Forest Garden: 64
9. The Permaculture Design Process: 77
10. PLENTY: A Streamlined Transformation-based Design Process for Everyone. 85
11. The Elements and Patterns of Transformative Aesthetics for Landscapes: 106
12. Benchmarks: The Transformative Landscapes Initiative Criteria: 120
13. Permaculture Zones for Aesthetics: 138
14. French Intensive Gardening: 151

15. **Some Starter Plans for French Intensive Gardens: 155**
16. 4 bed FIM Rotation 156
17. The Home Garden Veggie Circles Plan: 161
18. Nurse Log Patterns for Beauty, Ecology, and Productivity: 166
19. A Few Systems for Fertility and Soil Health: 170-182

20. **Ways to Start Gardens in the Holistic Natural System: 188**
21. The Transformative Double Dig: 192
22. Sheet Mulching Revisited. 196
23. Mike's "Good Enough!" Sheet Mulch for Aesthetic Landscapes: 200
24. Rough Mulching with Mulch Makers: 202
25. Hugelkulturs that are Cool and Cultured: 204
26. Grass-Proof Hugels: 205
27. Hugelsculptures, 206
28. Mycohugelkulturs: 206
29. Hill-gelkultur, or Hugelkultur Terraces. 207
30. Patterns for better Herb Spirals: 210
31. Herb Spiral Informed Designs: 213
32. Planting into Resident Vegetation: 216

33. **Patterns and Projects for Abundant Beauty: 218**
34. Meta Pattern 1: The Traditional Forest Garden: The Home Garden Pattern: 222
35. Realistic goals and expectations? 224
36. Meta Pattern 2: The Cottage Garden 231

37. Patterns to get the Cottage Garden Look 232
38. Meta Pattern 3: The Jardin de Curé: 233
39. Meta Pattern 4: The Post Wild Garden: 237
40. Meta Pattern 5: The Japanese Garden: 241
41. Meta Pattern 6: Novel Themes for Novel Times: 242
42. Terroir. 243
43. Meta Pattern 7: The Mosaic Landscape: 246
44. The Edible, Aesthetic, and Mighty Hedgerow, and/or Tapestry Hedge: 247

45. **Patterns for Sacredness: 321**
46. "Induction": 322
47. Wakefulness: 323
48. The Qualities of a Sacred Place: 324
49. The Sacred Grove: 324
50. Water, the Sacred Chalice: 326
51. The Mountain Protector: 327
52. Enclosure: 328
53. Flowers: 329
54. Biophillia: 330
55. Sacred Geometry: 331
56. Sacred Symbols: 332
57. The deep resonance of stacked symbols: 333
58. Visual Mandalas: 334

59. **Patterns for Garden Layout: 337**
60. The Fine Stopper Starter Pack: 338
61. Formal Layout, Wild Beds: 339
62. Hard and Soft Edges: 340
63. The Herb Hedge or Edging. 341
64. **Basic Patterns for Garden Color: 342**

65. **Patterns for Attracting Natural Beauty to the Landscape: 360**

66. **Aesthetic patterns for ponds and water: 364**
67. Patterns for simple ponds: 365
68. Edible Pond and Wetland Plants: 367
69. Water Guild Plants: 367
70. Patterns for Pond Beauty: 369

71. **Patterns for Beautiful Paths: 371**
72. Path Hierarchy and Orders: 373

73. Permeable pavers: 374
74. The Mown Grass Path: 375
75. The Mown Weeding Strip376
76. A Beautiful, Walkable Lawn Polyculture: 377
77. A Walkable Groundcover Polyculture for Shade: 378
78. Keyhole and Node Design: 380
79. Haphazard Mulch Mandalas: 384
80. The Walkable Stepping Stones Mix: 385

81. **Places to sit: 388**

82. **Aesthetic Guilds for the Beautiful, Abundant Garden: 394**
83. The Guild Matrix: The most important aspect of planting design? 400
84. The Beautiful Bountiful Scatter-Sown Polyculture: 405
85. The Aesthetic Edible Meadow Guild that Keeps Neighbors Happy: 409
86. Plants That Succeed Into Grassland: 414
87. An Aesthetic Three Sisters Guild: 417
88. The Beginner's Annual Garden Make-Over Guild: 422
89. The Beautiful Bountiful Mulch Strip Guild. 424
90. A Useful Ornamental Cover Crop 425
91. A Tree Guild in Cool Colors 426
92. Fruit Tree Guild in Warm Colors 428
93. Weeding Like a Jedi: Terminator Guilds 432
94. A Warm Color Terminator Guild: 435
95. The sunny forest edge community terminator guild: 436
96. Terminator guild ideas for dry, sandy sites: 437
97. Another guild for sandy sites: the Maine guild: 437
98. Beautiful Terminator Guild For Part Shade: 439

99. **Model Gardens for Inspiration**
100. An Ornamental Font Yard Food Forest: 441
101. A Suburban Back Yard Paradise: 442
102. Permaculture Edible Border Garden: 444

Mike Hoag's book is so inspiring; after 50 years of growing food, I am laying the foundation for growing a complex, beautiful garden that meets needs for beauty, friendship, rest, restoring the earth… as well as abundant food to share. Beauty and harmony are food and a thoughtfully designed garden can nourish us body and soul.
—Vicki Robin, *Your Money or Your Life* and *Blessing the Hands that Feed Us.*

Mike Hoag has written a long-awaited down-to-earth treatise on the topic of Permaculture plant guilds from a perspective that is both refreshing and easily understandable by both novice and professional alike. His presentation of plant guild examples brings a greater level of clarity to a topic of interest to all who are trying to enable a better planet for not just themselves and their descendants, but for all other residents of the "natural world"
—Bryce Ruddock, *Integrated Forest Gardening*

A true polymath, Mike Hoag is the most complete Permaculturist I have ever known (or heard of) and easily one of the best permies of all time. Humble, hilarious, and thoroughly holistic, this is the most significant Permaculture book since the Permaculture Designer's Manual.
—E. James Spielmaker, Hugelsculptures, HOBO, and Flintopia

Mike Hoag's permaculture masterpiece is humanity's rosetta stone between nature and society. I read it, and now I will lay it down in loamy earth, my hands performing its deeds, and I will be the happier for it.
—Noah Mandlin," the Permaculture Librarian," a student of permaculture for more than 30 years

Rich diversity, creatively, and heart-thoughtfully woven with smart design and beauty all throughout!
—Susan Opitz, Dancing Spring Garden

An absolute wonderful resource full of practical information on how to create that beautiful yet productive garden.
—Mothin Ali, *My Family Garden* YouTube channel.

Read this book to create abundant and beautiful landscapes that will make your life rich beyond measure. Other positive side-effects will include fighting climate change, building community and inspiring others to do the same. So, how could you possibly go wrong?
—Laura Oldanie, *Rich and Resilient Living*

The aesthetic patterns Mike is pulling together has some really great richness to it. Accessibly vernacular, but also something timeless and spiritually resonant in it all.
—Stician Samples, Permaculture practitioner

Part 1: A Manifesto of Beauty and Abundance

Chapter 1: Introduction, Growing Abundance and Beauty

Here's what always gets me: there will be ripe fruit hanging off the trees, berry bushes tempting us as we walk, fragrant herbs filling the air with their aromas, and fresh vegetables everywhere — and it's all of a quality you can't even buy at the store — but it's not all this amazing food that people most want to talk about.

It's the same everywhere I visit. For the last 20 years I've had the great pleasure of touring and working on hundreds of these radically new, ancient sorts of *transformative landscapes.* These inspiring gardens, homes, and farms care for people and for nature. They grow food for us humans, make a home for wildlife, save native plants, steward water, and fight climate change all at once. They show us that building a better world doesn't have to mean "tightening our belts," or lowering our standard of living the way environmentalists often lecture us. Instead, we can team up with nature to grow luxuries, delicacies, and delights in ways that make a richer world for all. We can grow paradise, without making life harder for others. And these gardens do all this do while helping us grow real wealth, freedom from the rat race, and right livelihoods — without destroying the planet.

In Gaia's Garden, author Toby Hemenway called them "ecological gardens," because in these edible gardens humans actually reconnect to their ecosystems as part of nature, joining in the bountiful feast of Mother Earth. We're not supervisors from above, but actually a part of the garden.

I've started calling them "transformative landscapes," because they transform who we are, how we live in the world, and have all of these positive impacts on society far beyond the garden gate.

Now, I like to eat. So, when I had the chance to really transform a landscape of my own, I thought people would want to talk about the unusual sweetness and flavor of my self-sown tomatoes, and the buttery smooth quality of my comice pears. *"How do I grow more amazing food than we can eat on a few hours of work per week?"* That's what I expected to hear.

But sitting down in the cool shade of a **forest garden or edible ecosystem** for a picnic of fresh fruits and herbal teas, it is not the physical sustenance that people always talk about.

What people always talk about is the way these gardens feed the soul. What captivates people is the *beauty of the thing*.

In these places, nature reigns. Just as in any wild ecosystem, it is the cooperation of all the species that maintain health and harmony. Except — looking around— you see nearly every plant is useful to meeting human needs. Here, human needs for food, fuel, building materials, and medicines are met off sun energy and ecosystem services, not fossil fuels and exploitation. And that is profoundly beautiful.

Probably the single most common comment I ever heard on hundreds of tours was that the visit was a transformative experience. I have multiple notes from church garden clubs and leaders telling me they had more of a religious experience sitting in the beauty of my home Garden of Eden than they've had sitting in their churches.

And that is the same experience I have had touring other great transformative landscapes. It's tough to put such an experience into words, but if I had to, I'd say that *it's like sitting in a vision - a revelation - of the kind of world we could grow together if we'd just take a break from trying to turn everything into money all the time. It's a vision of what is actually possible, and we can each build our part of that future right in our own yards.*

Talking to those hundreds of people who have transformed a part of their landscapes in this way, they all have said some version of that same thing. Yes, it's about the food. Yes, it's about caring for the earth, saving species, and fighting climate change. But the real reason for it all was the *beauty of it*.

A New, Old Approach

Many of us get into gardening for similar reasons: better food, better connection with nature. But that's not something that we see in the standard gardening approach sold in the big box stores these days. We want healthier, safer food but end up being advised to use more and more plastics and poisons. We want to connect with nature, but instead, nature becomes enemy number one, and anything that breathes or eats in our garden that wasn't put there by us is on the hit list. That standard approach has grown out of an old mindset that nature is indeed our foe, waiting to reduce our garden to "weeds," waiting perhaps to turn our mighty civilization back into a wilderness. (Hmm… now there's an idea.)

This mindset is that we are at war.

And that is what we see in the standard look of mainstream gardening: a battle against nature with poisons, lawnmowers, chainsaws, plastic sheets.... We even line up our hostas and petunias in strict battle formations behind bunkers of concrete landscape blocks, lest they be overrun or confused with the "wild" enemy, the weeds.

But the world has grown tired of the old war on nature.

Laying down the arms of fossil fuels, plastics and poisons against her, there's a rising movement forming a new alliance; liberating lawns, making pacts with weeds, jailbreaking our diets from the prison of Industrial Agriculture… We're jumping sides, going in cahoots with our landscapes to regrow a world of beauty, diversity and abundance for all.

When the wind rustles the trees, when the crickets sing and cicadas blow their trumpets, in the thundering silence of the stillest hour of morning, many of us hear that call, nature beckoning us to join her to regrow that world of abundance we deserve.

Changing Hearts and Minds

But of course, it can be hard to hear nature's call over the neighbors waging a full frontal assault with their lawn mowers, tillers, and leaf blowers. So our mission is to wake up our friends, family, neighbors, and communities to join us in the great reconnection with nature.

This isn't necessarily work for the intellect… it's difficult to convince people their side in a war is wrong. Talking to uncle Ted about his lawn at the dinner table is more dangerous than talking politics!

This is heartwork. And that is one major reason for this book. Beauty is a powerful tool. Beauty can move immovable internal objects. And when I say "beauty," I mean it in the fullest sense of self-expression, which can be pretty, or wild, or profound or, fun, funny, quirky, or adventurous. Beauty is simply about claiming our right to live in an uplifted way and declare who and what we are in the way we live. Creative expression is a deep human need. When we can fully present a vision of coexistence with nature with beauty, creativity, and expression, we won't have to lecture people or argue to change people's minds. People will line right up for this new world like it's Disneyland.

This book is about transforming landscapes in ways that produce food and care for nature, that also meet our human needs for beauty, meaning, and expression. It's a painter's palette if you like, for painting beauty onto productive, healthy landscapes, gardens, farms, and homesteads.

And nature's rewards for this work are great. By rejoining with nature and becoming part of our ecosystems, we have fresher, healthier food than most of us raised in an industrial food system could have imagined. Each day we wake up and find delicacies unheard of in even the best restaurants hanging from the trees, or just laying there right on the ground!

The sugars in apples, pears and mangoes begin to degrade and convert to starch as soon as they're picked, and that zingy, complex taste of fruit fresh off the tree or

Most materials on landscape aesthetics are just sales brochures for the latest corporate products and services. But for me, a landscape that isn't regenerative and supportive of a just society can never be truly beautiful.

What turns a landscape into a truly beautiful place? This image outlines the approach to creating beautiful places that we'll take in this book.

vine is a rare luxury reserved for the home grower. It's a difference you can actually feel. When you talk to groups of natural gardeners, you'll soon discover a common thread: the body rewards us with feelings of euphoria - and even gentle heady intoxication - for feasting on nutrient-dense, flavorful produce. Natural gardeners start to sound like mystics when they talk about their vegetables.

A Beautiful Way to Live

"We don't want to go on vacation anymore" is something I've heard repeatedly from these nature-centered gardeners who have successfully transformed their landscapes. "Why would we? We have created paradise right here!"

What I have come to realize is that most people who find this form of gardening with nature, rather than against it, come to it because of beauty. They had some sort of aesthetic sense of how they wanted to live, and in what kind of an environment. In some cases, they were drawn to a vision of being nurtured within their own home food forest paradise, surrounded by ripe fruits, and the scent of herbs.

Or perhaps they were shaken to their core, standing in a meadow at sunset, and realized they could bring that beauty home with them. Or perhaps they wanted to free themselves from dependence on systems they knew were harmful, take control of their own destinies, and create a support system that would truly set them free from the rat race. That beautiful vision of freedom lit the fire. Or they had an encounter with a wild rabbit, or dragonfly, sat watching a groundhog forage with its young, and vowed to lay down the lawnmower right then. It is nature's beauty that softens us from our warlike mindset, and beckons us to return home, to return to the forest.

Food and Natural Beauty

A typical daily yield at Lillie House, on just a few hours of work per week. Maintaining the ecological garden actually took less time than mowing the lawn that was there when we moved in.

While expert ornamental gardening books are very popular, and many are already enticing us to rewild our landscapes in various ways with native plants and naturalistic plantings, there are very few resources on the topic of beauty in productive, functional, ecological landscapes. As far as I know, this is the first. In fact, while most gardeners around the world do exactly that, in the Western horticultural tradition that goal is typically scoffed at. Vegetables are seen as the low-class residents of the plant world, and are barred from the ornamental garden.

When it comes to sustainable landscape features, we're taught a consumer mind-set approach. When we buy our homes, they come with an obligation to buy this and that new corporate products, or be judged for creating blight! We have to keep up with Jones' lawn and landscape blocks, but don't dare let the fruits and vegetables mingle with natives and flowers.

And yet, this is exactly what most people want: gardens that nurture their whole selves, body and soul. If they've started down that intuitive path, they tell me they actually feel guilty!

"I know I'm not supposed to, but I didn't have time so I just stuck the peppers right in this sunny flower bed."

Vegetables in a flower bed?! How scandalous!

"Well," I whisper "sounds like dangerously good gardening to me."

Part of my job as a gardening consultant and teacher seems to be Father Confessor for Gardening Sins. When people at parties hear I am a gardening professional, they'll suddenly look guilt-stricken.

"Bless me, Father, for I have let a volunteer tomato grow in the the rain garden and then… instead of pulling it out… I put *basil* around it!"

They've learned to expect a stern reprimand for their sin of planting an edible plant in their native garden. Little do they know I am more the wicked garden enabler than confessor.

People want to do something that cares for nature and uses their land well, and they must confess their guilt that they don't want to give up their grandmother's favorite daylilies, exotic fruit trees, or the vegetable garden. We want it all! And there's no good reason we shouldn't blur the lines and create gardens that meet all our needs and goals. Put those peonies in the pollinator garden! Grow monarda in with the veggies! It will be okay, I promise.

Even for experienced gardeners the world of horticulture seems filled with rules and shoulds and shouldn'ts and moral perils. This is no wonder since we live at a strange time where guilt and envy are the main fuels for the economy and the standard marketing procedure is to sell gum and deodorant by making people feel self-conscious.

"Your breath stinks, buy our gum!" translates to the landscape as: "Your lawn is weedy, buy our poison!"

The whole of what passes as a "lawn and garden" industry is about keeping up with the Jones' weed-free perfect lush grass. We make people feel guilty about having last season's concrete landscape blocks, or the wrong roses, or having squash bugs.

And don't get me started on poor souls worried that anything and everything is "invasive." Even a popular book on native plant gardening has a slew of scathing Amazon reviews for promoting "invasive!" species, because it advocates using a few naturally occurring color variants of native plants. These poor native species aren't native enough to avoid being attacked in the "invasive!" furor.

A client of mine confessed her guilt after being shamed by her extension agent for planting an "invasive!" pecan one county north of its historic native range.

Well, if there are such strict rules and mistakes are so grave, then the decision over whether to leave our dandelions or not is too important for regular folks, and the professional services of corporate landscape clergy are an absolute must.

$$$ cha-ching!

This is just the old pattern of marketing turning your landscape into another source of corporate profit. One famous prophet (or is it profit?) of this corporate clergy has said it should be illegal for homeowners to choose their own plants. Break free! You have the right to transform your landscape, and I

promise that if you look within, you have the ability to create healthy, beautiful spaces, evolved right into your DNA.

So this won't be a guilt book. I'm not going to shame you into doing this or that. I know you already want to be a good human while meeting your goals for beauty and productivity. I'm just going to give you tools for doing exactly that.

And this most certainly won't be another book about this season's trends in the big box stores, or the "next big thing" style of gardening. This book will give you a path and ideas for developing your own vision of beauty, making your landscape its own unique expression of place. You don't need HGTV, television experts, or big box stores like Home Despot to do that.

Transcending the Debate Over Invasive Plants

These days there is a great debate over whether to use only native plants, or whether we can and should also use some exotics, like the edible plants that make up most of our meals. There's no reason we can't have gardens rich in native plants and food!

Bill Mollison, founder of Permaculture, gave us these rules on choosing between natives and exotics. They're a good starting point:

1. Choose natives first. *So transformative gardens will always be rich in natives.*

2. Only choose proven non-invasive exotics *if there's not a native to do the job*. For example, potatoes come with a unique benefit that really only potatoes can provide.

3. Only plant unproven exotics if necessary for research, and plant them in controlled environments with care to not let them escape.

4. **Don't plant known problematic plants.**

5. Never interfere in healthy native ecosystems.

We should add to these that, in most cases, exotic plants necessarily increase biodiversity, ecosystem health, and habitat. Even with so called invasive plants, recent research has found only rare cases where a plant causes actual measurable decline in biodiversity or ecosystem health. And a recent major meta-analysis found that there were zero extinctions caused by invasive plants. Yes, we need to manage problematic species (many experts now agree that the term "invasive" is unscientific and impossible to define, and may cause land managers to make poor decisions) but we need to do so with smart, research-based strategies that don't do more harm than good.

And the most important, research-driven rule for managing landscapes is that it should be **PRIORITIES BASED**. This is the cutting edge of modern global landscape management practice, replacing the out-dated model of "invasive management." That means we need to base our actions on measurable, positive outcomes and an understanding that our time and resources are limited. These positive measurable priorities include things like: "increasing biodiversity; stewarding rare species; infiltrating water; sequestering carbon; reducing plastics, poisons, and petroleum consumption; and reducing dependence on the destructive corporate food system.

Removing problem species by itself is not a positive outcome. We can remove invasive species without having a benefit to humans or ecosystems, or we may even cause harm like erosion, disease or loss of biodiversity.

So, management needs to be considered in a holistic context of the positive priorities we have for the landscape we're stewarding. We can relax. Being too strident about invasive species might lead people to do harm, with no benefit.

The Transformative Landscape Criteria will help you set measurable positive outcomes for your landscape, including increasing native biodiversity.

Paths and Patterns for Beginners, Intermediate and Advanced

While this book is aimed at intermediate to experienced gardeners, farmers, and landscaping and Permaculture professionals looking to add aesthetic tools to their toolbox, it will also be helpful to beginning gardeners who want to create a beautiful, abundant yard.

This book was designed to be a complete path to having a beautiful, edible, ecological landscape.

Why? Because for me a toxic, destructive landscape can never be beautiful.

And I have seen many students and clients get "stuck" because they think they don't know enough, so they don't start. They just watch endless internet videos waiting to know "enough." So, for me, one of the most important things we can understand is what's "good enough." I don't aim for "the best garden ever." I aim for the "good enough" garden, which is far better than the perfect garden that never even gets started. So we'll root our transformation strongly in basic, beautiful natural gardening practices.

I've included some of the most practical gardening basics I know, things that have helped hundreds of my students have better, easier gardens. You'll find adventures for learning about gardening, and the basic techniques that will make it easier than you ever imagined. Unlike the standard gardening approach, you won't need to memorize a list of corporate products to kill all the natural foes. Research has shown that in healthy ecosystems the damage from pests and diseases will be minimal.

Getting Out of Nature's Way

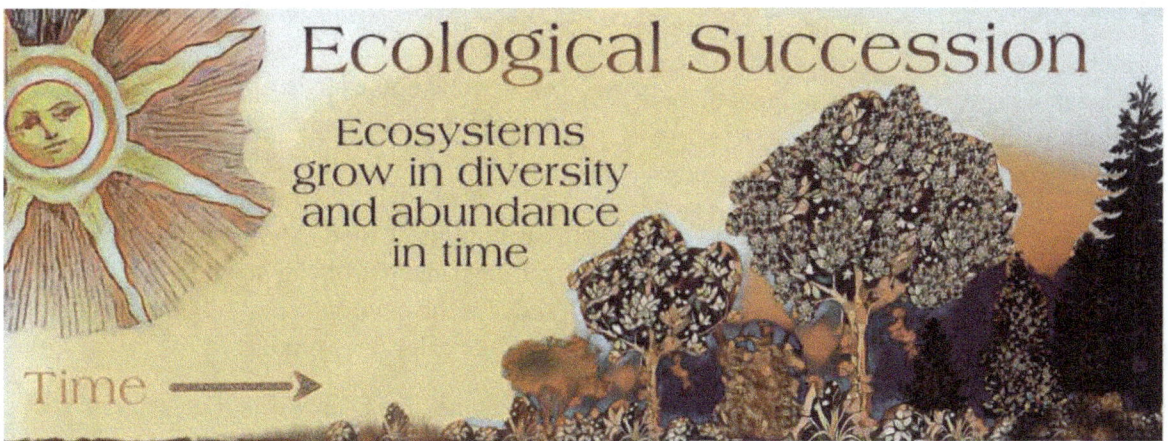

Because ecosystems catch and store energy, cycling it through living beings and increasingly complex relationships, they grow more diverse, abundant, and healthier in time.

This is counter-intuitive, but, lucky for us, the liberation of nature is the easy part. Many would-be landscape transformers are so moved to do good for this planet that they think they need to micromanage the whole thing. As if they need to beat nature into shape with brute force for its own good! And they end up exploiting themselves in the process.

The process of **Ecological Succession** means *every yard wants to grow in diversity, health, and abundance if we just get out of the way and let it.* Nature will grow soil if we let it. Each industrial landscape is already in the process of transformation, already becoming something richer, healthier, more diverse, more valuable, and more abundant. Each lawn is already on its way to becoming a diverse grassland, and ultimately a rich forest that will naturally steward endangered species, sequester carbon, clean water, and purify air for us. It actually takes a lot of resources and work to hold back this process, keep nature from her important work, all so that we can keep our landscapes frozen in time as lawn.

All we need to do is stop spraying poisons, wasting fossil fuels, and wasting time. Then the adventure of our co-evolution with our landscapes begins. As our landscapes become more vibrant, abundant, and diverse, we grow in wealth, health, and wisdom.

And each time a landscape is set free, we are all the richer for it. If you're reading this, you know that modern industrial landscapes—landscapes managed with the three "Ps" (poisons, plastics, and petroleum)—are an environmental disaster. While some appropriate lawn and grass increases the utility for people and the health of an ecosystem, the sprawling 90 million acres of lawn in the U.S. alone is now our largest irrigated crop, a huge waste of fossil fuels, a cause of climate change, and does nothing to assist in reversing the mass extinction event underway.

Meanwhile, our home and community landscapes provide us with one of our greatest leverage points for transforming our lives and the world around us. The liberated landscape can be an important tool for saving endangered species, maintaining biodiversity, feeding pollinators, and giving sanctuary to non-human animals. It can help us sequester meaningful amounts of carbon. It can help reduce the burden of our storm water systems, clean water, and restore aquifers.

And when we choose to use the home landscape to grow some food, we may be taking the most positive step of all. This is because the food system is arguably our number one driver of climate change, number one cause of mass extinctions, the leading cause of plastics pollution of land and water, and a leading cause of many persistent health problems. We interact with a very special leverage point when we choose to use our landscapes to reduce our dependence on the corporate food system.

The great challenge of creating a sustainable society is to meet human needs sustainably off of the surplus produced by healthy ecosystems and sun energy. And we can each participate in that revolution, right in our yards.

Because small scale home food production can be far higher in productivity than average industrial agriculture, 1,000 square feet of the home landscape devoted to food means a much higher acreage of industrial farmland can revert to wildlife.

And while home landscapes in urban areas are of low value to most wildlife, the highest best use of that marginal farmland in rural areas is to support wildlife. And because it's easier to grow food at home without plastics, poisons, and petrol, the home food garden reduces the environmental footprint of the average calorie we produce.

Of course, this is only true if we take ourselves seriously and grow food in ways that are environmentally friendly. But that is indeed what most people I meet today want to do. This is despite the warnings of experts, like my own university's "smart gardening" page on home fruit production, which says "growing fruit at home requires spraying." Real people don't want to grow food drenched with poisons or loaded with phthalates from plastics.

So this book will provide you with some tools to grow at home or in the market garden in ways that are easier than you ever thought possible—without plastics and chemicals. We'll show how people are meeting these goals by integrating the garden, making landscapes that have a place for people, and for the wild.

Winning Over the Neighbors

But while creating ecological landscapes may be easy, doing it in ways that meet human needs and don't make our neighbors hate us can be more challenging. In most cases, creating a truly regenerative landscape is as easy as selling the family lawnmower. Again, many people think it requires a herculean human effort and investment to make nature work, but really we just have to get out of her way.

But, this ecologically sound approach can often lead to unhappy neighbors or fines. And to be honest, while an unmown thicket may be a boon to lightning bugs and dragonflies, there's very little room for humans to interact with it, or meet our needs from it. In my 20 years of professional experience in this area, without space for humans, most "natural" landscapes do not last long. I have salvaged many native plants from expensive gardens that only lasted a couple years before the homeowners realized they actually need some space in their yards for themselves.

And the tools of conventional gardening—weapons developed in the long war against nature—are poorly adapted to helping us achieve our goals of cooperating with nature, of creating a nurturing place for human animals. Many conventional gardening books —especially those dedicated to aesthetics— are based on the poisons, plastics, and petroleum model. Most are geared towards the consumer culture lifestyle of buying carbon-spewing concrete blocks, plastic mulches, and plants that do little to care for the earth or people.

Up and down every street in every community, there are humans who hear nature calling us back to our rightful place as a beneficial species, but they have no idea where to begin, or how to do it in a way that will make their neighbors green (with envy, and new-found environmentalism!)

They need models and mentors to step up and lead the way!

Over 20 years in the fields of homesteading, natural farming, natural landscaping, and Permaculture, I have come to think that this one issue—how it is perceived as un-aesthetic or even careless of aesthetics—is the biggest limiting factor keeping sustainable and self-reliant lifeways from wider acceptance.

People assume that a healthy landscape has to be ugly, including many of us in sustainable living! "People just need to get used to things being wild and ugly," I've heard some of us say.

The typical stance of modern environmentalism is that "the party (of petrol, plastic and not caring about the earth) is over." Now we have to accept a lower standard of living and a more ugly, brutish way of life.

I completely disagree.

It is not that people have to get used to ugly, dirty lives and landscapes! We can create more beautiful lifeways and landscapes, and we can approach people and meet them where they're at as we attempt to do so. My projects, such as Lillie House and the others you'll see in this book are proof that sustainable landscapes can be productive and beautiful.

If your biggest fear is complaints from neighbors as you transform your landscape, you'll find a great deal of inspiration throughout the book. But I've simplified things with a specific set of patterns and tools we could call the "**fine-stopper starter pack.**" This set of patterns is designed to solve exactly this problem with the least amount of cost without increasing the time commitment of managing your landscape.

Recruiting Nature to Help

We'll go deep into designing **guilds**, which are gardens designed to act like natural plant communities. Because they function like plant communities, the plants actually take care of each other. Plants weed, fertilize, control pests, and so on, so that we don't have to do that work. A good guild can become very stable and require very little human input to maintain. I'll share some model guilds and overall food forest planting designs you can take and adapt.

And I'll provide you with a practical path for really transforming the landscape and meeting your goals.

"A garden is a grand teacher. It teaches patience and careful watchfulness; it teaches industry and thrift; above all it teaches entire trust.
— Gertrude Jekyll (2011)
'Wood and Garden: Notes and Thoughts, Practical and Critical, of a Working Amateur'
Cambridge University Press

Transformative Action for Meeting Your Goals

One of the most common problems clients come to me with is that they get stuck. They have a vision for their landscape, perhaps they completed a Permaculture Design Certificate course or paid a professional designer, and then they stalled out. Ten years pass and the landscape isn't too different than it was when they started. Meanwhile, people visit one of my new gardens and in two or three years they say "it's incredible, you've done more in three years than I have in ten!" I have heard this a lot.

These dramatic results certainly aren't due to my super-human work ethic. In fact, I think I put in less time to get that greater effect in almost every case. It's because I follow a *transformative* approach based on the key concept of banking my labor into transformation, like with long-lasting, self-maintaining guilds. This is a pattern that actually frees up free time as we go, by handing over more work to nature.

Most people will start with something like a big annual tillage garden. In year one, they start small and have decent success and feel like they have a few extra hours a week they could spend gardening. So in year two they double the garden. And since it's an annual garden, they must re-prep beds, plant all their baby plants, nurse them through their most vulnerable phase, harvest in fall, and then start all over again next year. All their free time is now used up!

So they will never get anything else done in the landscape, ever, without help.

And as soon as something gets out of balance —a new pest shows up, or perhaps they leave for a few weeks to go to their sister's wedding—the whole thing falls completely apart.

If we start instead by establishing self-maintaining ecosystems like **guilds, forest gardens, hedgerows, and edible meadows**, we put our energy in year one into a few gardens. In year two, those are mostly on auto-pilot, so we can move on to establish more guilds. In year three, those are now mostly taking care of themselves, and we're spending a lot less time mowing, so we actually have MORE time in year three than in year one, and we're getting lots of food out of it, too! We can put that new free time into even more permanent landscape transformation. It's like putting our work right into the garden bank.

This is what we mean when we say "**transformation**." Transformation is a change from one state, to a new state of stability. That means our actions are finite and have an end goal, so we can plan our actions as adventures which transform the landscape in a lasting way.

Transforming the World with Beauty

And that same type of transformative process is what we'll apply to our goals of creating beauty, waking people up, and transforming the world. Instead of using brute force, beating good ideas into people, and weeding out all the bad ideas, squishing all the bad people like squash bugs, in this book we'll try to create our part of a better world. If we get it right, others will follow us. That doesn't mean there won't be occasional bugs that need squashing, but there may be fewer if we start with this holistic approach. We'll transform people, yards, and institutions one at a time, and in this way we can transform the world.

"Setting aside time for the 'adventures,' self dates for fantasy and creativity, gives life a sense of adventure."
-Bonnie Ann Burnett

Adventure Step 1:. Visioning, Top Priorities

Our first step in transforming our landscapes is to begin a visioning journey. This book will guide you through that visioning process, helping you come to an idea of what your landscape can realistically be like. By reading and skimming I will hopefully be guiding you on that exploration. But as you read, there are a few top priorities to keep in mind that will have the biggest impact on the landscape. These are:

1. **Social uses in the landscape.** See the accompanying worksheet to help with this first and most important step. The energy that people bring will be the most important for your experience of the place, and for its longevity. How do you envision spending time with people in your outdoor home? Everything else I mention here just to put it in mind.

2. **Creating privacy.** Often people do not feel comfortable in their own yards while there are open views from neighboring yards and windows. The single biggest thing people can do to transform their landscapes is to create some feelings of privacy with well-placed plants. This opens up the yard for work and play. (See: **beautiful bountiful hedgerow.**)

3. **Framing the garden.** Along with a feeling of privacy, a feeling of a few outdoor "rooms" goes a long way to turning a yard into a garden. This is usually best done with hedgerows, bushes and trees.

4. **Framing views.** Once we have a feeling of privacy and enclosure, the next biggest impact on the quality of a landscape will be the views it provides. A view of a beautiful hill, a distant rustic shed, the house itself, a pond…. If there aren't naturally occurring views, we can make a few focal points. Even the best productive farms I have ever seen have done this well.

5. **Start getting to know the trees.** Tree selections will be one of the most important decisions you'll make. So, starting a list of tree crops will be important. Later, we'll have a whole worksheet to help you with this.

6. **Plan to obtain a yield early**! The single most common terminal mistake I see in projects is that they think only long term. They plant 10 acres of trees knowing they won't fruit for 5 years at best, thinking they can plant the understory later. For 5 years they are on the hook, tending to the garden with no return on the investment. Only extremely privileged people can make such a commitment, and most end up overcommitted. These gardens usually fail before they ever get a good yield. It is absolutely imperative to plan early good yields! This means a succession of annuals, fast fruits, berries, etc.

7. **Consultation?** If the site is new and unbuilt, I highly recommend a consultation with an experienced, certified Permaculture consultant to help site structures and main uses, especially one who's familiar with the goals and approaches in this book.

(Step 1) List Social Uses You Want in the Garden:

Social uses in the garden
- Outdoor eating/picnic/cookout area
- Shady living space for hot weather
- Warm/sunny space for cool weather
- Playground toys
- Outdoor office
- Outdoor workspace
- Dog runs
- Sunny space for some annual gardening.
- Family picture backdrop
- Tree house
- Firepit
- Natural swimming pool/water garden
- Outdoor "stage" for music, presentations, plays
- Spiritual spaces, sacred circle, meditation garden, drum circle, circle singing
- Outdoor gym, exercise space, functional fitness
- "Private" living space, where family members can be alone, study, meditate, take dates,
- Outdoor sleeping/relaxing space
- Outdoor games, horseshoes, baggo, croquet, frisbee golf, hide and seek, tag…

…in more public sites
- Weddings
- Art exhibits
- Outdoor theater
- Dance recitals
- Small, intimate concerts
- A biking/walking trail
- Frisbee golf
- Picnics
- Cookouts
- Campgrounds
- RV park
- Dog park
- Senior pictures and other portraits
- Vermicomposting near the fishing spot
- Athletic fields--yes, even these fields of grass could benefit from thoughtful design!

Chapter 2: The Transformative Power of Beauty

Chapter Key Points:
- We often dismiss beauty, but it is a powerful leverage point for personal and social change.
- Creating a beautiful landscape **"skillfully"** can be a key to a better world.
- A beautiful environment has many personal benefits, as well.
- Because of this dual action, landscape transformation is a powerful step to take.
- If we want to be effective, we can let this "solid base" of personal and community transformation lead our way.

As we look for ways to grow better lives, landscapes, and a better world, history has time and again proven that aesthetics—beauty—is one of the most powerful leverage points we can use. It's arguably the single most transformative point where we can consistently apply pressure.

If you were hoping for a book to just help improve your curb appeal, or stop your neighbors complaining about the appearance of your front-yard veggie garden, I hope you will get that need met from this book. But I would also like to share a perspective on the pursuit of beauty that I think will help you get the most out of your adventure, one that will help you bring some mindfulness and greater joy to the effort.

The more we put into something, the more we tend to get out of it.

So we'll be taking the task of beauty with some seriousness, understanding that making the world a more beautiful place can be powerfully transformative for ourselves and society.

Transcending the spiritual crisis.

Many of us today realize that the crisis of our time is not really an ecological or economic one. It is a spiritual one. Because we feel empty, many of us may look to fill the hole inside with consumer tchotchkes or superiority over others. Yet we can find deep spiritual fulfillment through connection to nature and community if only we choose to look there. For me, nature has been a great source of soul nourishment. And books and tools to help me connect with nature have always had a great spiritual weight. Whatever your own spiritual tradition (my tradition is simply to care for the Earth and all its inhabitants) I invite you to see the environment around you as sacred space.

Creating a Beautiful World is Serious Work

Yet, it's usually the last place most of us look to have a positive impact on the world. That's especially true of us committed activists and change-makers. In fact most of us would see beauty as a privileged distraction. Usually, we look to things like politics, the government, and laws to create change. The people who focus on aesthetics (artists, musicians, writers, gardeners) are dismissed as though they aren't very serious at all. Or, often, "serious" topics like politics and business are considered "masculine" (important) while things like creating beauty, creating a nurturing environment are considered "feminine" and unimportant. Even among a lot of us doing this work, things like grain production, energy, and housing are considered serious, while making things beautiful (or flavorful) receives almost no attention at all, or is considered frivolous or even counterproductive!

Yet, systems thinking pioneer and environmentalist Donnella Meadows (2008) studied how we can leverage meaningful change to complex systems like society and the economy, and here is the framework she came up with.

"Sustainability is a new idea to many people, and many find it hard to understand. But all over the world there are people who have entered into the exercise of imagining and bringing into being a sustainable world. They see it as a world to move toward not reluctantly, but joyfully, not with a sense of sacrifice, but a sense of adventure. A sustainable world could be very much better than the one we live in today."

— *Donella H. Meadows*

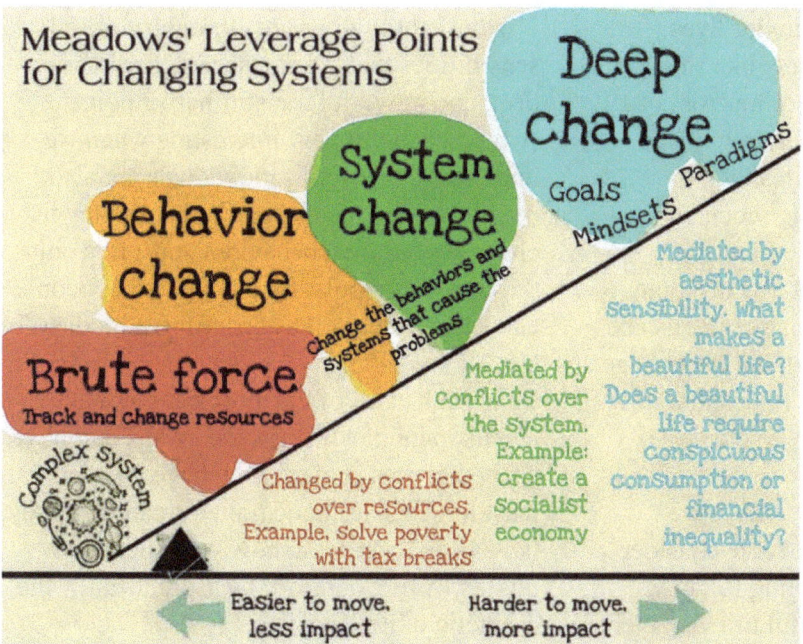

According to Meadows, these are the main places we can intervene to change complex systems like society (or our lives.) The points towards the bottom are easier to understand how to change, but have the least effect.

For example, if we want to impact poverty, we might pass laws to measure the impacts of poverty better. That way we at least better understand the symptoms, though to have an actual impact we might later have to pass laws or rules to shift some resources like food to those in poverty.

There are two important points to understand about change on these lower levels. First, these changes aren't necessarily easier to make, just easier to understand and strategize about. If we see poverty as a problem of who the system delivers resources to, that is clear what we have to change. We get the system to deliver more resources to those in poverty. Easy to see. That doesn't mean it's actually easier to do!

Which brings us to point two, change on that level is almost always a matter of conflict. If we want the system to deliver more resources to those in poverty, those resources have to come from somewhere, which almost always means getting people who have more resources to give them up. Even looking at our personal problems, change at this level is about conflict. We can start counting calories, but that means a conflict between wanting to eat that piece of cake and wanting to lose weight.

These are the "serious" places we usually look to impact important societal problems. But as Meadows points out, we fiddle with these low-impact leverage points while leaving strong leverage points in place. We give a small tax break to the poor but leave the tax incentives and corporate structures that keep generational wealth intact. That reliably creates castes with wealthy people—and inevitably—poverty. In most cases these small changes may even end up reinforcing poverty, because that is, after all, the goal of the system. The paradigm requires it.

Most importantly, we do not change the true, deep factors that create the systematic problem in the first place. In most cases, there's no "problem" with the system to fix at all! The systems are functioning exactly how they were designed to, if we look at these deep-level paradigms and goals. Inequality isn't a "problem." It's actually the goal.

By the way, we do the same in our lives, fiddling with low-impact points like stocks and flows (counting calories or macros) or maybe making rules for ourselves to break. Meanwhile we leave major lifestyle paradigms in place (like living sedentary lifestyles and eating corporate food.) Counting calories is easy, but low leverage. Creating a fitness lifestyle of good habits and natural food takes more effort, but the change will be profound and lasting. (Question: how does this same thinking apply to our soil?)

These deep change levels are within the realm of aesthetics, of what people consider good, beautiful, and worth living for, how they define a wealthy, beautiful life. Another example is the core assumption that a good life requires wealth, or requires a hierarchy to sit atop. So long as we keep putting wealthy people on magazines and celebrating them for hoarding resources, it doesn't matter how many anti-poverty initiatives we pass, the system will still reliably create poverty. If we can only feel wealthy in comparison to others lower on a hierarchy, we will need that hierarchy.

Another core assumption is that our lives require certain forms of conspicuous consumption and that these require exploitation of people and resources.

And yet another is that a beautiful life requires a war with nature, since nature is seen as "dirty" or "poor," or animalistic. This was justification for European colonizers, that colonizers were civilized, clean, and above lowly nature. Contact with soil, science shows, can cure depression and make us happy, but culturally we see the dirt as a sign of poverty or savageness.

It is only through beauty, through aesthetic sensibilities that we can access these very deep and powerful transformative points (which becomes very interesting when we understand that we as individuals are also complex systems, and aesthetics are key in changing our own behaviors, too.) It is only through these tools, for example, that people can learn to see soil as beautiful, not "dirty."

If we look at all of the most transformative events in human history, the events that have shaped human society, they have all been - at least in part - aesthetic transformations. They have been changes in how we experience quality of life, comfort, pleasure, beauty, the aesthetic of how we live.

We can trace these aesthetic changes in the move from horticultural villages to cities with temples. We can find it in jewelry as we advanced from stone tools to metallurgy. We can see it in the evolution of cultures over time, the kinds of houses we live in, the kinds of things we make with our life energy. Yes, some human advancements were technological, but even most of these also had accompanying changes in aesthetic sensibilities.

Looking at the world today, it is still aesthetic sensibility which most shapes society. Why do people drive huge gas-guzzling cars? An aesthetic sensibility. Why do we all own cars in the first place instead of mostly using public transportation? Aesthetic sensibility. Why do we live in ever-growing sprawling houses made of cheap energy intensive materials that contribute to climate change? Aesthetic sensibility. Why do we buy a never ending stream of made-for-the-landfill corporate junk we don't need? Aesthetic sensibility. Pollute the oceans with 7 layers of packaging on everything we buy? Mostly for aesthetic reasons.

Some of this is built into the way we see ourselves and others. Freud was the first to write about "the narcissism of small differences" in his "Civilization and its Discontents." As Freud pointed out, it is not the big differences between us we focus on, but the small ones.

The fiercest rivalries develop - not between very different peoples - but between near neighbors living very similar lifestyles. The focus on these small differences is necessary to give us identity, and may be woven into our biology to help us define our communities. For example, poor whites have been taught to hate African Americans living very similar lifestyles, rather than the ruling class elites hogging resources to themselves, who most impact their lives. And these differences, these conflicts and rivalries, are mediated by aesthetics (in the example, an aesthetics of "whiteness.")

And so, aesthetic sensibility may be a major leverage point in addressing these problems, too. Taking this example down to the level of our personal lives, we could see how we might do a better job of reaching out to people, including people, and finding new audiences by focusing some on aesthetic sensibilities.

Beyond theory and history, do we have evidence that changing environments can solve these difficult social problems?

It turns out we humans live in ecosystems, and like any other animal our behavior is dependent on our environments.

Environment has been widely linked with crime rates, violence, social equality or bigotry, war, pollution, and "the diseases of civilization." Yes, a beautiful, abundant natural environment has been found to reduce violence, reduce crime, reduce bigotry, and of course reduce war. While we struggle over resources in low leverage actions to reduce oppression, anthropologists have widely documented that equality (or the lack of it) is a predictable outcome of the type of ecology we live in and the healthy balance of our ecosystems.

Ecological economists have explained that inequality is a necessary feature of centralized agricultural systems, and bigotry is necessary to justify it. In fact, a great deal of research now strongly implies that the only thing that has ever significantly reduced inequality is changing the ecology it's nested in. As we'll see in the next section, many personal issues like addiction, emotional stability, IQ, and learning are also strongly impacted by environment, and these can become barriers to improving social problems as well. So the creation of a beautiful supportive environment is very likely key in addressing some of our most intractable social problems.

Skillful Means and Happiness

In many of the world's wisdom and philosophy traditions, (especially Buddhist and Taoist traditions) it is said that doing anything, especially artisan or artistic work, with true skill will lead to enlightenment and great happiness.

Because our true goal in any action is ultimately a happy life, if something doesn't lead us towards happiness, it must be unskillful. Right? So, a skilled gardener, for example, will carefully discard any actions that don't make us happy, and only keep those that "spark joy," you could say. If mowing the lawn or tilling the soil or covering the world in plastic don't bring us happiness we throw those out. If rushing around the market garden in a huff to sell another $10 in carrots for $3/hour doesn't make us happy, we throw that out and try to find a more skillful way of pursuing our goals.

And so it is said that any true master of an art will have arrived at a high level of contentment and satisfaction through their skill. Even just viewing the work of such an artist can spread that happiness. We can possibly share an experience of that enlightenment. We may see in the simple effectiveness of a great work of art that which is truly important, and that we can gradually throw out all the things that bring us frustration and anxiety in our lives, and keep only that which leads to happiness.

"The ultimate goal of farming is not growing crops, but the cultivation and perfection of human beings."
- Masanobu Fukuoka, One Straw Revolution

In this book, we will let this be our guiding ethic: to meet our landscaping goals, skillfully.

When it comes to making our lives better with skillful means, if we look at many of the factors that could most impact our own quality of life and our goals, and run that by Meadows' chart, we see that aesthetics will be a huge and powerful leverage point in a better, happier life as well as a better world.

Notes:
1. Donella Meadows, *Thinking in Systems*, Chelsea Green, 2008
2. Lara Cushing, Rachel Morello-Frosch, Madeline Wander, and Manuel Pastor, *The Haves, the Have-Nots, and the Health of Everyone: The Relationship Between Social Inequality and Environmental Quality*, Annual Review of Public Health (2015)
3. Rachel Morello-Frosch and Edmond D. Shenassa, *The Environmental "Riskscape" and Social Inequality: Implications for Explaining Maternal and Child Health Disparities*, EHP, (2006)
4. Walter Scheidel, The Only Thing, Historically, That's Curbed Inequality: Catastrophe, *Atlantic*
5. Nicholas Georgescu-Roegen, *The Entropy Law and the Economic Process*, Harvard press, (1971)

Transforming Your Landscape Can Help Transform the World

1. **Take Climate Action**: We can directly pull carbon out of the atmosphere and store it in the soil and plants.
2. **Help feed people**: gardens that integrate food can help grow local food sovereignty.
3. **Save endangered species**: we can provide habitat in our yards, and more importantly, reduce our reliance on the corporate food system that is the leading cause of ecosystem destruction and biodiversity loss.
4. **Steward water**: we can catch water, clean it, and sink it in to recharge our depleted aquifers.
5. **Reduce flooding**, by stewarding water we're reducing flooding.
6. **Increase climate resilience** by shortening long supply lines that are vulnerable.
7. **Reduce our consumption of limited resources** by growing sun powered resources like food, fuel, building materials, and medicines in ways that actually fight climate change.
8. **Learn sustainability**: we learn by observing nature's closed loop systems.
9. **Build social capital**: beauty, plants, and food are powerful touch points for building friendship.
10. **Support activism, art or other important work** through a healthy environment, stress reduction, savings, and financial support.
11. **Build community wealth**: by helping individuals, families and communities build real wealth and the means to produce the necessities of life. This can become the basis for a new sun-powered economy!
12. **Build political capital**: this social capital and wisdom grown in a garden can eventually translate to political capital.
13. **Fight corporate super-villains** by reducing our dependence on some of the world's most destructive industries, like the food industry, the leading cause of climate change and extinctions, and the plastics industry, the largest polluter of water and soil.
14. **Change hearts and minds**: changing people's minds and teaching them to value nature and the earth.
15. **Create viral change through the power of contagious acts of beauty and generosity**.

Are there other ways your action in the landscape can help transform society and the world?

Personal Benefits of the Beautiful Landscape

*Most front-yard gardeners upset over a ticket for a messy yard or for growing vegetables will attempt to solve the problem with low leverage points like petitioning their communities for rule changes or tracking information on biodiversity. Yet, if their neighbors thought their gardens were beautiful, there would be no problem to begin with. Beauty, and a sense of co-creating it with our communities, would be the better leverage point.

*We may struggle to build a real sense of community for ourselves, attempting to connect with people over politics, but we leave out the powerful cultural tools that real societies have always organized around, common story, art, music, beautiful place, and cuisine.

*A burned-out member of the workforce may purchase massages, vacations, or meditation retreats to try to fix their chronic stress and anxiety. But they miss the high magic of transforming their environments where they live and work into relaxing, beautiful places that call them out of their heads and worries, and into the sensory world of intoxicating sights, tastes and aromas. These everyday touch points of sacredness would be a far more powerful leverage point for a relaxed lifestyle than a stressful yearly vacation. (K. Wyles et al 2017)

*A vision of beauty has power. People and resources will be drawn to its gravity. I can say that 90% of the success of my projects like Lillie House were because people thought the projects were beautiful, and offered a beautiful vision for humanity. And communities and people with energy and resources naturally crowded in to see the beautiful view.

*Our environments have powerful impacts on us. A supportive environment has been found to:
- significantly improve IQ
- improve our health and fitness
- improve mental health
- facilitate cooperation
- lower criminal tendencies
- assist in learning
- facilitate community building
- improve mental focus
- be a powerful leverage point for breaking addictions
- assist in spiritual development
- help us transcend the scarcity mindset, so we can plan better and make better, more holistic long-term decisions
- build an environment of natural beauty that can directly make us "kinder, happier, and more creative," (Jill Suttie, Berkeley)
- create a beautiful natural environment that makes us feel more aware, more awake and present, more "vital and alive."

In many cases, such as with building new habits or breaking addiction, a beautiful natural environment could arguably be the biggest single factor in success. Patients struggling for years with addiction programs have reportedly been able to quit their destructive behaviors "effortlessly" by simply changing their environments. Research confirms that being in nature alone reduces addiction and stifles cravings.

And yet, as with societal change, we usually dismiss this powerful leverage point, to focus on "serious" brute force methods of willing ourselves and guilt tripping ourselves to change. These brute force methods deplete our willpower resources, while a beautiful environment has been found to restore them.

This is an ecological and holistic worldview, where we see our environments are likely our most important factor in determining our behavior. And this is why we start the design process by thinking about our goals and what makes a beautiful life. Our behavior impacts our environments, and our environments affect our behavior. "As above, so below."

So what do you want out of life? What do you want to be, and to achieve? It's likely that by creating the world you want to live in, you may be taking the most important step in shaping the kind of person you will become.

Notes:
1. Are Some Natural Environments More Psychologically Beneficial Than Others? Kayleigh J. Wyles, Mathew P. White, Caroline Hattam, Environment and Behavior, 2017.
2. How Nature Can Make You Kinder, Happier, and More Creative, Jill Suttie, Berkeley, 2016
3. The psychological science of addiction, Elizabeth Gifford, Keith Humphreys, Addiction 102, Issue 3 p. 352-361

The Transformative Power of Being Creative

Creative play itself, the kind we can engage to beautify the garden, has many personal benefits. Many of us have long understood the feelings of accomplishment that come with a completed work of creativity and felt that it helped them become better problem solvers, less stressed out, and better communicators. Now, research is backing these perceptions with data.

As the whole field of art therapy shows, art helps us have better, more stable mental health. It can both improve mental illness and prevent it by enhancing overall sense of well-being.[1] It can build our capacity for self-regulation and management of our own mental health.

Mindful creative acts such as we'll discuss in this book reduce cortisol levels and have wide-ranging health-enhancing benefits.

Creative acts don't just enhance our health, they make us better thinkers and problem-solvers. They have been shown to increase overall academic performance. A great abundance of research shows that creative activity increases innovative thinking and creative problem-solving in general.[2]

New research even shows that artistic action like beautifying the landscape actually helps us be more optimistic, and the cutting edge theory is that this is because it helps us to creatively envision possibilities, even envisioning a better world.[3]

According to author Girija Kaimal this is an evolved response. When things are darkest and conditions are bad, creative thinking allows us to imagine a future that is better, one that is worth continuing to work towards. And then, it can help us think creatively about how to get there, and even take steps to improve our surroundings, directly making the situation better. If we are creative, there is never a situation, no matter how dire, where we can't take skillful action to make things better. There is no finer version of this than taking action to make your own landscapes more beautiful.

"When you become truly wise, you can travel to even the deepest hell realm and make the flowers there bloom."
-Zen proverb

Even such practical-minded publications as Forbes and Business Insider have published articles on the benefits of creative play, ("Here's why you should be making art even if you're bad at it.") Because when we get down to it, it can make us more skillful at business, no matter what our business is.

And all of this allows us to step out of our roles as "consumers" and into our own lives as doers, builders, beneficial humans. A beautiful landscape isn't something we have to buy, it is our birthright as human beings to co-evolve that for ourselves. It's possibly the most basically human thing we can do.

Section Notes:

1. Suzanne Haeyen, Promoting mental health versus reducing mental illness in art therapy with patients with personality disorders: A quantitative study, *The Arts in Psychotherapy*, 2018, https://doi.org/10.1016/j.aip.2017.12.009
2. Smithrim, Katharine, and Rena Upitis. "Learning through the Arts: Lessons of Engagement." *Canadian Journal of Education* www.jstor.org/stable/1602156.
3. Brittany Harker Martin, The Artistry of Innovation: Increasing Teachers' Artistic Quotient for Innovative Efficacy University of Calgary, 2019.

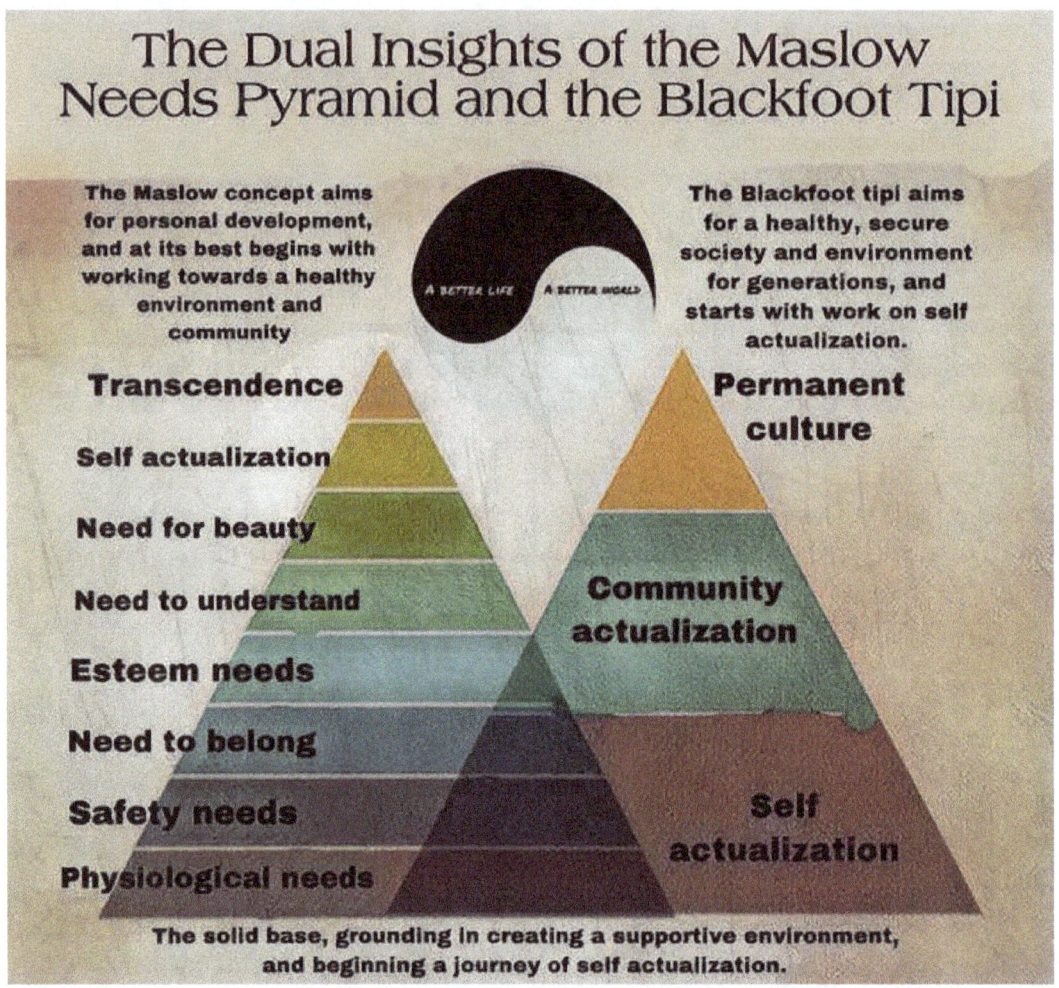

Where to begin a Project? The Solid Base: The Dual Insights of the Blackfoot Tipi and the Maslow Pyramid

When it comes to transforming the landscape, one of the biggest questions people get stuck on is "where do I start?"

Others jump right in, hoping to achieve independence with farming or homesteading, but end up stuck and burned out after a few false starts. Often, whatever progress they've made falls to pieces when life gets busy or some crisis hits.

To help my students and clients get off to a good, stable start, I have come up with the idea of the solid base. This is the foundation we build our dreams and landscapes on.

The solid base: a secure, beautiful, and enriching support system for ourselves.

Before people jump right in to farming, activism, homesteading, or even just a massive landscape transformation project, I always recommend people get a good start on building this solid base with some attention to their families, health, and so on. But, this also has implications that go beyond our landscapes.

I was discussing the role of creativity in building a better society, when I learned from my friends Dan Wahpepah and Stician Samples that the famous Maslow pyramid of needs was inspired by his visit to a Blackfoot reservation near Alberta, Canada, in 1938.

For those who aren't familiar with it, Maslow came up with his pyramid of needs because he was interested in helping people work towards "self-actualization," becoming the fullest, best versions of ourselves. To do this, he realized most people would need to have their basic needs mostly taken care of before they would be able to prioritize higher level needs necessary for this self-actualization work.

But Maslow didn't come up with this alone. He had been inspired by the high level of self-actualization he encountered among the people of the Blackfoot tribe.

This self-actualization means more than simply fulfilling our economic potential. In fact, it's widely recognized that the more self-actualized we become, the less we tend to care about material possessions. One who is comfortable and happy with themselves does not need a fancy car or numbers in a bank account to prove their worth. Self-actualized people are free to focus on playing a positive, but humble role in the betterment of the world.

Maslow wrote that by comparison, the Europeans he tended to meet were coarse and more concerned with money and watching out for themselves than with personal growth or the growth and happiness of their communities.

So, this Blackfoot tipi shows their goal is the creation of a just and sustainable society, a permanent culture, or a "permaculture," if you like. [1]

To do this skillfully, they understood that the individual must first reach a certain level of self-actualization. A coarse person only concerned with themselves can never contribute skillfully to the betterment of their community. They will prioritize their own needs first and then only contribute if it makes them look good or enhances their riches or power.

In the U.S., government policy seems to always prioritize making the wealthy and powerful more wealthy and more powerful first, and accomplishing anything is only a secondary goal at best.

A just and sustainable society is built on the foundation of self-actualized people.

Maslow defined self-actualization needs as "creativity, morality, the abandonment of prejudices, spontaneity, and acceptance of the self and of facts." These are the very meat and potatoes of what art is. Art is about exploring creativity, morality, and seeing and accepting the world and the self. And yes, that includes how we make art in our environments and surroundings. Around the world, this path of self-actualization involves exploring ideas, seeking the beauty in life,

---

[1] Teju Ravilochan https://www.resilience.org/stories/2021-06-18/the-blackfoot-wisdom-that-inspired-maslows-hierarchy/

seeking an aesthetic view of the world, listening deeply to the self-expression of other people and then seeking to express our own experience. These are all things in the realm of aesthetics and how we make our world and ourselves look.

To treat your surroundings as the palace of the gods and your body as the body of a god is old, powerful magic.

Some critics fault Maslow for making a personal accomplishment (self-actualization, or in later models, transcendence) the ultimate goal of his pyramid, rather than the creation of permaculture.

But Maslow perhaps understood that this path of self-actualization was possible in the Blackfoot nation because the needs of individuals were secured. In that kind of society, every child has their food and shelter needs taken care of, they have the love and respect of the people in their village. And so, they don't need to scrape and compete just to get some food, or a place to sleep. They aren't in danger of some scoundrel promising them a decent meal in exchange for participating in personally destructive acts like crime, theft, or violence, as often happens in less secure societies. In a secure society, no one will agree to defend the wealthy by doing violence to their community members just to get a little security, a paycheck, and esteem from the ruling class. In such a society where everyone's needs are met there is natural solidarity, and everyone has the space they need to pursue self-actualization, self-expression, art, spirit, community good, and true happiness.

But that is a very different society than we have in the Western world, where many of us, even children, must scrape and toil in poverty just to get our basic needs met. In our type of society, it is a privilege to have the opportunity to pursue self-improvement or community good. It becomes a vicious cycle where an uncaring society necessarily creates self-centered people who go on to perpetuate an uncaring society.

I'm reminded of an over-planted garden of scraggly vegetables, which will turn bitter and bolt in too much competition. Plants need space to get light, nutrients, and free soil in order to grow to their full potential.

"Ask not what your country can do for you, but what you can do for your country."
John F. Kennedy

So, some say that Maslow, understanding his audience of Western society, flipped the tipi upside down into a pyramid. They say he believed that the first step would be to move towards a society more like the Blackfoot, where people would have their basic needs met enough to pursue self-improvement.

And Maslow's pyramid meets people where they are at. In a society of people made coarse by the ecology of struggle and conflict over basic needs, the common goals would likely be personal accomplishment. Maslow offers a skillful way to seek personal happiness, one that begins with working to create a more just and sustainable society where people's basic needs are met. Maslow's hierarchy implies that if we want a less brutish society, we need to make sure people have the space to grow.

If there is insight in both the Blackfoot tipi and the Maslow hierarchy, we see that the path to creating a better world begins with a work to become better people. If we don't first develop personal wisdom, we risk imposing our faults and delusions on society (see Western culture's worship of greedy careless individuals.)

Which brings us to the solid base. We can find ways to work on both together. So paths of self-actualization that also help us invest in a healthy supportive environment and community may be our most powerful early steps. We will call this idea of taking steps to secure our physical needs while simultaneously beginning to seek self-actualization the **solid base.**

Yes, creating a transformative landscape as we define it in this book is necessarily working towards this solid base. It is powerful action.

Beauty in Abundance

But, if we're going to take our transformation of landscapes seriously and holistically, then we'd do well to make sure our plans really help in developing this solid base of security and self-actualization. But sometimes, this may mean we need to prioritize things like our health, our mental health, our family, and relationships as our first steps in transforming our landscapes. That's an important part of our work.

And when it comes to investing in positive community change, remember that one of the most important things we can contribute to others is to help them also secure this solid base, taking responsibility to ensure that the needs of our community members are also met.

This space by landscape designer Jeffrey Barnett creates a nurturing home environment

Some Practical Ways to Build the Solid Base in the Landscape:

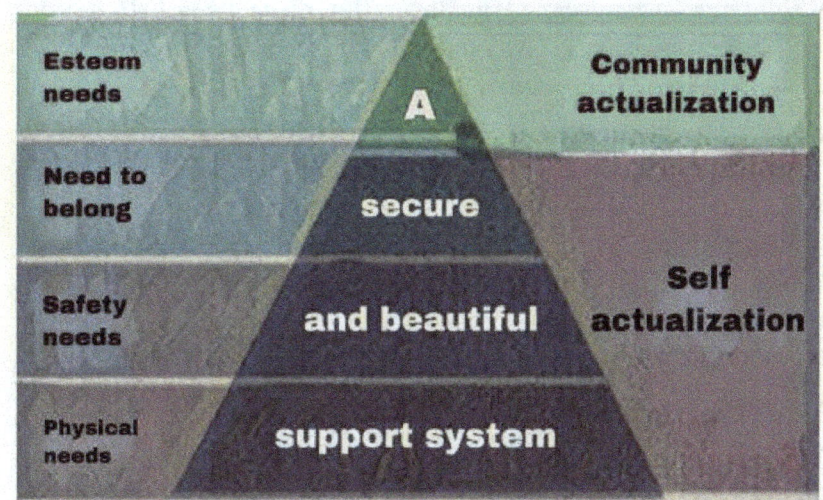

-Start building a support system of beauty and abundance in our landscapes.
-It's okay to prioritize our physical and mental health first. That's part of the transformation.
-Create a landscape that helps meet our physical needs and our need for security.
-It's also okay to prioritize just making something beautiful, if that's where we're at!
-How can your project feed your soul and be satisfying on a creative level?
-A transformative landscape should support people, not make more work for them.
-If you're sacrificing your needs for the project, it's probably not sustainable.
-If you're not having fun, you're doing it wrong.
-The earth doesn't need a martyr. It needs your healthy, free, creative energy.
-In each project, plan an element that will feed your creative self.
-Find ways to bring other people into the garden. It doesn't have to be perfect.
-Host tours. Get involved in permaculture or Transformative Adventures groups.
-Build an herb spiral.
-Make an ancestors garden.
-Explore spiritual patterns in the garden.
-Visit inspiring gardens.
-It's okay to hand more land over to nature. Not every inch has to be productive.

Adventure Step 2:
The Solid Base Self Assessment

On a scale of 1-5, with 1 needing improvement and 5 being very good, how would you rate yourself on each area of the solid base? This will help you set priorities for your project that help you secure this solid base. **A landscape transformation project can help develop all of these!** Once you have identified areas for growth, how can you take steps to improve?

Physical Needs: (rating 1-5)
Secure, safe, legal housing ____
Secure, safe water ____
Safe, secure, nutrition-dense foods ____
Basic caloric needs ____
Safe good clothing ____

Nurturing Surroundings
A beautiful, expressive place to live ____
Access to delicious satisfying meals ____
Expressive, respectable, clothes ____
that express who you are.

Safety Needs
Access to income or employment ____
Access to healthcare ____
Having a plan to become "FREE", ____
Financially Resilient and Economically (and Ethically) Empowered
Land access ____

Secure Surroundings
A unique set of marketable skills, ____
products or services.
A feeling of developing vocation ____
or purpose.
A place that instills a feeling of safety ____
A feeling of community safety ____
A feeling of knowing the neighbors ____
Secure food growing systems ____

Belonging Needs
Spend time with a few good friends ____
Have activities that engage you with ____
community
Contribute something of value to your ____
community
Have enough opportunities to connect ____
with other people
Have a feeling of identity or being ____

Self and Community Actualization
Have outlets for creative expression ____
Have opportunities for daily beauty ____
Have opportunities to encounter other ____
perspectives and views.
Have opportunities to share beauty and ____
expression with others.
Have opportunities for creative problem ____
solving and learning.

*It's worth saying that we have the right to invest in this solid base for ourselves, and that is not privileged or frivolous. I have known many activists who have felt that taking the time to invest in themselves is taking time away from the important work of activism. But this solid base is also the solid base to support and enable our best activism. And it teaches how we can help others develop this solid base for themselves, making our activism truly practical and valuable.

Investing in the solid base means investing in community—the best investment we can make. Photograph, PJ Chmiel.

Chapter 3: Models of Transformative Aesthetics

"We must tend our own gardens."
-Voltaire, Candide

(*Babour in the Garden of Fidelity, 1590*)

With it being such a powerful leverage point, it's no wonder that the most effective changemakers in history have often used a focus on beauty to support their goals. When we take responsibility for "tending our own garden," we join the proud tradition of the absolute best of humanity: those who have set out to quietly, and humbly create a better world right in their surroundings. Let's look at a couple of case studies for inspiration.

Trungpa, Shambhala, and the Magic of Beauty

"Magic lies in the state of mind of the artist."
—Chogyam Trungpa Rinpoche

Chogyam Trungpa was one of the first Tibetan Buddhist teachers to flee Tibet and come to the United States (U.S.) to teach. He found fertile soil for the seeds he had to sow, and his life in the U.S. was a flurry of accomplishments. He founded the first centers for Tibetan Buddhism, he translated dozens of Tibetan texts, he started Naropa University, and a school of secular warriorship, Shambhala warrior training, in addition to creating many renown and beautiful works of art.

In his book, Dharma Art, he talks about being "on the dot," connecting to the flash of inspiration, a seed syllable from which the work would naturally flow. "If you can envision a better world clearly enough in your mind, when you wake up in the morning it will be there."

It is said by many of his students that all of his actions and accomplishments began from this place of mind, that sense of clear elegance and truth. He would insist on getting this starting view right before working on any project. Everything was rooted in this beauty-vision.

For Trungpa, the blank page, or landscape, is "the magic mirror." The creative process begins with connection to the infinite possibility of this emptiness, from which all things emerge. It is in embodying this pure, empty state of mind that creativity begins. On this blank page mind, no preconceived notions exist. No distortions exist to our clear view of the world, just as it is.

Remember the theory of Succession?

We could see lawns as a wasteful problem. But in the lawn, all things are possible: a wildflower prairie or a rich forest is already there, waiting to emerge when we let go of our distorted attachment to the grass. As we look at the world around us, everything that is wrong, unsustainable, or unjust is an opportunity to build a better world. Each person we meet is a potential friend, if we stop putting energy into emphasizing our differences or our fear of rejection. Each group of people is a potential community.

There is a paradise planet waiting to emerge, as soon as we stop wasting energy maintaining this strange dystopia. This is the blank page view.

Trungpa's student, artist and teacher Jack Nihland recounted that when Trungpa was given a farm property to develop a Buddhist center, one of his first priorities was to paint the door, a seed syllable from which the entire site would grow. When he spoke publicly, he famously always wanted to make sure the space was beautiful and elegant. He took great care in arranging the flowers and art. When he met students, one of the first things he insisted upon was that they begin to take a sense of elegance and care about themselves, to live in an uplifted state. He wanted them to be mindful of their clothing, their postures, the sense of dignity in how they lived. To live as though they had potential within.

In Dharma art, he explains that all of life is dharma art, teaching art. Everything we do can be done with a sense of sharing this awakened, uplifted beauty.

When we live this way, we give a gift of beauty and sanity to everyone we come into contact with. Each meal can be art. Each conversation, literature. A goodbye, poetry.

And when we take that awakened sense to what we do, people tend to gather around it, to gather in and support that which is life enhancing and beautiful. It's no wonder that an activism that began with the seed syllable of beauty would lead to such profound accomplishments.

"Karmê Chöling, a Shambhala Buddhist meditation and retreat center and community in Barnet, Vermont (image from Karmê Chöling) Trungpa considered the beauty of the place a major importance to how it could transform the world and establish Buddhism in North America.

How to Observe a Site

In Permaculture, it is often said that the process of design should include a long, deep observation period. It's often repeated that this observation period should last a whole year or more. But this is also often misunderstood. In fact, one critic of Permaculture denounced the whole of the design system on this one point, saying that it was extremely privileged to wait a whole year before starting to garden. That was never any Permaculture designer's intention!

To clear this up, understand that we have 3 modes of observation we do for a site. We always need all 3.
1. **Interactive observation**. To truly observe a site, we need to interact with it. We will not notice very important things like heavy wind areas or wildlife sectors until we actually start trying to grow some garden. So, to observe, we must start planting some gardens and starting a good plant collection. This is an important mode of observation. It is just best if we wait a season before investing in major long-term elements.
2. **Directed Observation and Evaluation**. This is data observation, what is the soil like, how much sun and water there is, what plants are already there, and what kind of plants might like this situation.
3. **Open Observation**. This is the magic mirror, an attempt to get a true clear view of the situation without thought and evaluation. The idea is to try to get beyond any ideas we might have and be open to what is actually there. We often don't see opportunities because we're locked into our own ideas. But if we can find ways to observe with a true open mind long enough, sometimes a flash of insight will slash through our notions and we'll see the true opportunities before us.

The trick in this last one is to *just observe*.

Try this: Find a quiet, comfortable view in the landscape and sit comfortably. Schedule 20 minutes to an hour. Try to just relax and keep an open mind. If you start evaluating and thinking, say "there's time for evaluation later" and return to observing. If that is difficult, try watching the breath, too. With each inhale, try to imagine you are seeing this site for the very first time, as though you were just born and seeing for the first time.

The Ruler, the Radical, and the Tale of Two Tea Houses

Images, Top: Hideyoshi and his golden tea house. Bottom, Rikyu and his rustic tea shack. wikimedia, public commons

In "The Book of Tea" (1906, must-read for anyone concerned with gardening aesthetics) Okakura Kakuzō recounts the story of two famous tea masters who shaped the art of the Japanese tea ceremony, Sen no Rikyu and Lord Hideyoshi, who was ruler of most of Japan at the time.

To showcase his great wealth, taste, and power, Hideyoshi set out to give "the tea ceremony to end all tea ceremonies," and consulted with his friend Rikyu.

To showcase his power and aesthetic sensibility, Hideyoshi insisted that everything be special. The implements were rare and made of gold, and he had a gold tea house constructed to hold them.

Immersed in such luxury, Sen no Rikyu sensed a growing feeling of inequity in this tea ceremony, which he ultimately abandoned.

Afterwards, Rikyu gave up refined implements altogether, and built a tea house resembling a peasant's shack. Around it he created a garden that was simply like a natural woodland path. Instead of imported gold tools and fine pottery, he used simple hand-made pieces from local artisans. He created an aesthetic of simple elegance based on mindfulness, and connection to basic goodness, which became known as wabi-sabi. Wabi-sabi is an aesthetic that invokes the buddhist truths of life, impermanence and imperfection, a beauty one can connect to in moments of quiet isolation in nature. It is a beauty available to anyone, and it doesn't require money or educated "taste."

This tea house radical began inviting guests to his ceremonies of simplicity. It was a direct critique of the conspicuous consumption that the wealthy of society held over the poor. It insisted that even the poor could live an uplifted and dignified life of beauty, and that this simple beauty surpassed that gaudy finery of the rich trying too hard to impress each other.

The power of this aesthetic caught on, and completely transformed—not only the tea ceremony—but all of Japanese art. And according to most critics, it has become the most enduring and recognizable aesthetic of Japanese culture even today. This social revolution came from the seed syllable of simple beauty, shared over tea in a simple shack, in a simple garden, and went on to transform a culture.

Even in today's Japan—influenced by American greed and conspicuous consumption—there is still a sense of solidarity beyond classes, a sense that it is unaesthetic for the wealthy to be too conspicuous and that regular people deserve a sense of dignity. In Japanese schools, children clean up after themselves, so there is no need for a caste to clean up after them.

Today, this enduring tradition of wabi-sabi is even a spiritual path for some seeking self-actualization. Through this practice they develop an aesthetic view of the world, looking to see the inherent beauty in things, the beauty that grows with age and with the signs of life. The "wabito," someone on this spiritual path of aesthetics, treats life as art and seeks the beauty in everyday life.

A simple bamboo vase used by Rikyu

Patterns for Neo-Wabi-sabi in the Garden

This section comes with a warning against the idea that we can or should appropriate wabi-sabi into our gardens in exploitive ways. "Colonialism" is no longer a good look in the home landscape!

And yet, this aesthetic contains a powerful antidote to Western conspicuous consumption and spiritual emptiness.

So, I offer two routes to approach wabi-sabi ethically.

1. Do so with the help of a master in the art.
2. Unless you are Japanese, don't do wabi-sabi. Discover what this same aesthetic of impermanence, simplicity and tranquility looks like in your own culture, time and place.

If you choose the second path, these guidelines may help.

Grow an eye for the simple, for objects that invoke the tranquility of a long fall alone in the woods.

Use hand-made objects exclusively, with signs of flaws and marks of being hand-made. A machine-made plastic surface instantly shatters the illusion.

Use antiques, things with signs of age, old garden tools, found objects.

Select plants for beauty in the dormant season, beauty through the winter of life.

Use simple, rough materials.

Use sparseness.

Create a forest garden, with nurse logs, fallen leaves, woodland flowers, etc…

Edible plants for a coniferous forest garden

For many people, one of the hallmarks of the wabi-sabi I feel is the special solitude of a pine forest. Yet, most coniferous forests are low in edible plants. But it does not have to be so. We can create a coniferous forest based on edible plants.

Trees:
Monkey puzzle tree
Stone pines
Incense cedar
Some pines have edible cones.

Shrubs
Witch hazel
Contorted mulberry
Plum yew
Dwarf stone pine, pinus pumila
Blueberry
Serviceberry
Plum yew
Mahonia

Vines
Schisandra
Chocolate vine

Herbaceous perennials:
Tiger nuts, chuffa
Bamboo
Lemon grass

Ground covers
Strawberry
Creeping dogwood
Wintergreen
Lingonberry.
Forest sorrels

The Pre-Raphaelites

Wallpaper, William Morris, public domain

"It is only by labour that thought can be made healthy, and only by thought that labour can be made happy, and the two cannot be separated with impunity."
—*John Ruskin, philosophical founder of the Pre-Raphaelites*

Beauty in Abundance

In the era of early industrialization, government and aristocrat policies moved farmers off their land to free up labor for urban factories. The poor crowded into cities to compete with each other for underpaid factory jobs to pay for luxuries for the rich. As in Hideyoshi's time and place, the aristocracy competed to impress each other with the audacity of their conspicuous consumption.

In that context, the Pre-Raphaelite brotherhood (with some notable sisters as well) began making art that highlighted the beauty of a simple, spiritual life, and natural beauty. This return to simplicity impacted the arts, poetry, and fashion, and even music.

Inspired by this work, artisans like William Morris began making beauty accessible by making functional art like wallpaper. *They called for lives in which the thinking classes did their own labor, cleaned up their own messes, and the laboring classes had room for beauty, philosophy and recreation.* Morris' work within this Pre-Raphaelite framework is considered to have helped to establish the socialist party of England, leading to some of the most important social reforms we all enjoy today, such as weekends, sick leave, and public education. This movement even launched the **Cottage Garden** movement, the first European movement of gardening to consider ecosystem health, and social justice.

"Work," Ford Madox Brown (1863)

Hippies

Image: U.S. Department of Defense

No discussion of models for aesthetic revolution would be complete without discussing the hippies, both for the good and the bad.

In a context of rigid class and gender hierarchy reinforced by an aesthetic of whiteness and rigid conformity, the hippie movement utterly transformed society. And while the transformation was incomplete and followed by blowback, the change was lasting and meaningful.

Hippie aesthetics ushered in an era where conformity to whiteness was not so strict, where women were more free, where people could kiss across class, gender, and racial lines, where conspicuous consumption had less power and an understanding of environmentalism and a love of nature were finally acknowledged as worthwhile, important goals.

The hippies spread a critique of war, a critique of class, a critique of conformity, of religious oppression, and of mainstream white society to a much broader audience. So complete was the revolution that decades later, IT professional dads are still burning pictures of themselves in bell-bottoms so their grandkids won't know they were once hippies.

Of course, it's important to recognize there was blow-back, too. Perhaps this change wasn't lasting because it lacked the elegance of some discipline, the conscious decision to call everyone in, rather than calling people out?

Soon, those who found the hippie lifestyle unsustainable "sold out" by getting jobs, wanting to have families, wanting to have houses, and normal lives. Since there was no room in the hippie ideology of purity for these activities, most of the generation abandoned the hippie ideal for the yuppie ideal of even greater conspicuous consumption, and the Reagan/Thatcher era of excess began, with its aesthetic of flashy materials, gold chains, and yacht rock.

The Black Panther Party

"You cannot fight racism with racism. You fight it with solidarity."
—Bobby Seale

But the hippies are too controversial to talk much about these days, so let's move on and take a look at the work of the Black Panther Party.

Founded in 1966 in Oakland, California, by Huey P. Newton and Bobby Seale, the party's purpose was to patrol African American neighborhoods to protect residents from acts of police brutality. Famously, the BPP was so successful in calling attention to violence against black Americans that J. Edgar Hoover considered it "the biggest threat to America today" and brought the full force of the U.S. government, COINTELPRO (Counter-Intelligence) and its propaganda machine to put the party in place, to dismantle it, and shut it down with extreme force.

It's surprising to some people that a book on having a beautiful garden would have a section on the BPP. It's surprising to me that this is probably the first one that does. And perhaps it's awkward to have a white man writing on the topic. But honestly, I don't even know how to talk about having gardens or growing food without talking about the Black Panther Party. So, it's going to be a clumsy attempt, and there's no way to cover such a complex and controversial topic in a short number of words, so I'll ask you for some grace and benefit of the doubt as we try to generate some inspiration and thought.

First, let's face it: a lot of the "look" of modern landscaping is mostly about reinforcing class and "whiteness," right onto the neighborhoods where we live. It's a constantly changing code of visual cues that communicate who has power and privilege and who doesn't, who "belongs" and who doesn't, who "has taste," and who doesn't. And most importantly, who can afford to communicate that "taste" and who can't. It constantly changes so that those who aren't insiders (People of Color communities and working class whites) can never keep up with it. Most of the gardening industry, right down to our oversized lawns, is about selling this code to us, preying on our fear of looking poor, of not looking like we belong. The whole "suburban" look came about in an era of white flight and it's meant to communicate messages of class and race.

So if we don't think about these things when we start designing landscapes, we're likely to end up reinforcing these silent visual tools of oppression that may not match our values or intentions, or our vision of creating a world of equality for all. (And on the off chance that those words grate on your ears, I hope you'll think about why. Shouldn't we all want a world of equality?)

With the BPP, I think it's important to recognize these brilliant organizers understood all this from the beginning, and understood the power that culture would bring to building a movement. The party itself was born when it's two founders complained that in a college where a majority of the students were black, they never saw a black face in the history books. Culture was so much part of their thinking, that they had an official Minister of Culture, Emory Douglas. Under his guidance, his artwork, and the fashion of the Black Panther Party became immediately recognizable, again breaking down and confronting an oppressive aesthetic of whiteness in the culture at the time.

Artists have a way of instantly communicating essence. Things are made clear, almost like a language, and so art is a powerful tool to communicate with the community.
—Emory Douglas

Co-founder Bobby Seale has said, much of the impact of the party came from pictures of African Americans carrying guns and acting defiantly at a time when black Americans simply did not dare to stand up to white people, especially the police, under any circumstances at all. Others were inspired by a vision of black pride in a context where black children were taught to feel inferior in school. Others still were inspired by a vision for a beautiful world of freedom and solidarity that the Panther's eloquently explained, one where everyone had access to food, land, housing, medicine, and education.

I say to college students and to all people who want to see a new world and a better world that they should unite to form the type of power block that can defeat this racist power structure and put it in the garbage can of history where it belongs.
—Eldridge Cleaver, BPP Minister of Information, 1968

Secondly, I can't not talk about the Black Panthers in a book about people growing their own food gardens and having food security without acknowledging that those ideas have "deep roots in the Black Panther Party." (Cirrus Wood.) To remove this section would be erasing black accomplishments from the history of the food and environmental movements. The BPP started programs to feed children, to make "gardens in the ghetto," and food sovereignty programs that continue on today. Food was more than just about feeding people though, it was a cultural touchpoint and organizing tool. And the fact that black children were going hungry in wealthy regions highlighted that America has a human rights problem. And that message continues to resonate with food justice activists today.

...we had a lot of visuals that they (The African American community) could identify with. ...They were visual interpretations of the conditions people lived in. Inner cities, poor communities. Combined with revolutionary imagery. The people saw themselves in the artwork. They became the heroes. They could see their uncles in it. They could see their fathers or their brothers and sisters in the art" (Emory Douglas)

Despite the failure of the BPP to live up to its own vision, a failure which eventually included gender discrimination and alleged political violence against women members, we shouldn't dismiss the work of so many of its individual activists in establishing the concept of food and environmental justice in the movement today. It's a movement which spread in art, in food, and in gardens in ghettos. While the party itself ended decades ago, the vision and cultural impacts of its activists will continue to resonate deep into the future.

Some Cultural Dangers to Watch Out For on Our Adventure

As we look to transform our lives and landscapes, it's important that our actions don't have unintended negative consequences. Here are some pit-falls to watch out for:

Gentrification-caused displacement: In most cases creating beauty is a positive, but in some, it may trigger "gentrification," a process where property tax values soar with the "desirability" of the neighborhood to wealthier families. This forces out the historic residents, destroying communities and leaving people without housing access. Especially if we are new to an area, we may inadvertently impose an outsider's aesthetic that triggers gentrification and ends up destroying the culture that we moved to our neighborhoods for. This is also discussed in the section on **Terroir.** Gentrification is so important to discussions of landscape beauty that I've included a discussion and resources as Appendix A in the back of the book.

Appropriation: Since the technologies of sustainable land use come from around the world, there's a danger that we may unintentionally cause harm by not recognizing the intellectual and cultural property we are taking inspiration in. I hope that in this book I have given credit wherever due, and will be eager to correct any mistakes that have slipped passed. Wouldn't it be a shame to set out to create a beautiful space, and leave some of our guests feeling that their cultural heritage had been misrepresented or stollen? Again, I've included an Appendix B for more discussion on this important topic.

Unintentional exclusion: for me, a truly beautiful place should feel welcoming, and nurturing to all. I hope all find pleasure, comfort and friendship in my gardens. But aesthetics that imply gentrification, conspicuous consumption, class, oppressive gender roles, or appropriation may make some feel unwelcome or even threatened! Worse, we may even make our neighbors feel threatened in their own neighborhoods. The kindest compliment I have ever received in my garden was that it made all feel welcome. I was told that while many gardening ventures project whiteness, twangy guitars, and center rugged male farmers in overalls, my garden and whole project aesthetic avoided playing to those stereotypes and made all feel included.

Notes:

Meredith Roman. "The Black Panther Party and the Struggle for Human Rights." <i>Spectrum: A Journal on Black Men</i> 5, no. 1 (2016): 7-32. Accessed August 30, 2021. doi:10.2979/spectrum.5.1.02.
Cirrus Wood, East Bay food-justice movement has deep roots in Black Panther Party,
Garret Broad, The Black Panther Party, A Food Justice Story, Huffpo, Feb 2017.

Resources on Patterns for Culturally Relevant Gardens

To me, beautiful landscapes don't reinforce whiteness and class. Humans of every society and culture around the world have made gardens that communicated their values and their special relationship with nature. I hope the ideas in this book will inspire you to use culturally and spiritually relevant patterns, plants, and objects so that your landscape will connect you to your values and your ancestors. If you're reading this book and are inspired to search out more inspiration you can use in your garden, nothing would make me happier. There are too many options to mention, but I'd like to give a few resources:

1. Vaughn Sills, *Places for the Spirit: Traditional African American Gardens* Hardcover – Illustrated, August 31, 2010
2. Richard Westmacott. *African-American Gardens and Yards in the Rural South.* This book shows some patterns that are perfect for Transformative Landscapes, functional living spaces that also offered "solace." In a lot of ways, these look like Permaculture gardens or "home gardens."
3. Adriana Zavala, *Frida Kahlo's Garden.* For many people Frida is a cultural, national and feminist hero, and she was also an amazing gardener who integrated culturally meaningful plants with food plants in her landscapes.
4. *Buffalo Bird Woman's Garden: As Recounted by Maxi'diwiac* (Buffalo Bird Woman) (ca.1839-1932) of the Hidatsa Indian Tribe. By Waheenee, 1839-1932. Edited by Gilbert Livingstone Wilson, 1868-1930. For me, this book shows that hidatsa gardening was holistic, it wasn't just about growing food, but about growing a beautiful way of living.
5. Michael J. Caduto (Author), Joseph Bruchac, *Native American Gardening: Stories, Projects, and Recipes for Families* Paperback – March 1, 1996,
6. There is a whole literature on home gardens around the world, such as Home Gardens in Nepal, which could provide inspiration to people who feel a cultural connection to those places.

ART BY EMORY DOUGLAS, VIA WIKIMEDIA COMMONS

Punk, Sex, Fashion

SEX, WIKIMEDIA COMMONS

I can't not talk about punk. As a former punk kid, the politics and aesthetics of punk still shape my view of the world. My first real sustainability and Permaculture teachers were musicians Henry Rollins, Kathleen Hannah, Ian Makaye, and Jello Biafra.

Vivienne Westwood was a school teacher when her boyfriend started the shop Sex as a space for his band, The Sex Pistols, and asked Vivienne to make some clothes to sell. Vivienne took her radical politics and non-conformity and looked at it "as a way to put a spoke in the system."

Soon, her DIY aesthetic of imperfection became the look of the sex pistols and other punk bands followed. Eventually it became a revolution in the arts which has spread around the world.

Critic and publisher Leonard Koren has compared punk to wabi-sabi in its rejection of conspicuous consumption, and focus on "realness." It's an aesthetic of the way things really are.

Of course, I don't want to make it seem like Vivienne accomplished this all alone. She had inspirations, and co-creators, just as Rikyu built on traditions of Zen philosophy and the work of Japanese artisans. Transformation like this is never a solo operation, it's always co-evolution. Vivienne found inspiration in an already emerging subculture, and she joined in and added her creativity and energy. Again, what started with sharing DIY art in a small shop has spread around the world.

Notes: Vivienne Westwood, *climate revolution manifesto:* https://climaterevolution.co.uk/wp/The-Vivienne-Foundation-Manifesto/mobile/index.html

Plants for a Punk/Goth Edible Forest Garden

While we're talking about punk, one of my favorite garden designs was plants for a goth forest garden.

This set of plants and semi-stable guilds and useful ornamental oddities is designed to make all of Siouxie Sioux's Banshees scream at once and send Robert Smith back to the "Hanging Garden."

Black Star" Plant Choices

Woody Perennials:
Contorted mulberry,
Weeping rosa plum
Black lace elderberry
Black-tipped pussy willow
Black diamond apple

Perennial Herbs and Vegetables:
Black perennial kale
Richmond tree collards
Chocolate joe pye weed
Silver buckler leaved sorrel
Blood veined sorrel
Black daylily
Black alliums
Dragon's eye knotweed
Black columbine
Chocolate sage

Edible Annuals
Redboor kale
Merlot lettuce

Ornamentals
tulips, and lilies.
Light pink yarrow

Ground covers
Black and purple leaved sweet potatoes.

Guild Matrix:
Tropical, black sage
Walking onions, looking like little allium Medusas

Beauty in Abundance

What's Next? YOU DECIDE!

Cottage Core? Permacore? Solar Punk? What world awaits?

Just as we may have problems of gardening to solve, such what to plant and how, we still have artistic work to do in refining ourselves and our gardens.

While there are many great gardening systems, the aesthetic of cottage core (an internet phenomenon) has connected with young more people than any of them. A simple, perfect photograph of a garden can invoke a dream for hundreds of thousands of people—if not millions—around the world.

A single garden can include the blueprint for an entire ecological civilization. What kind of world do you want to see? How does that world start in YOUR landscape?

Is a beautiful landscape one based on corporate products bought at big box stores? Or does true beauty grow sustainably from the land? Can a garden invoke a post-apocalyptic paradise, a world that has recovered from insanity? Can a garden teach people that we could live like elves, sustainably among the trees? Can a garden help us envision a world of justice? Can we finally envision and learn to believe we deserve a sustainable, enlightened society, be experiencing it in a food forest?

A Garden Revolution?

I have often thought of Vivienne Westwood in her shop, of Rikyu in his rustic tea shack, while sharing a simple, direct experience of a beautiful world, sitting with guests, drinking tea surrounded by natural beauty, birdsong, fragrant flowers, and amazing food in my home forest garden.

Many guests have written to me that the experience of sitting in my garden for a few moments has changed their lives forever. Plants from my garden now grow in hundreds of gardens around the world. The guilds that I have designed and tested now grow in gardens around the world. The recipes I have made of sustainably grown produce are eaten at tables around the world.

If we can get this seed syllable right, if we can imbue our gardens with the core of a just and sustainable world, if we can perfect ourselves in these landscapes, and perfect our vision for what this world could be—what will we find when we wake up the next morning?

Lillie House, an urban food forest farm serves as a major case-study for this book. Many of the pictures in this book are from this project. It was created by myself and Kim Willis, with significant contributions by Melissa Malloy

Chapter 4: Designing and Transforming the Landscape

(How to use this book)

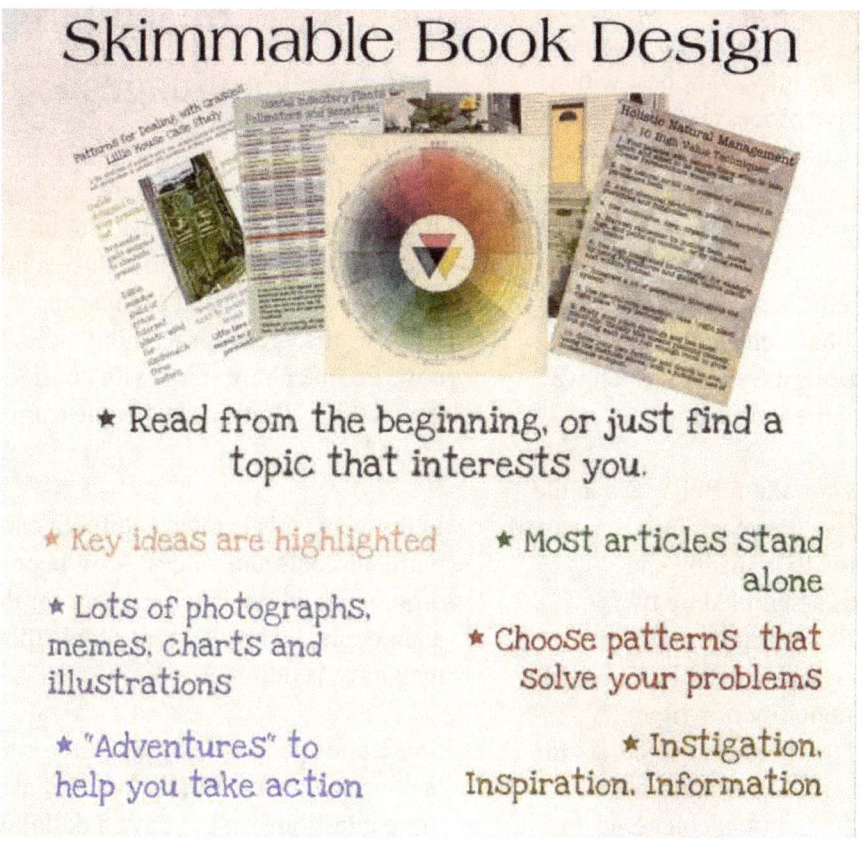

Hopefully you're fired up with the inspiration to transform your yard into a beautiful and abundant paradise. Great! Now, the problem many of us run into is: how do I actually do it, and most importantly how and where do I begin? Not surprisingly, this is where a lot of projects get stuck!

To get clear about how to get it all done, let's understand first that there are two basic strategies for how to proceed and where to begin. A lot of problems are caused by getting stuck in one mode or the other, as we'll see.

To keep things simple, let's say these two modes are **design** and **transformation**.

Design means we create a thoughtful plan up front, usually on paper. But even an idea in our heads is still a design. Transformation, on the other hand, means we go play in the garden, making useful, long-lasting improvements on what we have, until over time the landscape is gradually transformed.

Almost every good garden, farm or landscape is a mix of these two approaches, and that mix depends on the project and the temperaments of the people involved. Some people are just natural design thinkers who want to have a plan, while others prefer to fly by the seat of their pants and work more instinctively.

And, some types of projects are better suited to a careful up-front design, and others to a more transformative approach. For example, some home gardens may work best with a very transformative process that strikes while the iron of inspiration is hot. Professional spaces probably will benefit from more design to keep clients informed and happy.

The best method for designing sustainable landscapes that I have encountered is the **Permaculture Design System,** which we will introduce a little in this book.

As author/teacher Graham Bells says in the Permaculture Way, "Permaculture is a way of arranging your life to be happy and abundant." It was assembled by two Australians David Holmgren and Bill Mollison, and further evolved by a growing international community of writers, consultants, and teachers ever since. As for myself, I would almost certainly not be gardening still if I hadn't encountered Permaculture design, as the conventional approaches I was raised with simply did not make sense to me.

The Permaculture design system is so life-changing and powerful, that like most really impactful things it has gone on to inspire a movement, a community and a way of thinking, and in some cases, a global circus or even a religion or "worldview."

But in this book, we'll be using the word "Permaculture" (with a capital P) to refer to the original meaning, the *name* of a specific formal system for design. In the words of David Holmgren, "Permaculture is protracted thought and observation instead of protracted labor." It means we take the time to get things right before we begin making changes.

Design, Permaculture, and Pattern Languages

Permaculture has been covered in depth in other places, so we don't need to go into too much detail here. We'll introduce a little along the way. The end of this chapter includes resources for learning more. But for now, Permaculture starts with considering three ethics "Earth Care, People Care, Fair Share."

In my view, this is most helpful to check in with our goals and values. Most people probably think their actions care for the earth, and people, and well, most people think what they have is quite fair!

Sometimes people find these three ethics lacking in nuance. But the point of asking these questions isn't to have a definitive set of universal ethics, the point is to having something simple enough to get everyone involved in a project talking about these deepest values, and what they think it means to value them.

Next, Permaculture has sets of principles for getting clear about our goals, themes in design, methods in design, and a detailed design process we can follow.

For the purpose of this book, I'll refer to some of the principles as we go, including this one I consider the **uber-principle**. In my mind, most of the other principles reduce to this one, so it's good simple principle to keep in mind as we go about our work.

Solve problems by creating more skillful connection to nature and community.

Virtually everything we do as Permaculturists amounts to investing in connection to nature and community, since most of what we are doing is repairing the damage done to these connections by corporations, capitalism, and colonialism in the search for profit and power.

Of course, having lists of principles as thought tools is still very helpful, so I encourage you to explore the resources at the end of this chapter. But I'd like to keep things in this book as simple as possible, so as a brief check-in, this mantra seems to work well: **am I solving problems by creating more skillful connection to nature and community? Or am I moving in the other direction?** Am I solving them instead with corporate products, poisons, exploitation, or excess labor that come between our direct connection to nature and each other?

Finally, and most importantly for us, Permaculture is a pattern-based design system, meaning we look for useful patterns to solve our problems. Therefore, Permaculture has many "**pattern languages**" (libraries of useful patterns, ideas we can use like "raised bed garden," and "rain barrel") for designing different aspects of sustainable human settlements. You could say this book includes a pattern language for creating productive landscapes that are beautiful, easy, and ethical.

You May Notice this is a Very Strange Gardening Book

While most gardening books stick to the practicalities of how plants grow, how to fend off weeds and which plants to use where, we will be going to some strange and perhaps uncomfortable places in this section of this book.

These are the places that we students of art, creativity, and design have always traveled. Creativity is necessarily a strange, intuitive, mysterious, perhaps divine process. And so for those of us who have studied creative fields formally, these strange intuitive explorations usually provoke the most interesting discussions and the most growth. These discussions were certainly the most transformative to me as a designer and gardener.

And yet, they are almost never discussed in gardening books, even those having to do with aesthetics. Even Permaculturists like David Holmgren and Dan Palmer have suggested that Permaculture, a design field, should discuss design much more than it does.

And so, we will go to those places that creatives dare go. For those who wish to be professional designers, I believe these discussions are absolutely vital and overdue. But even for those who just want to make their own landscape more beautiful, I believe there is great value here. These design topics have given me a greater appreciation of creative works in fields beyond my own. I hope that even a skim of some of these topics will give you a greater appreciation for the genius of great places and landscapes.

Pattern Languages and Genius

The Pattern Language concept was created by Christopher Alexander in his book A Pattern Language as a way to make complex subjects easy to understand across disciplines. Alexander is an architect who wanted to make building more accessible to people without having to spend 10 years studying architecture. His solution was to create a library of useful patterns that any builder could look at, and choose the ones that fit the project.

For the purpose of a Pattern Language, **a "pattern" is anything that can be shared and replicated**. So useful patterns for gardening include "raised beds, rain barrels, mulch, drip irrigation," etc. These are simple, replicable ideas we can understand and apply to our own gardens. This book will include many patterns for a more beautiful, meaningful, and transformative landscape.

Pattern languages are powerful design tools. One study found that even a brief exposure to a pattern language for design dramatically improved the designs of amateur designers. So even just skimming through the patterns in this book is likely to help you create a more beautiful abundant landscape.

Pattern Languages help us organize a lot of information into a small space, so people who spend time with them can seem to develop an almost encyclopedic knowledge of useful information! This can be very different than the sort of random information we may accumulate by simply watching a lot of Youtube videos, or social media posts for example.

Pattern thinking makes us smarter, helps us identify what is actually important and useful information, and makes us more capable. Pattern thinking helps us organize, recall, and communicate large amounts of categorical information, not just "how to do stuff," but the truly important part: what to do, where and when to do it, and why. In other words, *what is wise*. People are often startled by the apparent genius of experienced Permaculture designers like Bill Mollison and Geoff Lawton, who seem to carry around encyclopedic knowledge on Permaculture topics. This is almost certainly from Permaculture training on pattern thinking. I know several non-famous Permaculture designers in my county who wield the same genius from their training. These professionals have spent a great deal of time talking to people about their problems, then exploring patterns to help solve them.

Featured Pattern: The garden growing wild

There's one problem that is at the core of every page of this book, which is this:

Can we have landscapes that are natural spaces but that also meet human needs for food, social space, meaning and beauty? Can we take advantage of the self-maintaining ease of wild nature without making the neighbors call code enforcement?

Yes, we can, if we have the right kind of garden. Christopher Alexander proposed a solution in A Pattern Language with his pattern the **garden growing wild.**

> *(Modern gardens) have none of the quality which brings a garden to life - the quality of a wilderness, tamed, still wild, but cultivated enough to be in harmony with the buildings which surround it and the people who move in it. This balance of wilderness and cultivation reached a high point in the oldest English gardens. (See* **cottage gardens***)*
>
> *In these gardens things are arranged so that the natural processes which come into being will maintain the condition of the garden and not degrade it. For example, mosses*

This is almost the opposite of what many of us who want healthier landscapes tend to think, which is to just let the landscape be wild and to figure out how accept and to make use of it. If we applied that approach to painting, it would be like leaving a canvas on the forest floor for a month and calling whatever showed up "a painting." Now there's absolutely a place for that, but if that were the only approach to painting in a museum, it would probably get a little boring and we'd miss opportunities for communication and expression.

"The garden growing wild" is a garden so well-designed and co-evolved that each year it grows into a high state of usefulness and beauty with little help. If it were painting, it would be like teaching microbes how to naturally grow the Mona Lisa in a petri dish. You just give them food and they pop out a whole museum's worth of master paintings. In the first approach we are considered separate from "nature," in the second we are a part of nature, we co-evolve as part of it.

With gentle, minimal tweaking, a garden filled with beauty and abundance just… grows, as naturally as any forest. But it will take a specific approach! We can't expect nature to grow a suburban lawn and rows of plants lined up neatly behind landscaping blocks.

So, that is what this book is all about. Every pattern in this book is about how to achieve this "garden growing wild."

Transformation through Adventure

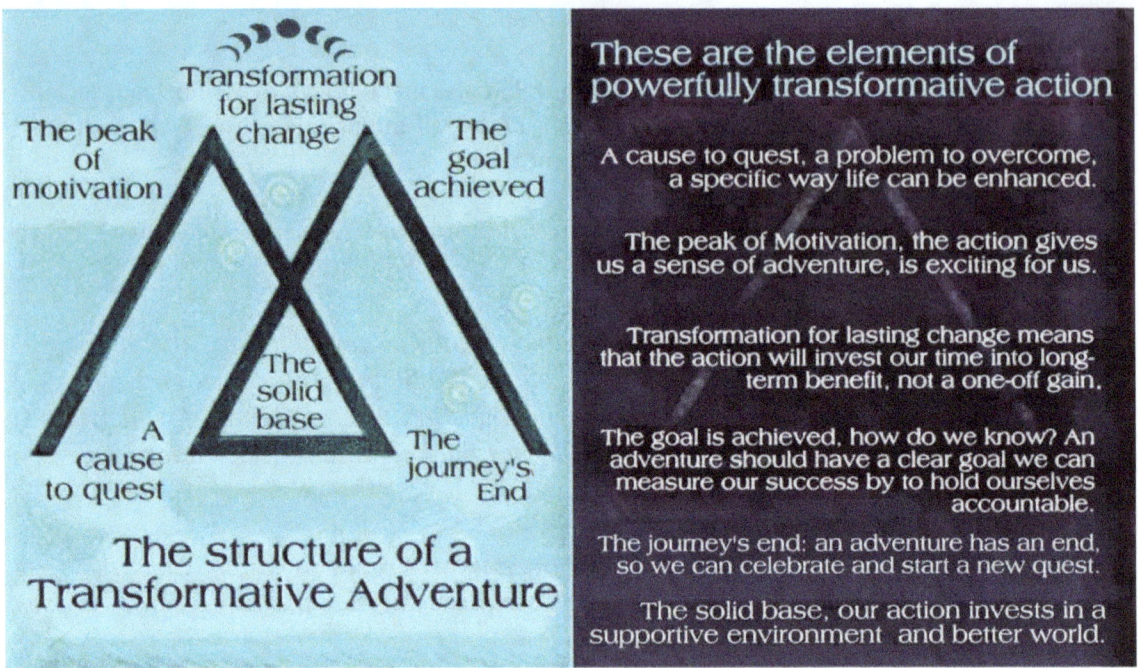

The six elements of a well-designed transformative action.

Our other mode of action is *transformation*. In this book, "transformation" means a change from one state of stability to a new state. For example, a manicured lawn is actually a very disorganized system. It doesn't have enough pieces to maintain itself, and so it depends on our constant work. If we stop mowing it, it stops being a manicured lawn. If we put in an annual tillage vegetable garden, we still haven't *transformed* it from this state. If we stop working it reverts to a weedy lawn. Transforming it means we put energy into organizing it, into a forest garden, for example, so that in this new well-organized state it mostly maintains itself. Now if we walk away and stop working it may lose some productivity, but it remains a forest garden and will keep producing. Transformation is lasting change. It's like money in the bank. We can put this same idea to work in every aspect of our lives.

After a decade and a half of work as a Permaculture activist, consultant, teacher, and designer, I began to notice a few important trends. First, that Permaculture often completely transformed people's mindsets. Especially the Permaculture design certificate courses I was teaching.

And second, that too often, it (disappointingly) failed to significantly transform people's landscapes! Many of my clients have been experienced gardeners and Permaculturists who were "stuck" in their projects despite their knowledge and studies.

This is similar to what Christopher Alexander discovered as the result of his book "A Pattern Language." People using his Pattern Language were still not creating architecture at the level he had hoped for.

In his next book, The Timeless Way of Building, Alexander explored why it was that humans in traditional societies could naturally evolve amazing places and architecture—the kind of places we like to visit on vacation—and yet armed with the concept of design, modern Westerners were not creating these sorts of places. Dan Palmer of the **Making Permaculture Stronger** project has also written on this, and has gone so far as to compare design to an inappropriate tree graft that didn't take, an imported set of tools from industrial fields with dissimilar values to Permaculture.

In my view, some people are just natural design thinkers. But for others, a design process turns into perfection anxiety and deadlock. For Alexander, the trouble was that great places evolved, naturally, through human interactions with place. Transformation. And that is a very different process than the plodding method of design and following instructions.

Just as the snail grows its shell and the beaver creates a dam, humans transform places through the act of naturally humaning.

Inspired to balance design with an action-based approach to transforming landscapes, in 2015 Kim Willis, co-manager of Lillie House, and I began offering programs based on experiential learning—actually taking action to solve problems, build real wealth, and transform landscapes. This was approach developed from my experience as a pedagog creating effective workforce training programs. Immediately, these *transformative Adventures* began to have a real impact. While students often take PDCs and never do anything in the garden, or work as Wwoofers (garden volunteers) and never learn how to transform landscapes, Adventurers in a TA program take monthly steps, build real plant libraries, and create their first model guilds that can spread around their yards. By the end of the program, a real process of transformation has begun, and the main tools for transformation have already been practiced and absorbed. Already, many Permaculturists have taken these programs and replicated them successfully in their own bioregions, creating beautiful landscapes, livelihoods, and relationships as they do. My hope is that this book will act as a program of Transformative Adventures for you.

Let's face it, some of us would rather skip the designing and go straight to the adventure of *doing*. And even for those who are designers, the PDC too often graduates activists who know what to do, but not how to get it done. Or if does, it gets done with brute force and hard, hard work that doesn't look much like "the easier way of gardening" we talk about. Following a combined approach of design and transformation, this book includes many patterns for design. But it also includes a series of Transformative Adventures which I believe will help profoundly transform your landscape, the way you see design, and the way you interact with your spaces.

If Permaculture helps us design better lives and a better world, Transformative Adventures help us make it a reality. If pattern languages make us genius problem-solvers, experiential action learning makes us highly skilled, experienced, and effective doers. Taking direct, short-term, adventurous action is key to lasting change in our lives: to establishing habits, learning foundational skills, finding discipline, connecting with motivation, and holding ourselves accountable for achieving our goals.

In the end, we will weave these two approaches of design and transformation together, giving everyone the best tools for their particular sites, projects, and personalities.

The research-based approaches and strategies of Transformative Adventures is built into this book. But for more information on Transformative Adventures, visit TransformativeAdventures.org.

Notes:

1. Christopher Alexander, Sara Ishikawa, Murray Silverstein, *A Pattern Language: Towns, Buildings, Constructions.* Oxford, 1977.
2. Christopher Alexander, *A Timeless Way of Building.* Oxford, 1977.

Anatomy of a pattern

As you read, you may notice that the patterns in this book usually follow a basic form, as recommended by Christopher Alexander. Since I want you thinking in patterns, it's helpful to know a little about them.

They start with **naming** the pattern and showing an example.

Next is the most important part, identifying **a problem to solve**.

Next, we use a few paragraphs clarifying the **context** with a few paragraphs on the major issues involved, and then state **the pattern that solves the problem**.

Following the solution, there will be some **discussion,** followed by other **related patterns**, and finally **citations and resources**.

The Permaculture Design Process

The best Permaculture is not about the right or wrong ways to do things. Bill Mollison won't rise from the grave and send "the Ghost of Soil-Loss Past" to rattle chains at you if you row crop your garden or till.

It's about our goals and designing to meet them.

This is actually a pretty transformative mindset! Most often we never really do consider our goals in a deep way.

So for example, what many people want is to escape the rat race and they hope they can make a living off the landscape, or at least reduce their costs. If they thought about it, they'd say their goals were things like:

I want to reconnect with nature, plants and animals.
I want to spend my days in a beautiful natural environment.
I want to have more time for friends, family and community.
I want to eat better, fresher, safer food and have better health.
I want to do something good for the planet, and Industrial Ag is horrible.
I want a simpler life with less stress, conflict, and pressure.
I want to cut free of the corporate system.

But rather than get clear about our real goals, we get lost in our preconceived notions, and get stuck on "farming" for example. We start learning about farming and take a $1,000 "profitable farming" course that promises we can "make $1,000,000 on an acre!!!" (Yes, there's a famous course that promises that in its ads. This is not a realistic goal.)

We get a loan to buy our teacher's brand name rototiller, tear up our acre of lawn, and row crop it. Inevitably, reality sets in:

Instead of connecting with nature, we spend all day at war with nature, killing any plant or animal that happens into the system, spraying poisons on the insects.

Instead of a relaxing natural environment, we are working with noisy, bad smelling machinery.
Instead of safer food, we are using lots of plastics.

Instead of more time, we have less, and the "profitable farming" guru keeps saying we just have to work even more hours to be successful. Like most farmers, we are working far longer hours than folks in the corporate rat race.
Like most other market farmers, we eat Pizza Hut all growing season, because we are too busy and tired after the long work hours to turn fresh veggies into meals.

Instead of feeling healthier, we feel worn down, like most farmers do, from long hours of repetitive labor in harsh conditions.

So we hire some cheap labor and get Wwoofers to pick up some slack and get back our time. Now we have conflicts and stress, and have to fire people. It seems people being paid less than minimum wage are unreliable and unhappy.

The million dollars hasn't appeared. Without rock-star status, customers won't pay the 4 times market prices for our produce. And unlike the celebrity gurus, we don't have an army of 100 unpaid interns who want to use our name to sell their own profitable farming e-courses.

And we're still doing as much sales and paperwork as we did in the corporate world.

Another study comes out showing that because of increased fossil fuel, fertilizer, and plastics use, small-scale intensive farming has a higher ecological foot print than industrial farming, and we don't even want to think about that!

This is actually a really common story.

If we had connected with our values to begin with, and done a good design phase, everything might have been different.

We could have started with a design to create a beautiful natural environment with space for flowers and native plants and wildlife all around them.

We could have built a small 10,000 foot no till garden and found we had the same overall productivity working a few hours a week with no wwoofers. It wouldn't have been optimized for the market, but the hourly wage for the work would have been far higher just harvesting for the family and selling the produce to immediate friends.

With our extra time not spent managing machines and labor, we start cooking amazing farm-fresh meals of exotic ingredients, which first turns into a value-added business, and later into a farm dinner and catering business fueled by all the excess produce.

Guests are drawn to come have amazing meals in this beautiful environment, with super fresh ingredients. A neighbor was an ex-publisher and another was an artist, and together you write and publish a cook book....

The garden requires no fossil fuel, few plastics, and no exploited labor. This is a landscape designed to meet our real goals.

At its best, that is what Permaculture helps us achieve.

"The greatest change we need to make is from consumption to production, even if on a small scale, in our own gardens. If only 10% of us do this, there is enough for everyone. Hence the futility of revolutionaries who have no gardens, who depend on the very system they attack, and who produce words and bullets, not food and shelter."
— *Bill Mollison*

Many of the author's other projects feature in this book as models, such as this small urban transformative landscape in Indiana.

Beauty in Abundance

What are Realistic Expectations for your Project?

On a few hours of work per week (2-5 hours on average over the season) we should be able to:

1. Grow a huge amount of food, and feel very food secure in year one.
2. Significantly reduce grocery bills by more than 10%.
3. With 1/3rd acre or more of decent soil, water and sun, we should be able to grow a complete diet, or better yet, grow extra high-value produce to share.
4. If desired, earn $7,000-12,000 U.S. dollars/year on top of home food production, by year three.
5. By year three we should be able to transform 90% of nearly any sized property into systems that will increase biodiversity, grow soil, sequester carbon and infiltrate water without further inputs, and then maintain that all on a few hours per week.
6. By year three you should be able to achieve the Transformative Landscape criteria in this book, making your property a truly beyond-sustainable landscape.
7. It doesn't require a lot of capital, you should be able to run a profit in year one and every year after.

Yes, the internet today is filled with unrealistic sales pitches about farming and gardening. No, I don't think it's realistic to expect to make $1,000,000/year on ¼ acre. I wouldn't be confident that you can even have a successful market garden business. Realistic farm profitability will depend a lot on your soil and local circumstances.

So what are actual realistic expectations for everyone?

I'm struck by these benchmarks Mollison gave on how to assess good landscape transformation:
Mollison's Benchmarks for design success (used for developing goals)
"These are the test of good design."

In a short time, 3-6 years, a successful design should yield:

1. A significant ability to reduce spending and save money (**10% pledge, FREE Benchmarks**)
2. Conservation and regeneration of soil, buildings, infrastructure, biodiversity, water.
3. Sustainable food production without further imports (after year three.)
4. A unique, preferably essential, marketable product for the region.
5. Viable income streams for residents if desired (a very realistic updated benchmark is $10-15k in part time production income.*)
6. Sound, safe, legal status for all residents. (No FBI Raids!)
7. A beautiful, harmonious and productive landscape without poisons or wastes.
8. Become a cooperative, information-rich part of the regional community.

Are these realistic goals for a project? With careful, explicit design, I'm proud to confidently say that Lillie House achieved these goals mostly by year three.

Even in year one the project achieved significant documented reductions in energy consumption, going from the bottom 20% to the top 5% in energy use in our region (according to consumers energy, documented at Lilliehouse.blogspot.com.) We grew a huge amount of food, including a complete hypothetical diet for 2, and even achieved a profit in year one.

By year one we had dramatically increased biodiversity, carbon sequestration, and soil conservation. By year three we were infiltrating most of the water that falls on the site, and we're becoming renown as a biodiversity hotspot and source of hundreds of rare native and heirloom edible plants. We used project income to make significant improvements in infrastructure and house.

Realistic Benchmarks for Landscape Transformation:

But what are realistic expectations for the work and rewards of transforming our landscapes? Some insist that transforming an acre will take 60 hours/week or more for the first decade, and that we shouldn't expect any yields until it's established. I think this is begging for failure. I know for a fact we can accomplish much more on much less work.

What can we accomplish without over-sacrificing? What can the average person who's not a gardening expert expect to get done on a few hours of work per week?

By year three we had reached full install and stopped importing materials.

By year three we had tested income streams and found a unique marketable stacked product for our region, the Landscape Transformation program. A replicable way for projects to find viable right livelihoods.

By year three we were beating the production income benchmarks with our design.

So *check*! I am still amazed by the beauty of this landscape! And the production it achieves on a few hours per week.

And my Permaculture design has certainly allowed me to grow very rewarding relationships in our region, perhaps more than I could have guessed. So yes, I believe these are realistic benchmarks for everyone.

These benchmarks are achievable in almost every climate on almost every piece of land at any scale. In fact, they are achievable if you don't own land.

Again, all this is achievable on an average of a few hours of work per week. I have seen people in many climates and many regions of the world achieve these benchmarks. In fact, these benchmarks are based on research of global human societies and what was done by people all over the world for all of human existence. My benchmarks here look similar to those given by other expert sources like Bill Mollison, John Jeavons and Peter Bane.

And yet, a strange phenomenon is that there are many people who believe it is realistic to make $1,000,000 on ¼ acre *if only they work hard enough* but think my benchmark of managing a landscape productively on a few hours/week is an unachievable fantasy.

My advice is these benchmarks are achievable, but there are certain caveats.

1. It takes the right goals. If the goal is to make $1,000,000 on a market garden, you're not going to be able to do that on a few hours/week, or 40 hours/week, or 100. Labor is known to follow the law of diminishing returns in agriculture, so I say you can realistically earn $7-12,000 a year on a few hours/week. But earning $20k on 40 hours per week is honestly tough and will require a lot of luck and probably volunteers. Other people say "I like to work hard!" If you just want to work hard for 48 hours/week, then that's perfectly fine! You don't NEED to reduce your work to a few hours/week. I'm just saying you can achieve these benchmarks without 48 hours/week. And you can achieve the same level of regeneration (soil carbon, biodiversity, water infiltration, pollution and waste reduction) on a few hours a week as you can with 48 in almost every case.

2. It takes a "Landscape Transformation" approach. That means investing those few hours/week in setting up perennial, low-maintenance systems. If you're putting too much time into annual gardening or tilling, that will all be extra work, that work will never go down, and probably won't increase yield much.

3. It takes experiential learning. There's a learning curve. So we need to start with projects that will help us learn what we need to know. That's built into the approach in this book. The adventures in this book will jumpstart your learning process while you transform the landscape.

4. It takes the right systems. If you've got 30 acres and you want 10 to be a market garden and 20 to be mown lawn, you won't get that down to a few hours work per week. You'll need to figure out how to put Mother Nature to work by finding the right kind of ecosystem to transition that land into.

A beautiful and highly diverse garden of annuals, and perennials by Michael Wardle, of Soil Savour Permaculture. This is an excellent example of a true transformative landscape that is beautiful as well as abundant.

PLENTY: A Simple Design and Transformation Process for Everyone

You don't need to be an artist or engineer. A simple pencil and paper sketch is plenty good enough.

Again, some people and projects will do best with a detailed formal Permaculture design process. I have seen many processes, but for me, the best I have seen is still the one Bill Mollison laid out for his students, based on his wisdom and experience. You can find a version of that on TransformativeAdventures.org.

But for those of us who just want to transform our landscapes, I offer a super-simplified design phase that looks like the way most people actually do it. This is also good for true professionals wishing to accomplish important things, but who just aren't design thinkers and who are more intuitive in their approach. This streamlined version should get many of the benefits of design without so much paper pushing.

The "Good Enough" Design Phase: Designing for "**PLENTY**"

1. **P**olish the mirror. Do your internal prep work. Make sure your health, mind, body, and family are prioritized first so you can see the project clearly in your mind.
2. **L**ook for inspiration, especially for big picture **meta-patterns.** Want to create a regenerative farm? Find models. Want a beautiful, abundant yard? Find an example to start with. This is the time for playing with the Permaculture principles or other design tools.
3. **E**lementary patterns. Make sure you "sweat the big stuff." Do extra thinking on siting major elements: water, major production areas, zone 1, social functions, animal systems, etc. before jumping into just changing things.
4. **N**eeds and goals. What steps will help establish a **Solid Base**? What are your most important goals?

5. **Themes.** Get clear on the big thematic things, get the aesthetics, production goals, and time commitment clear in your mind.
6. **Yields.** Let yields guide you on where to begin and how to proceed. I suggest beginning with the projects that will have the biggest real yields, the biggest positive impact on freeing up time and establishing your solid base.

Now that we've done "PLENTY" of design, we can feel free to start transforming the landscape a bit at a time. **Transformation** means that with each project, we're making sure it's sustainable and well designed. For example, so that it has built in fertility patterns and water collection. With a Landscape Transformation approach, we're looking for high-value, life and landscape-enhancing tasks we can accomplish with a durable, lasting impact.

As you can see, if you accomplish even half of these adventures, you will have utterly transformed the landscape in a lasting way.

Each task is undertaken with a sense of adventure, a sense of exploration, a sense of learning, and careful thought about self-sufficiency, sustainability, and the relationship to the whole. In this way, the landscape evolves one piece at a time. As it does, our wisdom and experience grow. Perhaps we shall make some mistakes, but some of these may be lucky accidents and nothing is perfect. And the adventures are generally arranged in an order where you won't be making permanent mistakes early on as you learn. You'll be building experience and wisdom before the stakes get too high.

Beginning Adventures to Transform your Landscape (most are in this book):

1. Start foraging by foraging your first meal. I always begin teaching gardening by foraging. Foraging changes your entire relationship with nature and the way you garden.
2. Commit to eating something foraged once per week all season long.
3. Create an **herb spiral,** or "herbs and greens" garden. Beginners are better off starting with perennial gardens instead of conventional annual vegetable gardens. Perennials teach us more, and are easier, meaning we'll be more likely to have successes we can build on.
4. Plant a fruit tree or transform a tree into a guild.
5. Design and install your first **mixed annual/perennial guild**.
6. Create a **scatter-sown dynamic polyculture garden.**
7. Create an **edible meadow**.
8. Transform a flower bed into an edible ornamental guild from this book.
9. Transform an annual veggie garden with the **annual garden make-over guild**, or start some **veggie circles**.
10. Create a **hedgerow**.

To learn more about these beginning adventures, visit *TransformativeAdventures.com*

Quick PLENTY Adventure Checklist:

(This checklist is based on the six key points of a Transformative Adventure. It can apply to a whole site project or an individual project like a guild or hedgerow.)

____ It's a good time to do this particular project, for me and everyone else.
____ I've looked at multiple patterns for inspiration, (not just one!)
____ I have thought about the major elements of the project.
____ I know where the water will come from.
____ I know where the fertility will come from.
____ I know how it will be maintained.
____ I know how any other needs for the project will be taken care of.
____ I am crystal clear about my own needs and goals are for this particular project and how much maintenance time I want to spend on it.
____ I am clear about what I want it to look like, feel like, what I want it to produce, how I want it to work, and how it fits with the rest of the landscape and the people involved.
____ I know what the costs and yields are, and I think this is the highest priority. No higher priority is likely to get in the way.
____ I feel fired up about this project!

Some Resources on Permaculture

It's hard for me to limit my recommendations to such a short list. I'll recommend more books throughout this book and stick to my favorite introductions here. A complete list of my recommended resources is available at TransformativeAdventures.org

1. PermaculturePrinciples.org. Learn about the Holmgren Principles.
2. PermaculturePrinciples.com. Learn the Mollison Principles.
3. Gaia's Garden, Toby Hemenway. A classic for home-scale Permaculture.
4. The Permaculture Handbook, Peter Bane. I recommend it for homesteaders.
5. Practical Permaculture, Jessi Bloom and Dave Boehnlein. A good introduction.
6. The Edible Ecosystem Solution, Zack Loeks. Beautiful art, compelling read.
7. Retrosuburbia, David Holmgren. A complete book on Permaculture.
8. Permaculturevisions.org. One of my favorite pages for learning about Permaculture.
9. The Permaculture Way, Graham Bell. I like this book for its simple explanations of the complete philosophy.
10. The Vegan Book of Permaculture, Graham Burnett. While vegans can build a complete veganic system with this book, Graham covers the subject in depth.
11. Permaculture Magazine. A magazine on Permaculture topics.
12. Permaculture Designer Magazine. Another magazine on Permaculture.

Chapter 5: How a Sense of Play Builds a Better Life and a Better World

Transforming the landscape gives us many opportunities to act creatively together with friends, family, and neighbors.

The most empowered way to respond to today's crisis of chaos is to see it as an opportunity: the unique adventure of our time is to transcend the delusions and systems that cause the chaos. It is not to respond with a dis-empowered, defeatist attitude of crisis, but with one of empowered, adventurous play.

Remember our twin tipis of personal and societal development: The path to personal growth involves nurturing a supportive environment and community, and the path to a just and sustainable culture requires us to grow as people. This path also links with our creative growth.

There is now a big body of research to back that up. Traits of "openness," including creative thinking, problem solving, sensitivity to our own emotions, aesthetic sensitivity, and curiosity all correlate strongly with empathy for beings outside our "in-group." That correlates to tolerance, and caring about people, the earth, and fairness. And not surprisingly, empathy makes us more creative and better problem solvers.

Meanwhile, an attitude of crisis keeps us locked into a scarcity mindset where we make bad, selfish decisions that prioritize short-term gains over long-term benefits. That scarcity mindset has also been found to make us less creative. Shifting to an empowered view of creative play helps us embody an abundance mindset where we'll make better long-term decisions that are less selfish.

Like our twin tipis, play works from both ends at once. In pattern play, we practice reaching out with humility to find different perspectives, experiences and answers. We practice finding the connections between things that is the hallmark of creative thinking. As we grow in this way we grow in empathy.

We understand ever more deeply that the best solutions are always holistic, that our well-being depends upon our community and environment. As we grow in empathy we grow as individuals.

It is no wonder traditional societies prioritized so much time for beauty, art, culture, ritual, design, symbolism, connection to everyday sacredness as paths to personal development. Thus we create a society of caring, empathic people.

By contrast, the standard Western approach emphasizes work, denigrates "wasting time" with creative play and beauty, promoting individuals who lack empathy for "other" beings, lack the creativity and openness to imagine a better world, and respond with revulsion when exposed to new ideas and ways of thinking.

So encouraging people to play creatively in the garden helps build a more empathic society. So, isn't it time we "roll down our sleeves and get to play?"

Notes:

1. Scott Barry Kaufman and Carolyn Gregoire, *Wired to Create*, TarcherPerigee, 2015.
2. David Greenberg et al, Personality predicts musical sophistication, *Journal of Research in Personality, 58,* 2015.

Chapter 6: The Permission to Get Weird

Painting, Rebecca Stockert. The Lunamoth has long been a symbol of transformation, and the feminine energy of the moon.

"Your guitar is not really a guitar. Your guitar is a divining rod. Use it to find spirits in the other world and bring them over. A guitar is also a fishing rod. If you're good, you'll land a big one."
—"Captain Beefheart," Don Van Vliet, American singer-songwriter

There is one aspect of the creative process, whether we're talking about design or transformation that rarely seems to get mentioned in the Permaculture world. And, most creative people I have met would call it the most important factor. This is where this book is going to get a little... *weird*.

Having studied creative arts formally in university programs and arts communities, I have found this was one of the most discussed topics.

Yet, this element is rarely discussed in landscape design, perhaps because we want it to be seen as serious business, not something exclusively for hippies. But I don't mind being seen as a weirdo, so I will share with you some of the experience of creative mind that was shared with me by my teachers in my creative traditions. Across traditions, wherever people are creative, you will find these same ideas repeated. In many ways, artistic traditions are less about the technique of making things than they are about cultivating the state of mind (sometimes called "flow" by scientists) of creation.

So, we may worry that it will be off-putting to "serious people," but you can't get around it, there is necessarily something mysterious in the process of creative design. Captain Beefheart's "Rules for Guitar" (quoted above) captures this mystery well. He goes on to say, *"If you're guilty of thinking, you're out. If your brain is part of the process, you're missing it. You should play like a drowning man, struggling to reach shore. If you can trap that feeling, then you have something that is fur bearing."* When we create, we can not simply think of the answer, the answer doesn't already exist! So, for many, it feels as if we are reaching to something, getting the solution from somewhere else.

"The true sign of intelligence is not knowledge, it is imagination." "Creativity is seeing what others see and thinking what no one has ever thought."

-Albert Einstein

At the moment of creation, we accomplish something more than what we have ourselves experienced. We transcend all that we are, and we become something more than we were. In some ways, the brain must misfire, there must be some wires that get crossed and things put together that had never been together before. Novelists often say they feel their stories come from somewhere else, another dimension perhaps. Sculptors may say they felt the finished works were already there in the stone, and all they had to do was carve away the extra pieces.

Creating something amazing, including a farm or home landscape, is never just a matter of knowledge or knowing the answers and best practices and putting them together "the right" way. There is no right way. Probably there are one thousand wrong ways that will be far better than what the experts think is the "right."

This is why I find that the best Permaculture designers are the new ones. There's a whole industry of internet know-it-alls dedicated to complaining about new Permaculturists just starting their adventure and filled with the divine spark of creativity. These new gardeners want to build gardens on the sides of buildings and food forests on roofs and all sorts of "wrong" things. Many of the internet experts even approach Permaculture as though it is all about right and wrong ways of doing things, rather than as a process for unleashing creative thinking about the landscape.

Not to sound too judgmental myself, but thinking too judgmentally is kryptonite for creative thinking. When we are young, or just learning, we may think that understanding what is "bad" is the most important information. We may think the more harshly we judge something, the more refined our own abilities and tastes must be.

In fact, when we are inexperienced we may even be drawn to someone spouting nothing more than an endless stream of criticism and gossip. "Wow! They must certainly have good taste if they hate everything!"

And there is some use to knowing enough to know what is good and bad. But mastery requires us to quickly move past that phase, because it is toxic for our own creativity. When we judge other people, we can become self-conscious and judgmental of ourselves. We can start to feel like others are watching and judging us, because our experience tells us that is what we are doing to them. Then we become hypercritical of our own creative efforts, which completely shuts down the whole creative process.

Creation requires an open mind that can put together things in novel ways without worrying that it is "wrong" or "bad." You may have heard of "no wrong answers" brainstorming for a reason, because this uninhibited thinking is the permission we need to even step into creative space.

Every great master of anything that I have ever met has been exactly the opposite of judgmental. The greatest opera singers in the world do not judge untrained voices (their egos don't need that!) They rejoice in hearing the natural sounds of humans singing in the streets for pure joy. The great architect Christopher Alexander didn't travel the world judging the accomplishments of those with less architecture training—he drew inspiration from them. The same is true of the great composers of the world. One might imagine great artists holding their noses as they look at the work of others, but instead we have stories of them rolling in ecstasy at the sight of the work of their peers. Most great artists don't just "see" or "hear" art. They allow it in, deeply. Everything can be an inspiration. They allow things to rock them to their very cores and transform them. They care less about a bit of botched technique and get right to the heart of what people were trying to say. They listen deeply without judgement. Instead of worrying constantly about what is bad and wrong, they are focussed on what is great, what can elevate us to new heights.

I have heard stories of Michael Jackson absolutely overcome with emotion, just from hearing a song or looking at album art. His mind was sensitively attuned to what plucks the heart strings, what moves the internal objects of our souls.

True creation exists at this connection of mind, reason; and soul, intuition; and heart, feeling. Would you rather work on a farm that feels like a factory or one that nearly brings you to tears with its beauty each day when you get up? Would you rather live in a house where the neighbors are glad you are conformist, or one where you jump around the front yard in delight at all the beautiful things to behold? Good design must be functional, but it must also connect with us on a deeper level, or we've missed 90% of it.

"Never sing, until the moment you feel you'd die if you didn't."
-Maria Callas, American Soprano.

Invoking Creative Mind

So, what can we do to cultivate this mind of creation?

Cultivate the creative mindset. That means understanding that there is such a thing, and that you can work on and acquire it as part of your journey. Wouldn't it be wonderful if this process of transforming a landscape also transformed you into a more creative, better person?

Get over judgement. If you are someone who is prone to negative thinking and judgements, and you want to be a good designer or gardener, my advice to you is to set a goal of growing beyond that into a more open minded approach. If you go to a museum or watch YouTube videos and your first thought is always "what is wrong" try to instead ask "what is good about this?" Give your own creative efforts the same benefit of the doubt. When you start doing this you will naturally be letting in more things to inspire you.

Believe you have the right of creativity. You exist as a unique expression of humanity, at a unique place and time, with unique sets of interests and experiences that only YOU can express. You have a unique weirdness inside that will delight other people. In school I was taught that the evolutionary point of life was reproduction. What a bunch of garbage. Nature doesn't care if you reproduce or not. You are here, from an evolutionary perspective, to be the most unique and truest version of yourself. Nature solves problems by evolving, by going off in many new directions to find new ways to solve the problem of life. YOU are one roll of the dice, one attempt to find a new, better, lasting way to be alive. If you follow your own unique genius to its end, then you will have answered the question and successfully fulfilled the meaning of your life. You have something worth *becoming*. Believe it.

Seek out new and challenging experiences. These give your brain more parts to play with. Creatives should always be learning new things. Seek out different perspectives, especially of those from marginalized backgrounds. Get out of your routine. Allow things to "change your mind." A brain that is constantly rewiring will be a creative brain filled with new ideas, and that affects how we approach the landscape, too.

Get into a routine, get into a ritual. Wake up at the same time. Make time to do some creative work every day or every week. Schedule time for the Adventures in this book. Routine takes some of the "mental load" off our lives and frees up brain power to be creative, rather than worrying about what to do next or how to plan our days.

Live with the time taken out. It's hard to be creative when we feel the ache of time's hook in our chests, ever pulling us into the future. There is an eternity in this moment, and all the time we ever need right here and now.

Set deadlines. Don't wait for inspiration, go harvest it. Do the work. Set goals and do it. Plan time to seek inspiration and study. Done is better than perfect, and often the permission to be imperfect that comes with "just get it done" will ironically promote greater creativity.

Don't worry about deciding if something is "right" or "wrong," look for ways to transcend right and wrong. Look for what is wrong in the right and what is right in the wrong. "*Out beyond ideas of wrongdoing and rightdoing, there is a field. I'll meet you there.*" (from Rumi's "A Great Wagon")

Chapter 7: The Nature of Beauty and Beauty of Nature

On one hand, I understand why many in the back-to-the-land/environmentalist subcultures — both rightwing and left—reject beauty and any sense of aesthetics in life or landscape. There's a sense that I want to be able to do what I want on my property. And I don't want my neighbors butting in, especially if they've got ugly, wasteful lawn!

What generally passes for beauty in our culture is whatever products the corporations are marketing this season, and that is it. "Beauty" is something you buy from Amazon to wear or decorate your house. So it makes sense that we reject that approach to aesthetics.

On top of that, with many people in the world going hungry, even going without basic needs met, it seems a serious injustice that we would waste precious resources and money to decorate our homes, landscapes, or bodies when others must go without food and shelter.

And so on the left, we see a rejection of the negatives of "mainstream Western society." And so, white, male "rewilding" teachers may wear torn rags, old moccasins, and a carefully crafted aesthetic intended to look like a rejection of aesthetics.

On the right, the same feeling about consumer society often results in a sort of utilitarian "roll up our sleeves and get to work" seriousness. Anything artsy is seen as no-good-pinko-commie propaganda. "There's a lot of work to be done!" That's also a natural reaction to American excess. Interesting that this rejection of consumer culture is a place where many people, left and right, appear to agree!

So these sentiments often come up as soon as anyone starts to discuss beauty.

But that stereotype is not how humans from vibrant, sustainable cultures actually behave. White, mostly male anti-colonial activists may intentionally cultivate a self-proclaimed "crusty" appearance they think looks "uncivilized." Here, uncivilized is used to mean a society which doesn't have the environmental, health, and inequality problems of a Western nation. Often, this means indigenous societies.

But modern-day Native Americans, for example, may not have the privilege to dress down in such a way, or they will not be hired for work, not be taken seriously, or may even end up "disappeared."

I learned this lesson myself as someone who dressed counter-culture while working with migrant farmer populations. My colleagues from native and immigrant communities politely taught me that they did not have the same privilege to dress in such a way. I found that I could still express my own rejection of consumer culture and mainstream conformity, but I could do it in a way that was not alienating to those who lacked my privilege.

Young, white, American Wwoofers (farm volunteers) around the world have become famous for this "riches wrapped in rags fashion." I have heard many stories of it being off-putting to the diverse communities they work with. In some American farm labor circles, you aren't taken seriously without a good patina of dirt on your worn overalls. We automatically raise up bearded "farming experts" who wear the standard holey overalls and country hats, even if they just started farming and moved from the city a year ago. We should start to be mindful that when we play to that stereotype, we are reinforcing ideas that exclude people from less privileged backgrounds and different cultures.

And it is a myth that "uncivilized" people had no aesthetics. In extant societies that remain largely horticultural, humans actually devote the majority of their time to cultural and spiritual expression, festival, art, etc.

"Work" to many sustainable cultures is a minor activity, only necessary for supporting things that are really important. I would like to see more of us "roll down our sleeves and get to play!" There's plenty of play that needs to be done in this dour world, and perhaps what people really need is to find ways to define their lives beyond their bank accounts. If you want to create "viral change," make living your art, and make your life something spectacular.

This is no wonder. The Elk in the woods does not shirk around to avoid looking magnificent. No! It steps forth onto the earth with its head held high, "I am a mother-___ ing ELK!" The Eagle holds its body and wings with magnificence every bit as spectacular as any gods in our mythologies. It is our very right as beings on this earth to behold ourselves with such dignity. It is our right to take pride and care in our bodies and our surroundings. It is our right as beings of this earth to claim our throne here, to celebrate ourselves and our lives, and to do so with creative expression and beauty.

Chapter 8: Transcending the Conflict Between Intuition and Reason.

As we seek to build vibrant lives and landscapes for ourselves, it is inevitable that conflicts will arise, such as the common conflict between intuition and reason. If we are a bit idealistic about the idea of community, it can cause a real crisis of faith when these first struggles come up.

We could also call this the conflict between secularism and religious or spiritual attitudes. I've seen this particular conflict send young community transformation leaders into instant burnout, or even their own crisis of faith. And I've seen it destroy multiple community projects that otherwise had tremendous potential. This comes up a lot in circles around gardening, where some denounce all "woo" while others want to plant by the moon and talk to their plants. We may see this conflict within ourselves in deciding how to garden or arrange the landscape.

I'd like to introduce you to a philosophy that has helped me transcend this barrier personally, and connect and include people who might seem to be in opposite camps. This tool is Ken Wilber's Pre-rational/Rational/Post-rational model for human and societal development. It shows why this particular conflict is such a common cause of passionate disagreement in our society today.

According to the model, individuals and society begin to understand their world through INTUITION and myth, and explain things like the thunder by imagining very loud super beings in the sky! Floods are the result of an angry river god.

But eventually, these answers fail our reason, so we transcend to a rational worldview.

Often when we do this, we reject intuition entirely as superstition, and may become quite passionate in trying to wake others up from their irrational beliefs! Science or atheism can become a religious dogma.

"Thunder is no longer the voice of an angry god... No river contains a spirit... no snake the embodiment of wisdom, no mountain cave the home of a great demon. No voices now speak to man from stones, plants and animals, nor does he speak to them thinking they can hear. His contact with nature has gone, and with it has gone the profound emotional energy that this symbolic connection supplied."

—Carl Jung, Swiss psychiatrist

Usually the opposite of a bad idea isn't a good idea, but another bad idea. So eventually, as Carl Jung pointed out, this strictly rational worldview fails us, too. Because it denies the powerful meaning-making and insight that we get from embracing intuition.

When this happens we are thrown back into a crisis and many people revert back to their pre-rational worldview, doubling down on Adam and Eve riding dinosaurs and so on.

Such people may try to impose their faith on everyone else using force, laws, and social coercion, because a rational worldview is a threat to their belief.

But there is another path, which is to transcend the rational worldview to a post-rational one, which embraces the benefits of reason and reclaims intuition.

This is not a rejection of science at all! But instead it is a full embracing of science, which has given us very good documented evidence that a spiritual path can contribute to happiness, that intuition can indeed give us insight, that practices like meditation and yoga have proven benefits, and that complex systems like ecosystems can exhibit "an emergent property" that sounds like spiritual mumbo-jumbo to those trapped in a "rational" worldview to the point where they actually reject this science. To reject these research based statements is actually a rejection of science in favor of dogma.

The cutting edge approach is to attempt to integrate rational and intuitive approaches, recognizing that each has its place.

This is the powerful emergent edge I have seen again and again in those people who are leading community transformation work.

People on this same path may identify as Buddhists, Christians, Muslims, Hindus, Jews, Taoists, New-Agers, Pagans, Secular Humanists, Atheists, or many other ways. As we develop a post-rational worldview we increasingly use science and reason to solve the physical problems of the world, rather than create policy based on what we think some god or another wants us to do. We just don't make reason a religion, or science an infallible god giving us commandments.

But we also become open to experiencing the intuitive, and using it to solve problems of meaning, community, creative expression, and problem solving. This is especially vital for any kind of good creative work or design.

For example, I have always secretly considered my compost pile a cauldron, in which I mix the strange alchemical elements of life, and make sacrificial offerings: herb, bone, ash, eye of newt. What strange dark forces lurk in the compost pile to appease the spirits of place and garner abundance in this year's crop? I've even thought of getting an actual cauldron to do my worm composting demonstrations with kids. The human intuition directly understands things like soil health, ecosystems and the biodiversity resiliency principle and communicates to us in a language we describe as "energies, chi, or spirit." Once we start thinking of gardening holistically, it can feel like we are trying to coax faeries and gnomes into the garden, ghosts into the soil.

There are some things which are best governed by intuition, with reason to help guide the way.

It is the integration of reason and intuition that is the cauldron of wisdom.

To a Western mind, this may sound as though white western society is leading this transformation, but this is mere ethnocentrism to assume that ancient indigenous cultures have no reason or science. In many cases, these societies have long ago transcended this crisis in their own ways, and have well-developed spiritual technologies and ceremonies to guide individuals through the experience. It is only that we in our young culture lack any spiritual or philosophical tools for making it through this hurdle to maturity.

I have found that most of the folks at the leading edge of transforming society are emerging into this last stage. Often they are well-versed in the sciences, especially the biological sciences of botany, ecology, physiology, and human health, but have also experienced sacredness in nature, or through their own bodies in a practice such as yoga. Many of us share a sense that the global crises we face today have a spiritual cause—a lack of meaning that compels us to seek "the meaning of life" in consuming the stupid corporate junk we know is destroying the planet.

This model is very practical in that it can help us see that others in our communities may be in different places on this spectrum, but that we're all on a similar journey of transcending this crisis of intuition and reason. That invites us to ask "what is valid about this person's experience?" instead of immediately rejecting it as wrong or assuming that we are exclusively right.

Instead of attacking the other's experience, once we consciously embrace a post-rational identity, we can invite each other to continue our transcendence along this path together and seek what is wise and life enhancing about each perspective. As we adventure towards a better world and better lives, if we seek to invest in both sides of the coin together, we are investing in wisdom and completeness.

And it can help us in creating a transcendent landscape as well. It can guide us to use reason and the best research-based approaches when it comes to the practical matters of gardening and productivity. And it can also help us understand that some Western science throws out good, real scientific practices when they don't reinforce the religious belief in infallible scientific progress. For example, many "scientific gardening" dismiss the large body of peer-reviewed scientific evidence showing plastics and poisons damage ecosystems and have significant risks to human health.

And finally, we can embrace intuition in the garden, too, and start listening to the plants and insects for their wisdom. We can learn to find sacredness in the soil and water, and in each being that comes to visit us in the garden.

The Magic Circle, William Waterhouse, 1886

Chapter 9: Ghosts

Adapted from transformativeadventures.org

There's a story that the Red Hot Chilli Pepper's guitarist, John Frusciante, used to tell that gives a great insight into creative genius. When John was recording his first album with the Chillis, they'd play pieces back, and he'd start yelling "listen to that! Do you hear the ghosts there?" John said that he heard ghosts and spirits all over their records. Of course, at this time, John was looking deeply ravaged by drug addiction, so people started giving him some worried looks when he'd mention ghosts. Understanding the danger here, Flea, their bassist, took John aside and said "buddy, not everybody can hear the ghosts, only we can." And so John stopped talking about ghosts… for a while.

Years later, a healthy, sober, and drug-free Frusciante started talking about ghosts in his interviews again. There's some footage of him sitting in a beautiful victorian hotel lobby where the band was recording and in it he explains why he always insisted that the Chillis record in such places. He points to all the fine craftsmanship of the crown molding, and to the obvious signs of age and wear, and he says "all these little things, these spaces, they hold ghosts and spirits. So when we record in a place like this, we get those energies inside us, too."

"How else could we get them onto the records?" he says.

―――――――-

That's one of the most beautiful descriptions of the idea of an "**emergent property**" that I've ever heard.

At some level of "complexity" the whole becomes greater than the sum of its parts. We call that an "emergent property." A human being seems something more than a big heap of cells, tissues and organs. A forest feels like more than a collection of trees. And when a song becomes more than a collection of notes, we say it's got "soul."

This is related to the concept of entropy, which is an important, ever-present ghost in the background of this book, haunting nearly every page when we're talking about ecosystems and communities.

Entropy is the idea that systems lose energy over time, and so they require continuous inputs of energy to be maintained. A car or a house falls apart, back into a heap of parts, and stops being useful as a car or a house. Once a complex system falls apart, it usually takes a lot more energy to get it going again than it did to maintain it. If an organism dies, it would take a lot of energy and technology to start it up again. If we let a relationship or community "die," it can take a lot of energy to start it up again. When we lose a healthy old growth ecosystem it is not easily replaced.

To some, John might sound crazy, but he's just speaking the language of "magic," "spirits," and "energies" that the human intuition has always used to speak of emergent properties.

When we walk into an old house, we see the amazing craftsmanship–work that few can replicate these days. We see the signs of age, and wear, and living. There's a large gash by the back door. Soot by the fireplace. And signs of care and love, too. The human brain is sophisticated pattern-recognition software–of course all of these things will speak to us, they'll whisper to us all their bitter-sweet stories.

Let's add a layer of complexity to this: artists in most every culture work in "movements" such as "Romanticism" and "Modernism." The emergence of these movements are always a response to what came before, and in turn, what had come before those movements, too. Each generation works together, developing and exchanging ideas and perfecting techniques… and because of this, they put so much more into their work than any individual ever could. A painting might be the work of a single painter, but if she was any good, she didn't just put her own soul into it–she got all the stink of her generation in there, too; their history, their struggles, their ideas, their philosophies of life, and what they thought of those who came before them. She knows her paint brush is a divining rod, with it, she can coax the ancestors into that canvas.

If we can allow ourselves to be open to it, we can intuitively "get" those layers of meaning, something that seems alive, that moves, speaks, and sings to our very souls: an emergent property.

Every couple of years, I try to visit Iargo Springs up on the Au Sable river, in the state of Michigan.

This place was considered sacred by the Ojibwa, who brought the sick here to heal in the curative waters. Since I was a kid, I have always felt the healing energies of this place. In the pilgrimage down the long flights of stairs the temperature can drop 15 degrees on a hot day. I felt enveloped by this coolness and the soft light through the high forest canopy. It was like entering into… some tangible and real force much larger than myself. The clear pools of water seemed to radiate with a deep stillness and the energy of the hills behind seemed comforting and safe.

I can come to this place and sit all day, and its energies get inside me and still my own. When I leave I am changed.

People of all cultures have recognized the *"genii loci,"* the "Spirit of the Place." Virtually every hill, valley, and settlement—any ecosystem with enough complexity—had one. These anthropomorphized versions of place were the first gods and spirits that we humans worshipped. It was often the shaman, the one who knew how to relate with these spirits, who sited new settlements, told where to plant the orchard or sew the grain crop, or planned the temple. In his designs, he'd bake brownies for ghosts, whisper to the ancestors, crack jokes to the genii loci to tempt them into the design.

How else would he get them in there?

And this has never gone away. In more intuitive cultures, people still speak of fairies and spirits in the woods, in the waters. This language is the tool they use to understand that there are some places that are curative, and some that are best avoided. People still build little nooks and crannies for elf-folk, gnomes, and dakinis hoping that if they attract the right energies, they can recreate the sacred feeling of a place like Iargo Springs.

Even in Western culture, urban ghost tours remain incredibly popular.

As a maker, gardener, and Permaculture designer, I have a deep desire to get as many ghosts of the right kind as I can into the human habitats I co-create with communities, ecosystems, ancestors, and the *genii loci* around me. The more inputs, the more voices and hands are included, the more likely it is the creation will transcend my own limitations and something… *more* will emerge.

To me, a garden that is overly managed by human goals, vision, and hands can never achieve that quality that Iargo Spring has. What we are seeing is relationship, connection, plants and trees dancing with each other in ways too subtle for the eye, but bold to the open mind. A garden of disconnected parts, flowers and vegetables planted into orderly rows may be pretty, but there is no room in such a rigid place for *genii loci* to hide.

While it's important to have our designs grounded in research-based approaches, it's equally important to find ways of speaking the strange language of intuition.

Beauty in Abundance

Perhaps you feel like your little slice of the world, your little humble yard is too insignificant to deserve this sort of attention. But even very small spirits need places to live, too. And we may need their presence far more than our worldview of reason will allow us to admit.

Despite all our technology, in this regard we moderns are no different than the thousand thousand generations before us.
There's only one door to this place—
you must find it in moonlight,
find a sincere desire,
put on the shaman's cloak,
beat the drum,
carve the antler,
seek the fae,
speak to ghosts….

When I become human
I will shed my clothes like leaves
Raise my branches high
under the solstice moon
root in the bare earth
And let the 10,000 birds within me sing

Eat wild roots and berries
Growl at machines
Howl at inequity
Snarl at bigotry
For how could such a noble beast
Homo sapiens
Wise man
bear a member
of his species oppressed
for profit or pride?

When I become human

I will finally be the equal
to my elder relatives
The kingdom of plants
The kingdom of animals
And I will wear the crown
Of a true citizen of the earth
The bear, the crow, and the oak
Will attend my coronation
And I will treat this mother earth
And each of her children
not as a kingdom to be ruled
But as a sacred throne
Of priceless jewels
To be served with care
Not a single gem to be tarnished
A throne that raises us all to the sky

When I become human I will fly

Chapter 10: The Elements and Patterns of Transformative Aesthetics for Landscapes

To get the most out of thinking artistically about our landscapes, it can be helpful to think a little bit about what that means. When we study art, we learn about "the elements of art," and there have been several attempts to break down "the elements of landscape design," too.

But, in my opinion, students get the most out of this not by just reading about them in a book, but through the adventure of actually applying them in the real world. And yes, I said the word adventure, so this will be one of our learning journeys.

But first, let's look at these elements briefly. I will give you both a conventional version and something of my own version you might use to think about landscapes. We'll discuss these at length in a moment. In visual art, the most commonly cited elements are line, space, form, shape, value, color, and texture. These are the things we can play around with in visual media.

But a garden is different in that it is a 3D object, one which we view from the inside, one which we view in the 4th dimension of time. In that way, a garden can unfold like a drama.

Line. These are the basic shapes we see in the landscape, paths, walls, etc. Are the lines straight or curved? Straight lines may be more formal and have an immediate appearance of tidiness to most people, especially Westerners. Meanwhile, curved lines are more naturalistic, more relaxed, and tend towards making us feel more at ease. In feng shui, the wisdom is to make paths to the door curved and meandering, so that visiting guests will not approach with a business-like, dehumanizing attitude, but must walk

through the garden past pleasurable and relaxing sights and smells before arriving at our door. Hopefully, they arrive with a softer attitude. In addition to being straight or curved, lines can be "hard," or soft, having a broken or disturbed texture.

Space; defining it, enclosure or not. What is the shape, size, feel of the space? In a way, all gardens are defined by a sense of enclosure, or a frame, or the lack of one. Perhaps that enclosure is formal like a wall or hedge, or simply the frame of trees or perhaps a neighbors' yard. See the patterns on enclosure in the chapter on garden layout.

Views and Vignettes. Just as forms and shapes on a canvas are elements of a visual art, views are one of the major objects we experience in the four-dimensional garden. Are there major views in the landscape? Great gardens are often based on great views. And great garden designers are thoughtful about bringing the guests to those views. A bench placed in such a way that we sit down, then notice the mountains in the distance, or perhaps a hidden rock cairn dedicated to the ancestors. This can make the most of a view.

Moon gates and trees framing views can also help call our attention. Long paths leading to beautiful views can make a garden an event. Have you ever traveled up a mountain or hill then arrived at the top to a view of the entire countryside? Even a small urban garden can take us on an experience, through a winding path with the discovery of a fountain, an altar, a beautiful shed, a tea house, or picnic table waiting at the destination. For Permaculture teachers or entertainers looking to create powerful landscape tours or pleasure walks, such a feature turns each garden tour into an adventure.

Theme. What does the garden say? What is its style? Is it modern? Does it invoke the natural beauty of a bygone time? Or does it help us envision a future that balances the best of the modern and of nature? We'll discuss garden themes in a few chapters.

Symbolism. Does the garden contain symbols or symbolic meaning?

These can give a garden special depth. In fact, the potential for a landscape to fulfill us on the spiritual dimension is one of its most powerful potentials.

We'll discuss this in the chapter on **patterns for sacredness**.

Transformative Landscape in Vermont, by Stician Samples

Color. How is color used? Are there color patterns? Are the colors high-contrast and energizing or soft and soothing? Are they warm or cool? How color is used is one of the most immediately noticeable features of a beautiful landscape, so we will have a whole section on **color theory later**.

Context. What is the context of the landscape, a suburban yard in the midwest? A tropical farm? Does the landscape harmonize or contrast with the context? Does it do so in an appropriate or inappropriate way? Thinking about context can be one of the most important elements when it comes to whether a landscape causes us social harmony or conflict with neighbors. Living in a suburban subdivision with a home-owners association and having a country farm-like aesthetic with animals running wild and long grasses is likely to cause issues. In a city core, a bright, lively landscape might be popular with neighbors. In a historic neighborhood, an eccentric garden might be more welcome if it fits the historic form of the house (we'll discuss matching the house to the landscape in a later section.) Terroir is a pattern will address related to context.

Mark-Making. How was the garden *made*, and what does that communicate? Was it thoroughly "moonscaped" by tilling and spraying before planting? Or did it evolve in cooperation with nature? A "wild" garden of native plants using industrial materials, plastics or heavy moonscaping to refine may still communicate a message of a very managed landscape.

Materiality. What materials are used? Industrial materials from big box stores? Found and repurposed objects? Environmentally friendly materials? What does this "say?"

Texture. this can refer to the plants and materials. Are materials rough and rustic or smooth and space-age? Are plants large-leaved or small? More importantly, are the plantings filled with blocks of similar-sized plants (a smooth texture) or are there big contrasts in the size and leaf shape of plants? As with color, high contrast can be energizing, while low contrast gives a more relaxed and relaxing appearance.

Scale and proportion. What is the size of the space and the size of the elements within it? In a very large space, small items can get lost, while a large focal point can unify the space. In a small garden, a large object can be overwhelming. Traditionally, small spaces use small plants, but a small space with very large plants and elements can have a bold impact!

Unifying features give a landscape an identity. For example, a color scheme, a repeated material, or a repeated charismatic plant will give the whole garden a character. The more unique the unifying features are the more unique the garden itself will be. For example, outlining all the beds in a garden with rocks (see the pattern, lithic mulch) or all the paths with crushed white shells will give the garden a put-together identity, but it will not necessarily be a unique one. Either may be your goal! Lining all the beds with bowling balls or using paths made of metal factory parts as stepping stones would give the garden a more unique identity. Repeating alliums in a garden can be enough to give it identity. Repeating an unusual plant like cardoon would likely give a garden an unusual, eccentric character.

Rhythm, Repetition, and Rhyme help us talk about how elements are repeated to form patterns in the garden. Rhythm talks about how frequently we see an element or color, and rhyme talks about near misses. Color echoes is a pattern we will discuss in the chapter on colors, which means a color repeats through a garden. Usually perfect symmetry in a garden is impossible because plants just don't behave enough, so intentionally making slightly asymmetrical "near misses" can be better. Instead of repeating the same plant or same color on both sides of a path, it can be interesting to put it slightly off or to use a slightly different 'rhyming" color. This will give the garden a feeling of balance without making it look too tightly controlled by humans.

Time (Anticipation.) An element that is unique to gardens compared to other arts is that gardens exist in time. We may experience a garden as we experience drama or literature, with a theme, or even a plot. Through the season flowers bloom then die back. We can use this to create different characters at different times, or even create a sense of anticipation. We may also encounter an element of time as we walk through a garden. A walk through a garden may take us on a story, hinting at hidden views ahead, revealing secret spaces and nooks to explore. We may experience mystery or excitement or induction into relaxation as we experience a landscape.

Non-visual sensory elements include smells such as, taste, sound, touch, and temperature. Our experience of a garden may include the heavenly aroma of a lilac, or the sound of rustling seedpods. We may feel the mist of a fountain on a hot day, or walk into a cooling grotto for refreshment.

Adventures in Landscape Elements

The first adventure is very simple: take this book to one of your favorite spaces of natural beauty. Identify some of these elements of landscape art at your favorite place. How are they used?

For an advanced set of adventures, next do one element at a time. Go for a walk in a beautiful place or a garden, and focus on only one element, like plant textures. How does this one element impact the beauty of the place? This is the type of exercise great artists engage in as they develop their craft.

Plants and Patterns for non-Visual Richness in a Transformative Landscape

Plants for an aromatic experience:
In this list, I'll be focusing on the best options for a transformative landscape, meaning the plants I'll recommend are easy, often multipurpose, and beneficial to wildlife. For aromatics, I'm looking for plants that fill the garden with perfume that can be appreciated from far away.

Annuals: Petunias, tobacco, heliotrope, sweet alyssum, moonflower.

Perennials:
Lavender, in dry weather can fill the garden with aroma.
Mints, in mass can be quite aromatic for the passers by.
Tulsi, in tropical gardens can grow as a perennial and be amazingly aromatic.
Valerian is perhaps the most aromatic perennial I know of, and on a good day it can fill the whole garden with a heavenly and complex aroma of cherry pie and earthiness.
The evening dame's rocket is named for the perfume it puts on in the evenings. Quite a pleasure throughout the garden.
Joe Pye Weed has a vanilla-like aroma that can be tinctured, and it is a pleasure to stroll near it on a good day.

Tropical Perennials: Passion flower, plumeria, angel's trumpet, Ylang ylang, jasmine, coffee, oleander, Hoya vine.

Shrubs and Trees:
American mountain ash, can scent a landscape for hundreds of feet in all directions and is one of the most aromatic plants I've personally experienced
Lilac, a classic, but most people don't know that it has edible uses.
Fruit trees, a number of fruit blossoms have a heavenly aroma. Plums and cherries can be particularly nice, while pears (having a "mushroom" smell) might be unwelcome in the aroma therapy garden.
Rugosa roses. We've all heard to stop and smell the roses, but some rugosas will have a smell that will stop passers-by in their tracks. They also have the largest and best quality edible hips.

Patterns for Sound:
The most common elements of sound in a garden are man-made, such as wind chimes. I have been in a garden that was designed as a walk through harmonizing wind chimes at different scales and it was quite mesmerizing. But natural elements can also provide sound, such as birds, and yes, even plants. Poplar trees, and grasses, and dried seed pods can be especially musical. Water features can create a mesmerizing sound. Frog ponds and grasses to host crickets can make a summer garden evening a musical experience.

Flower and Plant Shape for Texture, Rhythm, Rhyme, and Repetition

A good painter creates visual interest on the canvas by using different sorts of pressures, different sorts of brushes and strokes.

In the garden we use different forms of flowers and plant shapes to create effects. So it can be helpful to think about what kinds of flowers and plant shapes we have to play around with.

Flowers, duh. This is the one shape that every garden needs to have. It is the shape people most associate with a "flower." Without this, we run the risk that people won't know that there's a garden! Sounds crazy, but it is true. A garden can be filled with beautiful flowering plants, but people will say it looks like weeds. I have even had someone say to me that a flower garden in full bloom could have some flowers in it! But add one of these flower-shaped flowers to the bunch and people say "oh! It's a flower garden!"

Bells, such as the bellflowers.

Points include the round flower heads of most alliums. These give a punctuating pop to the flower-scape that is almost a feature in itself.

Umbrels have a distinctive flower shape that we most associate with the carrot family, and they can give a hazy, cloudy feel to the visual field. A nice effect is to repeat these at different levels.

Spikes. These tend to draw the eye upwards. These include lupines, baptisia, foxglove, and even anise hyssop and liatris.

Sprays are often associated with weeds, because these are small flowers that haven't

been cultivated for flower size. But usually these can have the biggest visual impact of color in a harden. A big mass of Turkish rocket flowers adds a big patch of gold to the canvas. Bee balm gives a big bold patch of scarlet. Sea kale gives us a big patch of white.

The key to using these is to create variety, repetition and rhythm in the garden. A bed of all "flower-shaped flowers" can look unnatural and boring. Adding in some other shapes increases energy and interest, and makes a planting look more natural.

We can also play with plant shapes in the same way. A bed with only large-leaved plants like hostas can look unnatural. Mixing in other shapes will usually look more interesting, healthy, and more natural to the eye. But if the goal is to create something that looks very human-managed, using all the same texture will have that effect.

Some plant shapes to consider:
Large-leaved: squash, fuki, hostas, sunflowers, comfrey, burdock.
Finely textured leaves include marigolds, apiaceae plants, musk mallow, baby's breath.
Compound leaf "starbursts," walnuts, sumac.
Architectural forms. Cardoon, artichokes, pruned tomatoes.
Lobed leaves. Compass plant, tomatoes and potatoes.
Round leaves include nasturtiums, fuki, pillea, and violets.

A mix of textures can be as interesting as a sophisticated color palette. We can play around with staggering large leaves and fine leaves of different textures, as well as foliage colors. The blue leaves of borage, cardoon, and sea kale have an interesting effect here next to variegated corn.

Chapter 11: The Power of Transformative Landscapes

Given the problems of the world today and how often we are told by politicians and "thought-leaders" that we are powerless to confront them, it's no wonder so many people are looking to their landscapes as an important place to take direct action.

All over the world people are waking up to the fact that the better world we want to see starts right in our own yards, our own homes. We are tired of living with destructive landscapes, lawns, and destructive food systems.

Many of us are replacing our lawns with native plants, to help steward native biodiversity. Others are growing our own food, recognizing that the food system is a major driver of global inequality, the largest driver of biodiversity decline, and habitat loss, and the largest single driver of climate change. The best of these "foodscapes" look to grow food in ways that are "regenerative," recognizing that home gardens can be just as damaging to ecosystems, if not more so, than industrial Agriculture. We are ditching the petroleum, plastics, and poisons paradigm in favor of working with nature in ways that are beyond sustainable.

But the most inspiring landscapes I have seen are those that are looking to use the land in ways that are truly societally transformative. These home yards, landscapes, farms, and gardens aspire to divest from the industrial food system, build local food justice and sovereignty, steward biodiversity, give sanctuary to wildlife, clean and sequester water, regenerate soil, and sequester carbon.

While many people recognize a pet problem like biodiversity, what these land stewards have recognized on some level is that the true danger of our predicament today is one of conflicting priorities. It is not just that there are some stupid greedy humans somewhere taking careless action that harms biodiversity, or drives climate change. It is that we as a global community have a system that creates a legitimate set of difficult priorities to navigate.

This tangled web of peril is made of many threads spun on our landscapes.

Food: We must continue to feed a global population of humans amid crop losses from climate change, rapid soil loss, resource limitations to fertilizers, fuels and water, and the decreasing effectiveness of pesticides. If we cannot keep up productivity (which has been at a plateau since the 70s) then billions of humans will potentially suffer.

Biodiversity: But we are also amidst the fastest mass extinction event in global history, and the major cause of it is habitat loss driven by converting land to agriculture. The extinction deficit theory suggests that extinction often takes 100 years or more after habitat loss occurs, so it is likely that if we fail to rapidly restore land to wildlife habitat, a cascade of extinctions is sure to happen in the future. The fate of these species is already sealed unless we take action now to return the land to habitat.

Climate Change: Yes, climate change has been driven largely by the food system, but at this point it would not be enough to simply stop polluting, we need to pull carbon out of the atmosphere. The best technology we have for this is to set aside land for sequestering carbon. Yet, in most cases, setting aside land for carbon crops may not do anything for biodiversity or food, making those problems even worse.

So we find ourselves with a three-way bind. If we want to feed people using the systems we currently rely on, we will need to convert even more land to industrial agriculture. If we wish to preserve biodiversity, we need to quickly convert industrial farm land back to habitat. And if we want to fight climate change, we will need to prioritize land for that purpose. And we need to do this in a context where global connectivity is awakening us to historic systems of oppression and inequality, where people in poor nations may no longer accept being poor so that people on the other side of the world may lead lives of extravagant luxury.

And that is the power of these Transformative Landscapes. They help meet human needs without causing these problems to begin with. They meet human needs while preserving or even enhancing habitat for wildlife and endangered species. They grow food in ways that sequester carbon, steward biodiversity, and care for water. And these systems are compatible with a world of justice and equality. In a world where people in rural India can speak online to the wealthy in North America, we can no longer accept the inequality in consumption we have grown used to. And since we already consume the world's resources at an unsustainable rate, there is no path to equality except one where those of us in the wealthy world consume much, much less.

Yet, this does not have to mean a decline in our standard of living, not if we start growing our own luxuries off of sustainable sun power, right in our own yards. We can move towards equality while simultaneously growing better food than we can buy in any import stores. We can surround ourselves with beauty and meaning that exceeds any we could jet off to for a weekend. We can grow wealthy in our connection to nature and community.

And as we'll see in future chapters, most of the world's food and nutrition already comes from small-scale sustainable systems like we'll create in this book. It is only folks in the Western world who are missing out, relying upon an industrial food system that produces less food, of lower quality, but creates an oversized carbon and ecological footprint.

And this is why we call these landscapes "Transformative Landscapes," because they are the single most powerful leverage point we have for undoing all of these problems before they even begin.

So what specifically does it mean for a landscape to be Transformative?

Benchmarks: The Transformative Landscapes Recognition Criteria

The purpose of the Transformative landscapes Recognition Initiative (TLI) is to promote land management methods that are shown to have a measurable positive benefit to ecosystems and society.

These solutions are accessible at any budget, yet are proven to be effective and high impact.

Recognition is available to any type of landscape, including residential, farm, commercial, or public. And while the methods for accomplishing them are easy and accessible, these criteria require proven methods that make greenwashing impossible, so TLI recognition intends to be the gold standard for demonstrating socially and environmentally sound landscape management, that can't be bought or faked.

Transformative Landscapes use proven methods that measurably create positive societal outcomes. By following these criteria, landscapes will meet the following goals:

Help feed and care for people
Protect water
Sustain global forests
Protect native biodiversity
Provide insect and pollinator sanctuary
Fight climate change
Use energy efficiently
Build soil
Use materials responsibly and sustainably

Recognition signs are available in four different levels:

Four Levels of Recognition

Aspirational: Anyone can apply and receive aspirational recognition and sign

Transformative Landscape Recognition: (achieves 70% of the criteria)

Monarch: (achieves 70%, including all required Monarch criteria)

Luna moth: (achieves 90%, including all Lunamoth criteria.)

Aspirational Criteria Overview:

Anyone is eligible to apply for a sign that shows what your landscape aspires to. This allows us to communicate what we hope to accomplish when we stop mowing, build water catchments like swales, or hugelkulturs.

Criteria: has applied for recognition and submitted a plan to Transformative Adventures or a local Landscape Transformation professional.

Full Transformative Landscape Recognition Criteria

Learn more or apply online at: TransformativeAdventures.org

I. Societal impact:

—Grows some food, medicines or materials for human use on site.
—Reduces the site stewards consumption of corporate products by 10% or more. (For Monarch level.)
—Reduces consumption and produces an excess to provide for other people through sales or donation. (For the Luna moth level.)

II. Water:

—Catches and infiltrates water for at least 60% of site.
—Uses water-wise holistic techniques.
—Has plan to catch and infiltrate approximately 90% of rainwater on site. (Monarch)
—Also includes water recycling procedures, greywater system or other methods to treat site water.

III. Pollution:

Fossil Fuels: No aesthetic mowing.
—No (or rare) fossil fuel mowing on sites under ¼ acre.
—For larger sites, mown grass is limited to pathways and used recreational areas.

Carbon Sequestration.
—40% forest where applicable. At maturity, polyculture tree canopy (minimum three species) will cover > 40% of site for sites with >30 inches of precipitation. (Submit site plan map)
—Healthy diverse polyculture grassland, savanna, prairie or other native ecosystem for climates where forest is not appropriate, including holistic management of grazing systems.

Full Transformative Landscape Recognition Criteria

Learn more or apply online at: TransformativeAdventures.org

IV. Sustainable materials:

Concrete. The concrete industry is responsible for a massive amount of pollution and atmospheric carbon. Limited use of new concrete pavers, bricks and landscaping blocks, less than 10%. Use of recycled materials wherever possible for landscape features, retaining walls, and pathways.
-No new concrete materials, or where necessary, low-carbon sustainable materials or re-constituted materials are used. (Monarch)
Plastics. Agricultural and landscape fabrics are the fastest growing cause of plastics pollution of water and land, and are strongly associated with micro-plastic and phthalates contamination of food.
—No new plastics pledge.
—Adopt a plan to reduce or eliminate plastics.
—Achieved a documented reduction or elimination of plastics in the landscape. (Lunamoth.)

V. Biodiversity:

—Use of multiple plant community or ecosystem types: forest, meadow, prairie, intensive gardens, wetland, hedgerows, etc.
—Integration of wildlife habitat features, such as rockeries, brush piles, bat houses, native pollinator habitat.
—Season long blooms, something is flowering at all times in all parts of the landscape.
—Provide wildlife food sources: a variety of nuts, seeds, fruits, native flowers, that are shared with wildlife.
—Provide multiple water sources: birdbaths, ponds, water features.
—No spraying of pesticides or herbicides. (Organic sprays acceptable for production purposes only.)
—High biodiversity site: >200 species/acre, or includes endangered species. (Monarch)

VII. Soil creation: Many of the above criteria are also proven to contribute to soil creation, but these additional factors are special indicators of soil creation.

—Minimized annual mechanical tillage, including the use of "tilthers" or harrows.
—No synthetic fertilizers.
—No monoculture plantings.
—Minimal bare soil. Use of mulches or permanent ground covers.
—All fertility, compost, organic fertilizers, and mulches are grown on site or from a waste stream. (Monarch)
—Submit plan to grow fertility on site. (Lunamoth.)

Adventure Step 3: Water Wise Design

Image, forest fire sky, Victoria Gunn, Permaculture designer.

Sun, soil, and water are the most important keys to a successful garden. A plan for water wisdom is absolutely essential. With good water design, in many climates it should be possible to eliminate watering work and irrigation altogether and still have high yields. I haven't had to irrigate one of my gardens in nearly a decade, and have had yields that far surpassed conventional agriculture high yields. As climate change rages and aquifers drain, this becomes even more essential.

And, for those who are interested, to achieve the higher level Transformative Landscapes Recognition, it is necessary to submit a plan for dealing with water on our sites.

As Permaculture designer Brad Lancaster says, always "plant the water first." When you make a new garden or bed, know where the water will come from and what systems you'll use for it.

Our goal should always be to utilize as much rainwater as possible before using fossil water. And we should try to avoid city water whenever possible, since such systems are almost never sustainable. I like to create a whole design sketch for the water systems I'll use in the garden. If a site is designed to catch and store over 90% of the water that falls on it, then it is entirely ethical to use well water (unless you are in a depleted aquifer) as you will be recharging more water than you are withdrawing.

Our ethical objective is to catch, clean, and infiltrate as much rainwater as we possibly can, and turn it into productivity. When the rain falls on our land, it becomes our responsibility to care for it and not flush it out to the storm water system, where it will turn into a damaging pollution problem. This "stormwater" starts its life doing damage to gardens, landscapes and home foundations, then it goes into the drains where it costs taxpayers money for upkeep. In most Northern American cities, flood water systems were designed to move water from wealthy white neighborhoods to low-income neighborhoods, which is why whenever it rains, these neighborhoods flood. Next this water pollutes our waterways, doing damage to biodiversity, and making our waters unswimmable and unfishable. Finally, this waste water makes its way out to the ocean, causing large ocean dead zones. Meanwhile, we're plumbing fossil water from aquifers that will take tens of thousands of years to recharge so we can water our lawns and gardens.

And this is why I recommend having a whole site sketch of water systems. See the **Lillie House Case Study** for an example of a water plan for the Transformative Landscapes Recognition.

Systems for A WaterWise Site and Gardens

(Use these in your site water design plan.)
Water-Saving Systems (patterns in bold are included in this book.)

1. **Good Permaculture zone design: our best tool for water savings.**
2. **The holistic gardening methods:** Many elements in this system transform soil into the best possible water-saving system we have. Mulching, no-till, polyculture, French Intensive Gardening, good crop selection all save water.
3. **Nurse log mulch.**
4. **Enclosure: wind breaks, hedgerows, forest gardens.** These provide shade and protection from desiccating winds.

Water-Harvesting Systems:

1. Beds designed to catch their own water, such as the **beginners' annual garden make-over bed.** These are especially useful in Permaculture zones 1 and 2.
2. Infiltration basins, like rain gardens. Best in zone 1 or near buildings or parking lots.
3. Permaculture Swales. Swales are ditches that are perfectly level on contour, so that they catch water instead of flushing it out into the ocean. There is a great deal about digging swales in the Permaculture literature and online, including resources for getting them right at TransformativeAdventures.org.
4. Mini-swales and water direction trenches in zone 1.
5. **Nurse log passive swales.** Good in any zone.
6. Contour plowing, this simply means that if we're going to till, make furrows on contour.
7. Keylines. A keyline is a single, plowed furrow with the intention of slowing water down, spreading it out, and soaking it in.
8. Terraces, such as the hugelkultur terraces in this book. Terraces turn a hill into a series of nearly level water collecting beds.
9. **Trees and Forests.** Forests are the greatest water-harvesting tool we have. Most research shows that forested landscapes can catch and store nearly 100% of the water that falls on them, without swales or other catchment features. Even young treed landscapes can catch a huge amount of rain. By year three, apples and oaks will catch and infiltrate enough water to fulfill our ethical obligation, probably more reliably than swales or terraces. These are useful in any zone, but especially at the exit points for any last runoff water. For fulfilling the Transformative Adventures requirement, a mini-forest, or hedgerow at the final exit point for water will be considered to catch and store as much water as is reasonably possible. Any area with trees or bushes at 90% canopy will be considered to catch 100% of its rainwater. The tree systems in this book are considered to be water-harvesting systems, including **hedgerows, tree guilds, forest gardens, etc.**

Patterns for Harvesting Rainwater
Arranged by slope, cost, and effectiveness

EARTHWORK	Slope	Benefits	Costs	Notes
Infiltration basin, net and pan, rain gardens, banana circles	Flat to keypoint	Can be used on flatland.	Can be dug with shovels	At 8" depth, use about 250'sq for every 1000 SF of catchment. .
Contour Plowing, furrowing, beds on contour	Flat to as steep as equipment will allow	Proven effective, can be used in most garden contexts.	If already tilling, no additional cost	For annuals, requires yearly tilling. Long lasting effect.
Keylines	Pastures or orchards on slight slopes as steep as equipment allows	Used for pasture or sometimes for orchard establishment. May be dependent on site conditions as some research found no benefit.	Inexpensive if access to a chisel plow is available.	Requires reinstallation every few years.
Contour logs, passive swales, contour felling, nurse logs, log erosion barriers, log terraces.	Everywhere you can secure a log	Useful for all crops. Long lasting holistic benefit, infiltrates water, sequesters carbon, reduces disease, increases fertility,	Low cost, if wood is available. Moving wood is required.	Excellent value and usefulness. Long term effect.
Swales	up to 15% grade, used to establish TREES	Primarily tree establishment systems, but in landscapes with depleted groundwater and large rain events, may help replenish groundwater.	Special design and tools for a short lifespan.	Biggest effect is in the first 3-5 years in most climates.
Terraces (on shallow slopes this might just be "regrading."	15% slope (or less) and above	Very long lasting. Useful for all crops. Useful on all soil types. "Micro Terraces" for trees.	Design and installation is costly	Long term good return on investment.
Forests (and prairie, and meadows, and other natural ecosystems, no mow	Everywhere, often used even on the sides of steep terraces	The ultimate rain water harvesting feature. Catches and stores nearly 100% of water. Long lasting. productive, provides habitat and generates fertility.	Very low cost if planted with seeds and cuttings	Best ROI and long-term effect.

Beauty in Abundance

Patterns for Harvesting Water on Large-Acre Flat Land

People see all the cool Permaculture water-harvesting techniques for slopes and feel disappointed they're on flatland. The thing is, all those techniques are to keep the water from flowing down the hill, a problem we don't have on flat land.

But in a heavy rain or on poor bare soil, the water will still form sheets and flow away, so we can do things to keep the water in place, help it soak in, and then keep it where the plants need it. This last bit is accomplished by preventing the sun and wind from boiling it off, and improving soil carbon and structure so the soil itself becomes a water-holding device.

Some techniques:

- **Permaculture zones.** This is still our most important water-saving feature, but not so relevant on broad acreage, so I'm assuming you already did other techniques near the house. Which means I'm also assuming you can't get acres of mulch delivered and put in place.
- **Grow in guilds and polycultures.** This builds soil and infiltrates water. Tilled bare soil can have infiltration nearly as low as a concrete parking lot.
- **Make mulch/compost basins** and plant trees densely around them. These create underground mini-lakes near the trees.
- **Nurse logs in general.** Nurse logs are one of our best research-based approaches for keeping water near crops. They tag-team with fungi to create super soil systems that fertilize crops. Nurse logs are also proven to increase plant and beneficial insect biodiversity.
- Combine those two and form **mulch basins** out of **nurse logs**.
- **The haphazard mulch mandala** in the chapter on paths is a great water-saving feature.
- For tree establishment, **plant "sacrificial trees"** densely around crop trees, then chop & drop them for "rough mulch." You don't have to run this through a chipper, just hack them with a machete. (I get seeds from pioneer species at local parks, part of the Miyawaki technique.)
- **Nurse log water harvesting works.** We can even create "net and pan" (you can search online for it) or large 30/40 system basins with nurse logs. (These are 30*40' catchment circles or rectangles with nurse logs and an infiltration basin on the slightly lower side.)
- **Use broad-leaved plants** like squash, gobo, and catalpa, which conserve water and build soil.
- **Use smart watering design** for emergencies, like the five-gallon bucket plan (poke holes in the bottom of five-gallon buckets and fill them with a second bucket.) Five gallons is an optimal weekly watering for new trees if there's no rain. With nurse logs and mulch, three gallons will do.
- **Hugelkulturs** of different kinds can be effective, though I usually consider these inappropriate for broad acreage.

We can stack many of these techniques together and create systems that are more effective than swales, and that are also holistic, building soil, sequestering carbon, reducing weeds, increasing plant biodiversity, and fertilizing plants while saving water.

Chapter 12: Gardening Beautifully with Holistic Natural Management

"I want my gardening to be a gentle art."
—Toby Hemenway, author and Permaculturist

To create truly beautiful landscapes, we must have beautiful ways of gardening. We have to abandon the old war with nature and find ways that recruit nature to cooperate with us.

This section will give a few gardening patterns and techniques that will help you kick plastics, poisons, and petroleum in the landscape, and instead start gardening in a beautiful and gentle, yet effective way.

Once we are clear on our goals to have a truly transformative landscape, and we need to know what to do and what to prioritize.

The big problem is that in the world today, there are a thousand videos on everything you might consider doing. You can find out instantly how to do anything. But there is very little interest in how to know to do.

How to Know What We Need to Know

If we want to grow food, heal ecosystems, get healthy, or interact in pretty much any biological system (which includes human

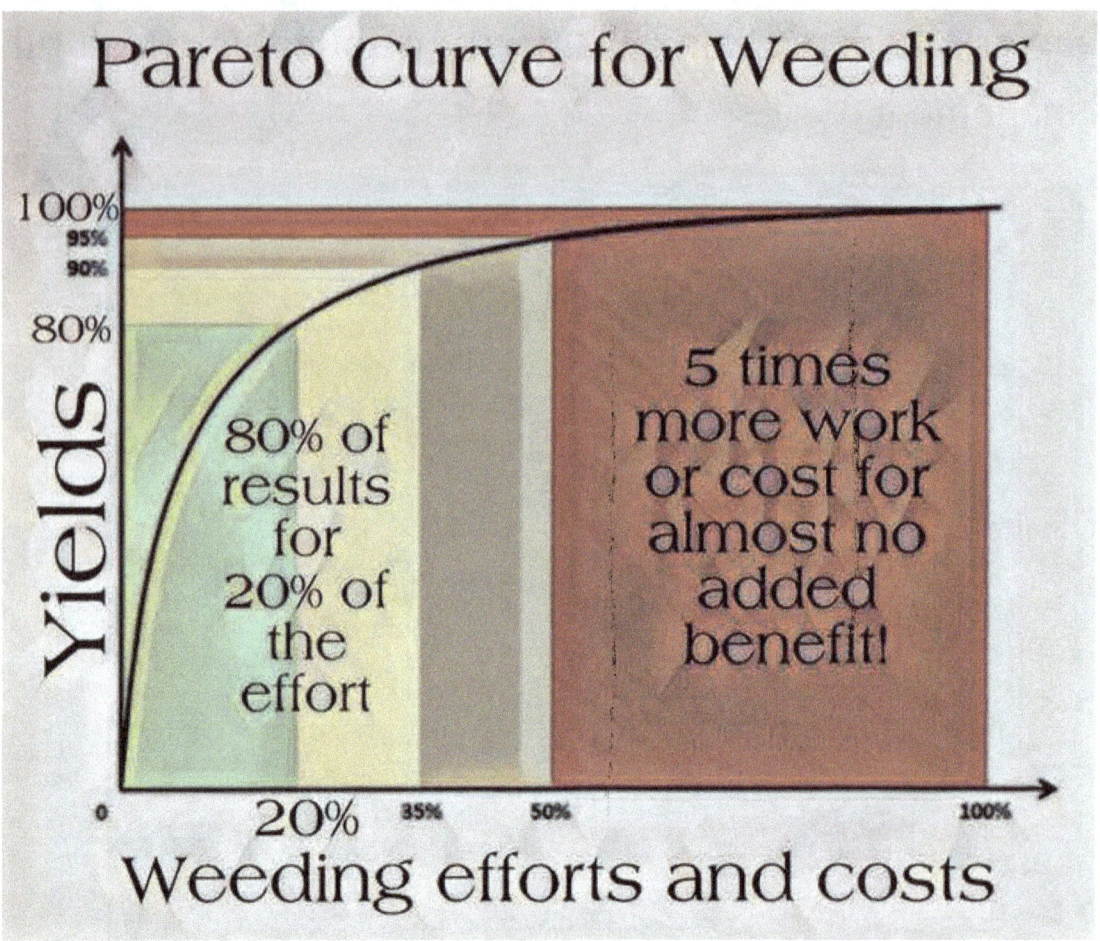

communities) probably the single most useful thing we can understand is the **80/20 principle, or the Pareto curve**.

Simply put, we get 80% of the results (change) for the first 20% of our work (inputs.) This has been a key in agricultural science for pretty much as long as there has been Agricultural science, and it's the theoretical basis for Integrated Pest Management (IPM) and new alternative approaches like Holistic Management and, yes, Permaculture.

Research shows we get an 80% reduction in pest damage from our first 20% of pest control efforts. After that we get "diminishing returns" for the same level of pest control. Same for weeding, watering, fertilizing, etc.

So, the process of REAL, VALUABLE education for most of us is to learn enough to know what that 20% of truly transformative information is. This was part of the thinking and design of the original Permaculture Design Certificate course, to give "regular people" a selection of 20% powerful info on a variety of topics necessary for designing a human settlement—including food, passive solar energy, community organizing, money, etc.

But in academia people are rewarded for mining deeper and deeper truths well past the point where there's any practicality, so it's easy for us to fall into the same trap! That level of knowledge is useful for pure knowledge, but I don't personally need to know the maximum air speed of an African swallow to enjoy watching birds.

For example, in Permaculture it's easy to fetishize the minutia of soil science, and for hundreds of years agriculture graduates have been peddling "scientific agriculture" products like rock dust, designer microbial tea concoctions, and humus (which later are found to not even exist!) [2]

But none of these expensive products appear to give much if any improvement at all over what French intensive gardeners were achieving 150 years ago with composted manure and their 20% info of scientifically optimized plant spacings, minimal pest control, and scientific watering procedures.

When it comes to achieving sustainable high productivity that also stewards biodiversity and ecosystem health, it would be difficult to improve much on the self-maintaining traditional forest garden systems, slash-mulch systems, and so on that indigenous peoples evolved around the world, and from what I understand they accomplished all that without too many PhDs and microscopes.

If you want to become a great, knowledgeable practical gardener, here's my advice as someone who's been at it for decades now: follow the Pareto curve in your studies, find the things that will have the biggest Return on your Investment (ROI.) Don't overwork to be perfect on one thing, work to be **"good enough"** in a lot of areas of the garden.

Know that general principle that soil microbial diversity and abundance improves plant health and productivity. Then, let PhD's fuss over identifying the various 10,000 organisms in a teaspoon of soil while you learn which SYSTEMS will most encourage microbial diversity and abundance in your garden. Focus on the 20% that will get you 80% of the benefit— and for the least possible cost. Hint: most of the time research has shown that's probably just using polyculture, perennials, and organic mulches, which works better than fancy teas and inoculations. Done. So the thing to really study is systems for growing with deep mulches and polyculture, not memorizing the names of 28 microbes in somebody's signature tea.

If you want a good return out of your garden, then stack 20% approaches: do the 20% most impactful actions on pests, the 20% for water, the 20% for weeds, the 20% for fertility, and so on. That probably looks like our Transformative Adventures "Holistic Natural Gardening" management system.

If you want a good return on life, IMO, it's probably good to stack 20% efforts there, too, do the 20% you need for a healthy body, the 20% you need for a healthy environment, the 20% for stable finances, the 20% for a vibrant community, 20% for an optimized diet, the 20% for a healthy mind…. That's "holistic life management" I suppose. It looks a lot like our **"solid base."**

---

[2] The contentious nature of soil organic matter, Johannes Lehmann, Nature, 2015.

Holistic Natural Gardening Patterns

Following the Pareto curve, there are ten high-impact holistic management techniques we can put to work in our gardens and landscapes. Each of these is so high powered because they have holistic effects, meaning they impact many different elements of garden performance, they impact many different Pareto curves.

For example, mulching can virtually eliminate weeds, provide habitat to beneficial insects, dramatically improve water use, improve plant health, increase soil carbon better than adding compost, and even provide all the nutrients the garden needs without added fertilizer. So, by doing just one step, mulching the garden, we are replacing a whole lot of work we would need to do otherwise, work that would never be as effective as mulching. For example, there is an abundance of research showing that fertilizing plants causes many problems, including increased pest and disease issues. Meanwhile, 4 inches or more of organic mulch can provide all the nutrients we need, without causing these problems. And in multiple studies, mulch alone improved plant growth and productivity better than the fertilizer!

Because people often doubt that mulch can replace fertilizer, let' look at the numbers. In the Grow BioIntensive system, compost paired with good crop rotation has been demonstrated to create sustainable yields over time while growing soil health. In that system, one inch of compost is added to the beds annually. This is the standard recommendation for organic vegetable production. One cubic yard of good quality finished compost contains eight lbs of nitrogen, which will be released over multiple years. One cubic yard of loose green grass clippings weighing 325lbs will contain 4% nitrogen, over 12 lbs of nitrogen. Fallen tree leaves will contain between 1% to 3% Nitrogen, giving oak leaves as much N as cow manure per weight. 4 inches of mulched leaves will at minimum have as much N as an inch of finished compost. Gardens with consistent, deep mulch should have all the nutrients they need to maintain high productivity.

The same effect applies to watering. Good deep mulch can mean that in some climates no additional irrigation is necessary. While irrigation may cause problems in many crops, such as powdery mildew or blossom-end rot. Mulching instead, removes the risks associated with irrigation.

The same multi-faceted benefits apply to all of these ten powerhouse techniques.

Holistic Natural Management
10 High Value Techniques

1. Find balance with nature. Know when to take action and when to let nature lead. (Create Permaculture zones.)

2. Use natural no-till (no poisons or plastics) in permanent beds.

3. Avoid chemical fertilizers, plastics, herbicides, pesticides and fungicides.

4. Use continuous, deep, organic mulches.

5. Harvest rainwater by putting beds, nurse logs, and paths on contour, or by using swales or basins.

6. Use high integrated biodiversity, for example, by using Polycultures and guilds, native plants and wildlife habitat.

7. Integrate a lot of perennials throughout the system.

8. Use careful crop selection, take "right plant right place" very seriously.

9. Study good plant spacings and use them mindfully. Usually this means planting densely, but giving each plant just enough room to grow.

10. Grow your own fertility and mulch on site, using the methods above, with a judicious use of homemade compost.

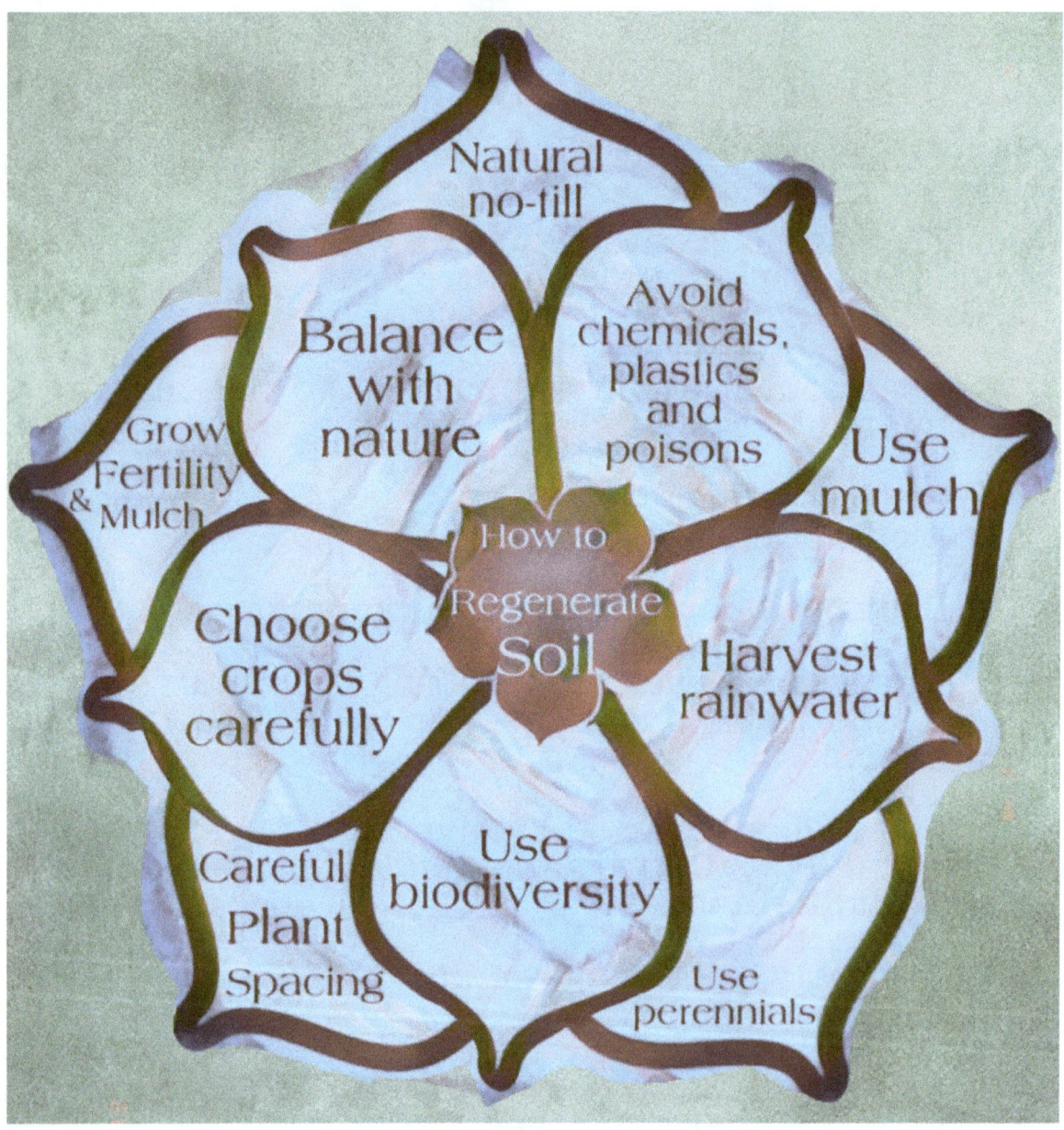

I. Find Balance With Nature

This is number one on our list of holistic natural management techniques, because it is our most important technique for an easy, productive garden. Understand, this isn't just a vague aspiration, it's a detailed and highly practical method.

When it comes to weeding, this is our number one tool. When it comes to pest and disease management, this is our number one tool. When it comes to irrigation, this is our number one tool. And when it comes to keeping labor down this is our number one tool. Get the idea?

At its most basic, the idea is that smart humans create a gradient between human-managed systems, and nature-managed systems. We know when it is worth our while to put our energy into managing things like intensive gardens, and when to hand over the reins to nature, such as in forest gardens.

For all of human history, right up until the invention of oil power, knowing where to put our energy and where to hand things over to nature has been the most important technique humans used for all of those problems. And yet today, most gardeners never even think about it. But most of our garden problems are caused by not thinking about it.

The image of the conventional garden looks pretty much exactly like most gardens and farms I visit, even most regenerative and Permaculture farms.

I can think of one case in particular, where I visited a regenerative farm for a consultation and the farmer began to tell me about his day.

"First thing in the morning when I get up, I hop on my quad (all terrain vehicle) and head out to the livestock."

See, the quad was absolutely necessary. It was a long narrow property with the house right in the middle, and all the livestock was on one far end of the large hundred acre property, and the market garden was on the exact opposite side.

"Then, it's back to the house, and after breakfast, I head out to the market garden to work. There's hours of weeding work to do every day, so I'll usually spend all morning there. After lunch, it's back on the quad to go check on the workers by the livestock, then I quad back to the veggies to work for the afternoon. Then it's back on the quad at night to make sure the animals are locked up."

See the pattern? Back and forth, back and forth all day on that quad. When it was time for meals, they usually didn't have any vegetables, because you'd have to walk all the way out to the garden to get them!

When I spoke to his workers, they told me they had the same day, back and forth, back and forth all the work day. Only, they didn't have quads! One worker said he thought he spent 60% of his paid work day walking back and forth.

Now, if you've got all the vegetables a ten-minute walk from the house, with no water, you are going to need irrigation. And, if you know anything about irrigation, you know irrigation hoses leak. This is why Permaculture founder Bill Mollison said "if you spend your money on mulch instead of irrigation, you'll get twice as far and be twice as happy for it."

When I was there the irrigation hoses had multiple leaks. What happens when you go back and forth, back and forth all day on quads along a leaky irrigation hose? The whole thing was turning into a mud pit. They had a real live moat surrounding their house.

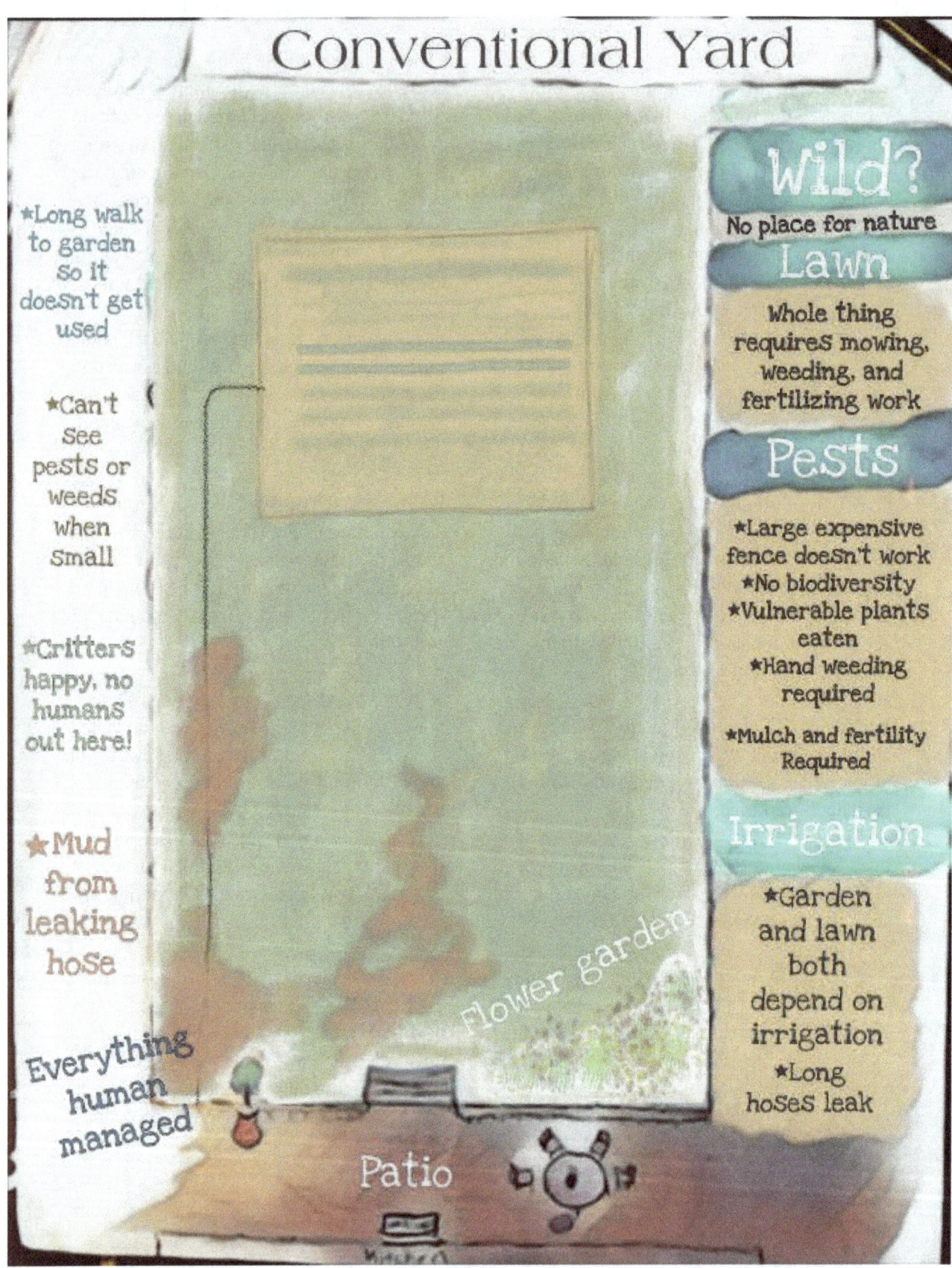

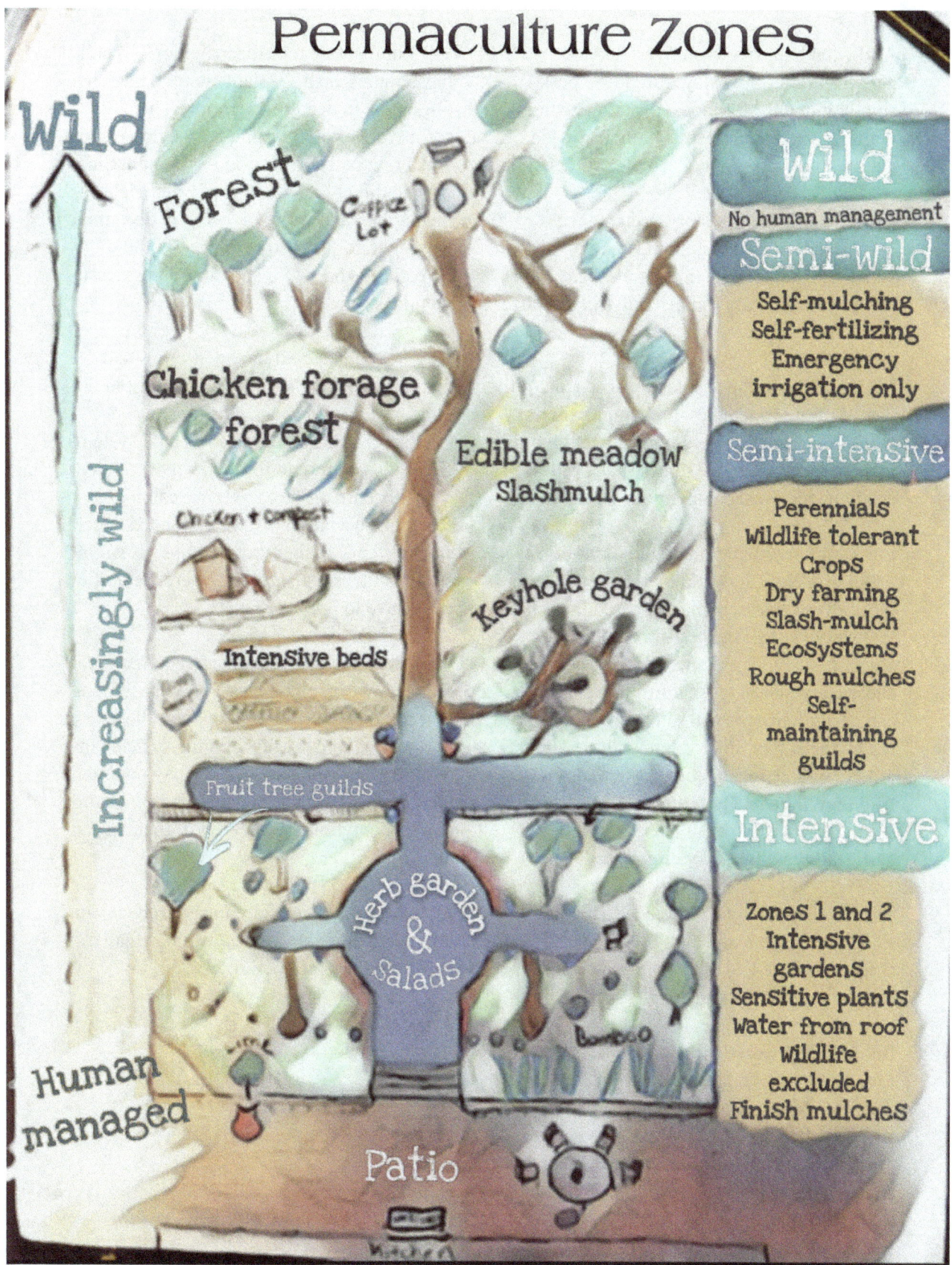

And what happens when livestock and gardens are far out away from those pesky, smelly humans? It's a pest fiesta! Messing with electric fencing was a daily job, trying to get the thing to reliably keep the critters out.

Now, when I tell this story I often have homesteaders say, "this sounds like my yard!" So, here's how finding the right balance with nature can instantly fix all those problems.

First, let's go to the vegetable garden and pull out all of the vegetables that require a lot of work. That includes tender greens that require lots of water, are prone to pests, and fade easily under weed competition.

Let's stick those in a small garden right by the house. Now, it would be impossible to fence off a three-acre market garden, but now that we've got all the pest-prone crops in one place, we can use a tighter planting design and we can shrink those down to a space we can easily fence off with little cost. Because it's by the stinky, smelly, human hole, the critters will be reluctant to go there anyway. And they'll be less persistent about getting around a fence. Now, an intervention like garlic spray will be very effective.

The number one tool for weeds? This farmer said "I can't deep mulch my whole three acres, so I have lots of weeding to do! Deep-mulch gardening doesn't scale!"

But now that we've shrunk the size down, we can easily do a deep-mulch garden for these weed-sensitive plants, reducing our annual weeding work to nearly zero. Since we can grow the mulch right near our beds with **mulch-maker plants**, we don't even have to work to cart mulch around.

The number one tool for irrigation? Luckily these same sensitive plants tend to be the ones most in need of irrigation. The most sensitive we can put right by the down spout and a rain barrel with flood irrigation. Since we can direct the whole roof water right to the garden, and we're growing in deep mulch, we might not have to ever spend time irrigating! And, we'll spend no time messing with leaky hoses.

Now we can take all the other crops and clump them together by their needs, too, in good systems for them. The most weed-tolerant crops can go into very wild systems where they can fend for themselves. The crops that are prone to deer we can try to fence off, because now we only need a much smaller fence. Other crops can be grown in systems like **slash-mulch**, where we don't even have to prep the beds.

Now, with the growing space required greatly reduced, we can let the rest go wild. With a few small tweaks, we can convert those ecosystems into useful foraging systems that require almost no care. These may not be the most productive, but they require almost no work, so anything we get out of them is just extra.

In this way, we can dramatically reduce our labor, irrigation, and costs, while actually increasing our productivity.

Permaculture Zones

An excellent system for designing a better balance with nature is the Permaculture Zones system. This system breaks down these levels of human/nature balance into five different zones, as seen on the graphic in this chapter. I've also included a list of my crop selections by zone.

In a later chapter, we'll cover an approach to Permaculture zones specifically for garden aesthetics.

Beyond the practicalities, this type of system, helps us better define our relationship with nature. It means that in every garden, there is some space reserved for nature, and that there is a balance. We don't have to go to war over our whole properties. We can leave some for other species to thrive. With this balance better defined, we actually start to welcome and enjoy the presence of wildlife, because they are part of our design, not enemies to our poorly designed gardens. Blown up over a whole culture, this becomes a society where there is room for other species, indeed, other species are actually welcome.

Related Patterns

On the whole concept, we have patterns for **mosaic landscapes, Permaculture zones for aesthetics, patterns for garden layout and paths, and crops by Permaculture zone.**

On the human-managed side, we have **intensive gardens, Holistic Intensive Gardening, ways to start beds for holistic natural gardens, sheet-mulching, and double-digging.**.

On the Nature-Managed Side, patterns are included for **food forests, agroforest systems, edible meadows, edible hedgerows, slash-mulch, nurse logs, and terminator guilds**.

Permaculture Zones for Aesthetics

The key to applying zones to create beautiful landscapes is to understand that we have limits on the time and resources available. So we need to use our resources where they will have the biggest impact, giving the impression of an impeccably managed and maintained landscape.

If we're frustrated because the landscape's too messy, it's because our zone 1 is larger than we can handle.
Zone 1: Approx. <1,000 Square feet of mulched garden beds, and 1,000 SF of weekly mown Lawn.

Zone 1 Aesthetics: Can be "tidy" formal gardens, geometric designs, straight lines, annual vegetables and flowers, herbaceous perennials, clean hard edges, relatively weed-free. Can use compost to maintain "garden soils" that can grow most plants.

At this scale, it should take a few hours/week of maintenance to mow and tend the zone 1 gardens. Larger properties may have smaller zone 1 areas, and very small properties may have larger zone 1 areas.

This zone includes the most visible areas, such as near the driveway or areas near the road frontage, near home entrances, main window views, and main outdoor living areas. These can have intensive gardens, small mown lawns, raised beds, and **"hard edges"** that are maintained with an edging spade or borders.

Zone 2 aesthetics: more nature-managed. Wilder. **Cottage gardens, Jardin de curé**. Google: Gertrude Jekyll, Piet Oudolf, Cambo potager, wild, and post-wild plantings. "Romantic." More herbaceous perennials, trees, and shrubs. No bare soil, tight plantings. Use of "guilds," hedges, hedgerows, forest gardens. Mostly **soft edges**, curved lines. Or "formal layout, wild plantings." In ornamental landscapes formal edges may be maintained by hardscaping. Mown grass or hardscaped paths.

Zone 3. (Only in professionally productive landscapes, production gardens, nurseries, farms.) Usually <1/3rd acre for most families with a full-time worker. Aesthetics are usually a secondary consideration. Interplantings, herbs, and flowers, companion plantings, signs of care at entrances, garden gates, etc.

Zone 4: Semi-intensively managed ecosystems. Usually < 3 acres for full-time homesteading families. Primary tasks: intervention care for principle crop plants, mown or grazed paths. Aesthetics: wild nature, woodland, ecosystems, hiking trails.

Zone 5: Wild nature. Tasks: possible annual path maintenance. Aesthetic: wild nature, woodland, ecosystems.

I. Use Natural No-Till; No Plastics or Poisons.

This is another important pattern because it impacts virtually every aspect of the garden's performance.

In the holistic natural garden, we rely on biodiversity to do our work for us—on beneficial insects to do our pest control, on microbial diversity to keep diseases in check. As soon as we use plastics or poisons this natural system breaks down, and we are dependent upon more poisons and plastics to keep things working.

Better to not get started in the first place.

No-till is a powerful technique, which has been demonstrated to improve soil fertility, completely stop soil loss and erosion, enhance biodiversity, improve soil texture, keep plants healthier, and even nearly eliminate weed-germination.

Patterns for Natural No-Till in this Book:

I've included a whole chapter on **starting gardens the holistic natural way**. Most of the techniques in this book are to help you be successful with an natural no-till approach.

III. Integrate for an Easier, Better Garden

Here's why we say "integrate rather than segregate."

There is human tendency to want to create "pollinator gardens" and "rain gardens" and "native plant gardens," and vegetable gardens," and to keep them all separated and segregated into different parts of the yard.

But nature doesn't think that way. Nature doesn't know what the heck-pie a "pollinator garden" is. Or a "vegetable garden."

Those distinctions don't make any sense to fungi or birds, or insects or the plants themselves. Basil doesn't think she's an "herb," she thinks she's an F-ing GODDESS.

So when we prioritize those made-up categories we necessarily do so at a loss to other priorities, like ease, function, beauty, and productivity.

Let's say you make an "herb garden" of typical "herb garden plants." That's all fine and good, except an herb garden is not a naturally occurring ecosystem (although with the right soils and conditions one can create a natural ecosystem with a high quantity of aromatic herbs, such as occurs in certain hillsides in Mediterranean Europe.) Nature doesn't think in "herb gardens," she thinks in ecosystems and plant communities. So, this herb garden will be missing some of the necessary components to make functioning plant communities, and you (the overworked gardener) will have to supply the work that would otherwise be done by plants in a functioning ecosystem.

Or let's say you make a "rain garden" by the downspout and only put master-gardener approved "rain garden plants" in it. The first problem is that a "rain garden" is not a naturally occuring ecosystem, it will be missing important niches, and you will find plenty of non "rain garden plants" will happily volunteer in those niches as "weeds." A second problem is that the nice, wet, rich soil in that basin will be ideal habitat for other plants you may want to grow, like the wetland vegetables skirret and marshmallow, or the herb valerian. These COULD fill those open niches making a more stable, low-work plant community. If you have the idea that these "don't belong" in a rain garden then you'll have to move them someplace less appropriate, and work harder to supply the conditions they would have naturally had in the rain garden.

Or you make a "vegetable garden" in the sunniest, richest part of the yard. Some of those veggies are natives and would be absolutely ideal in "native gardens" except that we don't think they should go there. And some native plants, flowers, and herbs would help fill out the function of the veg garden making it closer to a self-organizing plant community, except we insist they shouldn't go there. Instead, we segregate them off to other parts of the gardens where they will do less well, and without the help they would have received from some of the vegetable plants. We rob both the veggie garden and the other plants.

If you really, really, really want to have a pure "native pollinator garden," that's okay, but understand you'll be sacrificing other goals and working harder to prioritize that human-centered goal.

In the integrated garden, or Transformative Landscape, we follow nature's rules instead of senseless human distinctions. We start by creating stable plant communities that best meet our priorities of ease, productivity and beauty, instead of prioritizing rigid made-up categories. We start by listening to what *really* cares for the plants, the earth, and people.IV. Use Deep Organic Mulch

We've already discussed how deep organic mulches can superpower the garden, saving water, building fertility, hosting beneficial organisms, defeating pests, eliminating weeds, and so on. People always ask "which mulch is best?" The answer is, of course, "it depends." We could go into a great deal of depth on which mulches to use and where, but let's shoot again for a "good enough" approach.

Mike's Simplified Mulch Rules:
1. When you have access to clean materials, use straw mulches or other herbaceous materials to mulch annual vegetables.
2. When possible, use woodchip mulches for most trees and woodland plants.
3. When possible, grow appropriate mulches in the garden, right next to the crops if possible.
4. Use a finish mulch like cocoa hulls or mown grass in high visibility areas.
5. If those aren't possible, just use whatever bulk clean organic matter you happen to have around. Use it wherever it's needed. It will be more than "good enough."
6. **Only use clean materials. Many agricultural wastes, lawn wastes, and straw may be contaminated with persistent herbicides, which can kill the garden for up to ten years or possibly longer.**

Later we'll visit patterns on mulching including **finish mulching, sheet-mulching, and nurse logs.**

V. Plan Gardens That Harvest Rain Water

If we want to make resilient, healthy, low-work gardens, then planning and creating gardens with built-in irrigation is another key point. In fact, it's so important that I could write a whole book on the topic. But of course, many other books already exist to cover it. Many books on Permaculture will be helpful in designing a whole site to be truly water-wealthy, abundant and drought-resistant. "Water for Every Farm" by Mark Sheppard is a milestone on the topic, and "The Designer's Manual" by Bill Mollison is still one of the most comprehensive treatments of the topic.

For the purpose of this book, I'm taking the approach I've developed for Transformative Adventures. Instead of worrying too much about having a comprehensive site water program on paper, we will make sure that every garden system we make, every new garden, every new bed has its own water plan built in, along with a fertility system, and pest and disease systems. If we build a garden out of self-sustaining elements, we can work a bit at a time and come up with something that will be "good enough!"

Related Patterns:
Throughout this book there are patterns for good water design, but some of the most important include the new garden design elements of "**deep mulching,**" **nurse logs, hugelkulturs, French Intensive gardens, food forests, hedgerows, edible meadows, mosaic landscapes, Permaculture zones, the Permaculture Garden Make-over Guild,** and some of the other guild designs.

VI. Use High Bio-diversity for Fertility, Pest and Disease-resistance

The key concept here is is polyculture, growing many plants together, instead of monoculture, growing only one crop in a bed or field.

Speaking of multi-function elements with holistic impact, biodiversity is a powerhouse.

High biodiversity has been demonstrated to dramatically reduce pest and disease issues, increase productivity, and even generate fertility without fertilizer, compost, or Nitrogen fixers!

The drawback of high biodiversity, is that it is a high-skilled technique. One has to have good information and skill to design plantings that pack in biodiversity without reducing plant health or productivity. Some YouTube videos show high biodiversity polycultures where plants are so overcrowded that only small misshapen vegetables are produced.

That is 100% not what I advocate for. If that's the result we're getting, we need to thin more.

I advocate for systems that allow us to use high biodiversity while maintaining good plant spacings, and so actually increase crop quality and productivity. But that means we can't just cram a bunch of plants together, we have to know what we're doing.

Much of this book is about creating plantings that will help accomplish this goal.

Related Patterns:

Of special help will be the sections on **French Intensive planting** designs, which show actually successful high-productivity gardens I have used for many years. The sections on **scatter-sown- polycultures, and the aesthetic guilds** will all help. The **herb spiral and the Garden Make-Over Guild** will be good for beginners.

VII. Integrate Perennials, Including Woody Perennials, Trees, and Shrubs

Perennials are the long-term thinkers of ecosystems, always investing in soil and connection. Because perennials cultivate long-term relationships with microbial communities, and continuously explore the soil with their roots, they are the great builders of soil structure. They will generally be the great builders of insect communities that will protect your crops. And of course, trees are the greatest of these perennials. Trees in the garden bring new levels of ecological health and abundance.

VIII. Use Careful Crop Selection

If you want to have an easy garden, plant easy plants. Done. If you plant things prone to pest, disease, and weed problems, you will be doing a lot of pest, disease, and weed management. Some of this relates to zones, which help us grow more needy plants where they can get extra attention, and allow the more resilient crops to do without.

Here are some additional tips on crop selection:
1. Grow organic seeds: they have not been dependent on spraying.
2. Grow heirloom, open-pollinated varieties. These, too, will usually be more resilient, because they are less cultivated for a particular quality.
3. Grow hybrid "disease-resistant" varieties ONLY when the particular disease has been problematic and other attempts have failed.
4. Grow purple varieties, especially of brassicas. These tend to be far more disease-and-pest-resistant.
5. Save your own seeds. Keep seeds from the best performers, as this will localize your crops to your unique pest and disease pressures.

IX. Use Careful Plant Spacings

Generally speaking, this means tight spacings. In French Intensive gardening, this is called the "greenhouse effect," in which the tight plantings create a zone of moist, rich soil, conserving water.

But that does not mean too tight! Ideally we want plants so that they are barely touching one another at maturity. If we over-plant, we can harvest plants as they grow, always keeping plants barely touching, or thinning them when they begin to touch.

Alternatively, careful spacings CAN mean shifting to extra-wide spacings. In times of drought and heat, I often mass harvest my gardens, leaving only the tough drought-tolerant plants like tomatoes and peppers. These are now spaced extra-wide and I mulch very deeply to conserve water. So the idea is to be mindful of spacings, giving plants what they need.

X. Plan Gardens to Grow Fertility and Mulch

One of the most important things we can do to create a truly sustainable garden is to grow our own fertility, and our own mulch.

Not only does this ensure that the garden is sustainable, it means that there is always a source of mulch nearby, and that the garden fertility will never be a problem.

In the chapter on Transforming Soil, we will discuss **patterns we can use to design fertility into the garden**. Several of the **patterns on polycultures and guilds** will help to design mulch makers right into every part of the garden.

Adventure Step 4: Zone, Sector, Microclimate, and Use Analysis

Our next adventure will be one of observation, directed thought and evaluation.

We'll use a couple of tools from Permaculture: zone analysis, sector analysis, microclimates, and uses. Start with a sketch of the site or a print off map.

Zone Analysis was discussed in this section, and talks about finding our correct balance with nature.
-Where are the areas you naturally visit each day?
-In the areas closest to the house, where are the places with the best sun, best soil, and best water access? These will be our most important opportunities for high-productivity gardens.
-Are there opportunities for productive gardens further away? How will they be used?
-Are there areas you rarely visit or that are difficult to access?

Sector analysis talks about other opportunities and energies. On your map, mark:
-any entry points for deer or other wildlife.
-good and bad views.
-flood or fire risk areas.
-areas with full sun.
-areas with shade.
-existing trees.
-areas with good water access.
-areas where people will see your property most, this is an advertising or beauty sector.
-aspects: North, East, West, and South-facing areas. What are the opportunities here?
-areas close to the house that could easily be enclosed.
-frost pockets or other microclimates.
-any other energies or opportunities.

Uses: next sketch out a rough idea of what each part of the property will be used for. Allow this to change and develop as you move through this book.

A good Permaculture design course should go deeply into these topics, especially microclimate design and crop planning.

*What's the best aspect (slope, side of the house) for a garden?
Answer: it depends. We can use all the microclimates to our advantage, matching the crop to the conditions.
The image shows an herb garden to the hot dry west, where those conditions will make potent, fragrant herbs that are easier to harvest. The south has tomatoes, representing crops that require long days and a long season for full sugar. The east has a garden of tender annuals, which will be slower to bolt and appreciate the gentle heat and protection from frosts. The north has forest gardens, sheltering the home and gardens from the winds.*

Garden Aspects:

South
Warm, dry
Long season
Longest days

Crops that require ample sun for best sugar content and ripening:
Tomatoes, peppers, melons, high-sugar fruits

Crops that require season extension.

Crops that are at a premium in early spring: strawberries, asparagus, honey berries.

East
Gentler heat, moist, shorter days, later spring, least frost risk.

Tender plants that appreciate the gentle heat and moisture and don't require maximum sun. Brassicas, perennial vegetables, peas.

Crops that are prone to bolting: Radishes, lettuces, cilantro, carrots, fennel, etc.

Crops that are susceptible to spring frosts: apricots, peaches, citrus, persimmons.

West
Hot, dry, windy, shorter season.

Plants that thrive in hot dry conditions, like herbs: oregano, lavender, thyme, rosemary, bay, germander, medicinal herbs of most kinds.

Annual and herbaceous vines to shade house from unwanted sun through western windows. Yet, these can die back in winter to allow potent western sun into the house: squash, hops, groundnut, cucumbers, luffa, etc.

North
Cool, wet, cold winds, latest spring, Earliest winter.

Forest gardens for shelter from cold winds.

Tree crops that can reach up high for sun, making aspect irrelevant.

Perennial crops that don't require long seasons. Shade crops grown under trees and will appreciate the water conservation.

Crops prone to spring frosts, since spring will come later, after the coldest weather has passed, though it will be more prone to frost than the Eastern aspect.

Crops that require more chill hours: apples and pears in sub-tropical locations.

Some microclimate opportunities, listed from those with a strong effect to those with less effect. The greatest effects are those from the largest scale, like major lakes, valleys, or mountains. But we can still take advantage of small opportunities.

The background image shows several opportunities at Lillie House:
Enclosure, location in a thermal belt slope, the strong thermal mass of the house, a frost channel that drains frost away from the sheltered microclimate, and dense plantings. Lillie House also uses small ponds and stacked stones. We get the greatest effects when stacking microclimate features in this way.

Some Microclimate Patterns to Consider:

Smart design can extend the season and hardiness zones and improve yields. At Lillie House, a combination of multiple climate control patterns consistently extended the growing season a month or more each year. In one case, a very early frost ended the growing season in our whole region in September, but we were still harvesting tomatoes and peppers in December.

Enclosure: the strongest microclimate factor is enclosure, such as with hedgerows in the home garden pattern. Enclosure protects from winds and frosts, and has been found to increase crop growth and yields by up to one-third.

Windbreaks: well-designed windbreaks (like a tapestry hedge) reduce wind and can increase crop growth and reduce heating costs. Ideal windbreaks slope gently upwards and let ample air filter through to decrease turbulence and create an air buffer. A windbreak that is one-third evergreens functions well in winter. Poorly designed ones (such as the common row of dense evergreens) can increase drafts, increase heating costs and plant damage. (See image.) A windbreak upwind of a planting can help establishment of the whole site.

Aspect: the slope or side of a structure the garden is on has a significant impact. Trying to have a lettuce farm on a western slope, or apricots on a southern slope may be doomed to failure without special design.

Thermal mass: objects can absorb heat during the day and radiate it out at night, reducing temperature swings. The larger and denser the object, the better. The strongest thermal mass is water, so lakes and ponds have a large effect. Stone walls or even individual stones may also have an important effect, protecting tender plants from frost.

Thermal walls: equator-facing stacked stone or clay walls will have a strong thermal mass effect, and trees espaliered against them may buy as many as two hardiness zones of protection depending on crop and site specifics.

Urban heat island and lake effects: these are simply some of the largest thermal mass effects.

Frost pockets: since cool air sinks, valleys, small depressions, or structures can collect cold air and cause frosts.

Thermal belt: this is a strong frost pocket effect. Wherever there is a valley, cold (heavy) air sinks, creating a blanket of warm air on the higher elevated hill-sides around the valley. Such places are very beneficial for tender crops.

Frost channel: since cool air drains downhill, we can move frost away from tender plants by creating mown paths away from crop areas. We must choose whether to have paths on contour for water collection, or frost channels. Clever design can incorporate both.

Frost pockets occur in any low spots where cold collects, especially where there is morning shade.

Well-designed windbreaks offer significant protection from wind and frost

Chapter 13: Holistic Intensive Gardening

Dynamic Polyculture at Lillie House, throw-cast then selectively thinned.

When it comes to growing produce in ways that are beautiful, there is one form of gardening that has a reputation for creating gardens that are both extremely productive and gorgeous: French Intensive Method (FIM.) It also teaches the gardener true advanced gardening, just from practicing it. In this book, I'll present a holistic version using modern methods, developed by teachers and students in the Transformative Adventures Cooperative.

This old, evolved set of techniques uses planting designs with precise, tight, non-row spacings and smart interplantings – all to achieve the highest possible productivity and quality. The style called French Intensive Method, (FIM) was pioneered in French gardens, famously loaded with produce and flowers so they end up being abundant in beauty as well as bounty. And FIM naturally teaches us about true sustainability, companion planting, soil building, plant spacing and size, and producing top-quality produce.

FIM is one of the major things that gives my gardens, including the garden at Lillie House, a distinctive look. Conventional gardeners find it incomprehensible, or even "impossible." Gardeners often recoil at seeing these dense plantings without bare soil, despite them being the research-based optimal spacings for superior produce and total yield.

I have learned a great deal from FIM, and the simplified systems taught by Alan Chadwick (BioIntensive French Gardening) and John Jeavons (Grow BioIntensive.) The book "How to Grow More Vegetables" by Jeavons is still one of the best books on vegetable gardening ever written, and would make a fine companion to this book. Add our Holistic Natural Management techniques for greater ease and sustainability, and you gave Holistic Intensive Gardening.

To come right to the point, it's absolutely revolutionary to understand how these methods optimize the "return on investment" of a garden or farm system.

First, FIM gives the highest possible yield per square foot of any system. Consider this: like historic FIM gardeners in the suburbs of Paris, gardeners in peer-reviewed research have achieved yields that are four to six times the best conventional yields, and in some cases over ten times. So, the FIM gardener can do on 1/4 or 1/6th an acre what a conventional market gardener using a tiller and planting in rows does on 1 acre or in some cases, four acres.

Of course, this requires more work, more design and more fertility management. But — here's the key—NOT PROPORTIONALLY more.

In other words, it will take significantly less time on average to manage 1/a quarter or even half an acre using holistic intensive gardening than it would to manage that acre conventionally. It's realistic to grow *more* produce on a 1,000' holistic intensive garden on a few hours per week than you could grow on a third an acre of tilled row crops or "plasticulture," and that would be a full-time job.

And it will not require a tiller or imported, unsustainable fertilizers. And finally, quality is often higher, and so is profitability. So, while it will typically take a couple of full-time workers to manage that one-acre farm, one person could get the same (or better) outcome from a quarter acre under holistic natural management.

This leaves three-quarter-acres which can be managed very "extensively," by handing management over to nature, using other patterns in this book, like edible hedgerows, edible forest gardens and edible-meadow type systems, or possibly small livestock. The best of these are traditional, evolved patterns with long-established proven viability and management techniques. All of this can add significantly to yield, while helping to maintain fertility sustainably. NOW, we're using good energy-efficient design! And it's also just good math.

This *intensive natural polyculture by Michael Wardle, of Soil Savour Permaculture, is both beautiful and productive, just like the best FIM gardens.*

Some Rules from French Intensive Management

Holistic Intensive Gardening is a type of gardening that requires knowledge and experience, which are best learned by practice. This book aims to get you started on that adventure with some starter plans.

Beyond the planting plans, there are some key points, which I've taken from Jeavons, Chadwick and Aquatias, one of the first to write about French methods for an English-speaking audience.

1. Grow in permanent paths and permanent beds. Ideally, these should be double reach, allowing us to reach into the center easily without ever stepping on the beds. Approximately 4-5'. The **keyhole designs** later in this book will be even more space efficient.

2. Intensively manage soil. This is done with compost, manure, deep organic mulches or the **fertility systems** presented in our chapter on soil.

3. "The Greenhouse" (Chadwick) – tight plant spacings with no rows. "Close plant spacings, as found in nature" (Jeavons.) Starts are spaced tightly in a grid-like

formation, rather than rows, with naturalistic spacings so that there is no soil visible at maturity and leaves brush each other. Research by ecologists has demonstrated that plant cooperation in such conditions outweighs competition, helping to maintain optimal growing conditions in the top soil layer and the atmosphere under the plants. This is probably why FIM systems are so productive, sustainable, and healthy.

4. Use Polycultures. While Jeavons and Chadwick reduced this a little for their American audience, intercropping was a major part of the French tradition, and one of Aquatias' four principles. This maximizes utility, yield, use of space, and garden health for home and small market-garden systems.

5. Include many beneficial plants, aromatic herbs and perennials kept in the garden over a long period of time. These are traditionally in every bed and near every crop.

6. Grow your own fertility (Jeavons) or source it smartly and sustainably (Chadwick, Aquatias.) The patterns in this book will help you to do this.

7. Use open-pollinated seed, rather than hybrids, to enhance seed security, diversity, and self-reliance. Better yet, save seeds.

Getting Started

Holistic Intensive gardening is a method that trains expert gardeners. This is one of its main benefits. But that expertise takes time to develop, just as the soil develops over time with holistic natural management.

Beginning gardeners may want to start with Bartholomew's "Square Foot Gardening" program, but also try creating some FIM beds. Jeavons' How to Grow More Vegetables is an excellent place to start, with resources for spacing and companion planting, as well as sustainability.

The patterns in this chapter should help you get started with some basic designs you can adapt as you gain experience.

Yes, Holistic Intensive Gardening methods take some extra knowledge and design time. But the rewards are phenomenal.

Some Starter Plans for Holistic Intensive Gardens

(See images *For a smaller garden, use circular Holistic Intensive beds as in the pattern: **Home Veggie Garden Circles.**)

Garden Overview

I recommend that these French Intensive style beds be used in the context of a zoned garden with different levels of wild and human management. In such a garden, we'll use wilder systems to grow plants that can thrive in wilder management. Such a garden will use forest gardening elements, hedgerows, and perhaps a 3 Sisters garden for corn, beans, squash, cucumbers and melons. Difficult melons and cantaloupes might be grown on trellis systems. Potatoes could go in a **slash-mulch system** or be grown in containers. And we will have some French Intensive beds for the vegetables that will thrive under tight management.

A 4-Bed Rotation

The simplest way to get started, once we have pulled out other vegetables into wilder systems, is to use a 3, 4, or 6 bed rotation. A 4 bed rotation will use a solanaceae guild, majoring in tomatoes, peppers, possibly eggplant, and their companions. A brassicas bed will grow summer cauliflower, broccoli, and Brussels sprouts, and we can add a bed of peas and fall brassicas or a scatter-sown polyculture bed. The 4th bed will be perennials and light feeding self-sowers for a rest year, or possibly a cover crop.

The Solanaceae Guild

This planting majors in tomatoes and peppers, or possibly eggplant, and ground cherries. These are planted north to south at garden spacing. Tomatoes go to the shaded side, planted at ½' to 2' (pictured at 18".) Peppers are planted at 1'. Basil is planted at 1' and marigolds at 6" in a diamond pattern. Carrots are scatter-sown and thinned in any openings. This bed shows a fortress planting along any grassy edges to prevent grass from overtaking the bed.

The Summer Brassicas bed: The salad bed can be seen with large summer brassicas staggered down the middle in an 18" diamond grid, a fortress planting on the outside edge, and beneficial plants filling in spaces. Ideally, purple varieties are used to discourage pests.

An Ideal Cover Crop can be seen in the chapter on a useful ornamental cover crop.

A Peas/Fall Brassicas bed can be seen in the chapter on Home Veggie Garden Circles.

The Beautiful Scatter Sown Salad Bed is found in the chapter on Guilds.

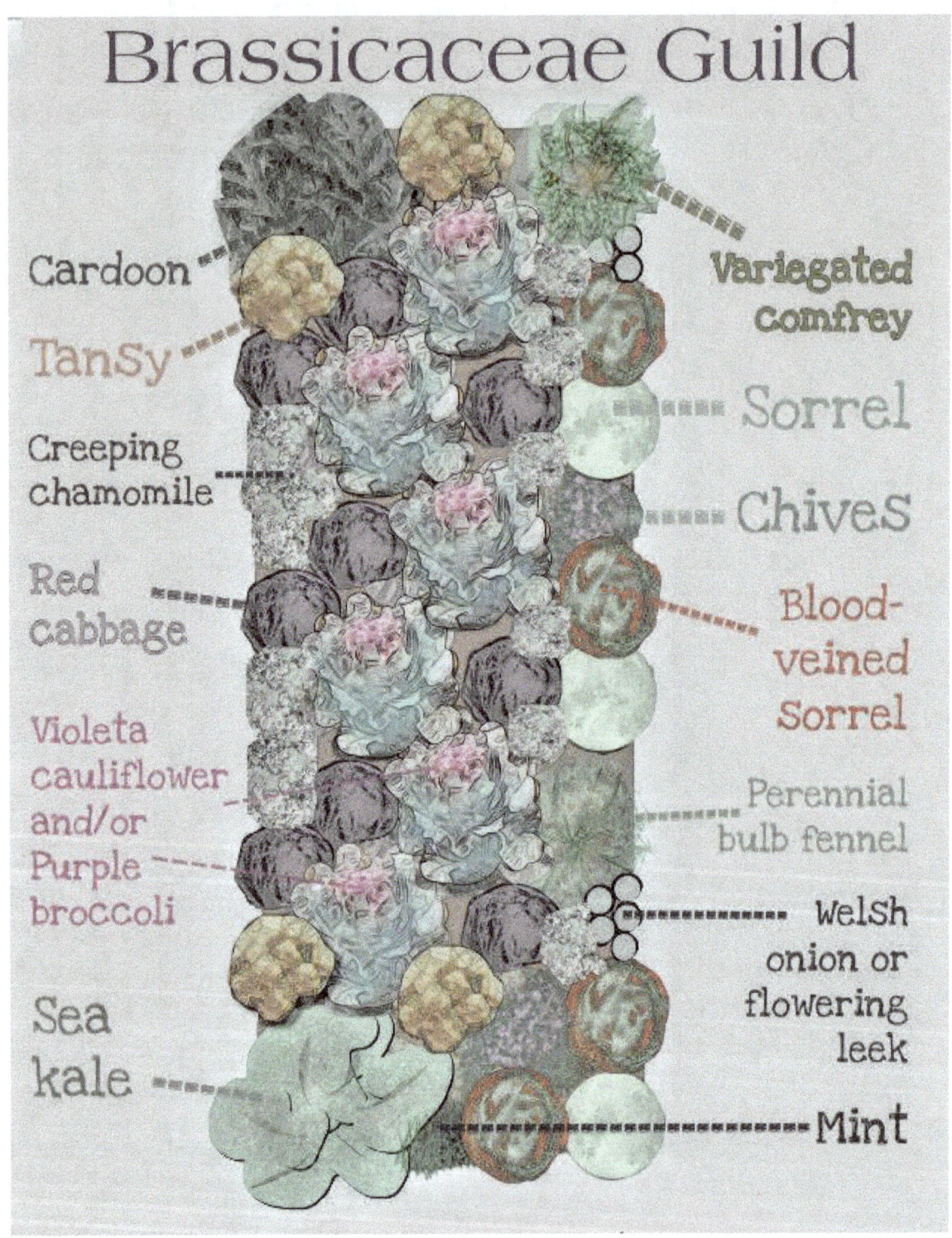

The solanaceae guild in spring at planting. This year, the rotation was in a bed with sea kale. The tomato spirals are visible in the back.

These guilds at peak season. Solanaceae guild bed is on the right. Brassicaceae guild is to its left.

The Home Garden Veggie Circles Plan

Perfect for home landscapes, these little veggie garden circles can be integrated into guilds, forest gardens, edible meadows, or pretty much anywhere.

One of the biggest problems I encounter for smaller-scale home gardeners is the balance between growing some vegetables in a way that makes sense, but is also not too much of a time burden. In most cases, for busy people on a small scale, forest gardens, hedgerows and other extensive gardens just make more sense. On a small scale, home gardeners feel like the time required to start and maintain a veggie garden often isn't worth it.

Here's what I mean: for a few hours a week one can manage 2-3000 square feet of intensive garden and grow most of one's one vegetables, a lot of calories, and have a lot of produce left to trade and gift. But a tiny garden of one-hundred square feet can often take about the same amount of time and have less than 10% of the yield. So, there's a minimal scale.

The solution I've come to and implemented with clients is to grow vegetables in high-yielding **holistic intensive** plantings, integrated into self-maintaining edible landscaping guilds. Nestled into a low-maintenance landscape with plenty of mulch-makers around, and surrounded by a set of stepping stones and walkable, useful ground covers—these small, circular, deep-mulch beds practically take care of themselves. All that is required is planting in some starts and then topping off the mulch each year.

Everywhere I have used this design, I've been happy to see homeowners using these annual guilds every season.

You can see a model of a real design that has functioned well in a small urban yard in the pattern "[An Ornamental Front Food Forest](#)."

The Design

Most often I use three beds to give a very simple rotation plan, with beds for a solanaceae guild (tomatoes and peppers,) a guild for spring peas and winter brassicas, (the cabbage family plants,) and a bed for a **scatter-sown salad bed**.

When there is room for a **forth circle**, I add the **Useful Ornamental Cover Crop**.

When **5 circles** are possible, I will create a mini three sisters guild with 1 mound of Japonica corn or sweet corn, and surround it with cucumbers, melons or squash. With **6 or more circles**, I just start repeating these circles with slight variation.

The Tomato-Pepper Circle:

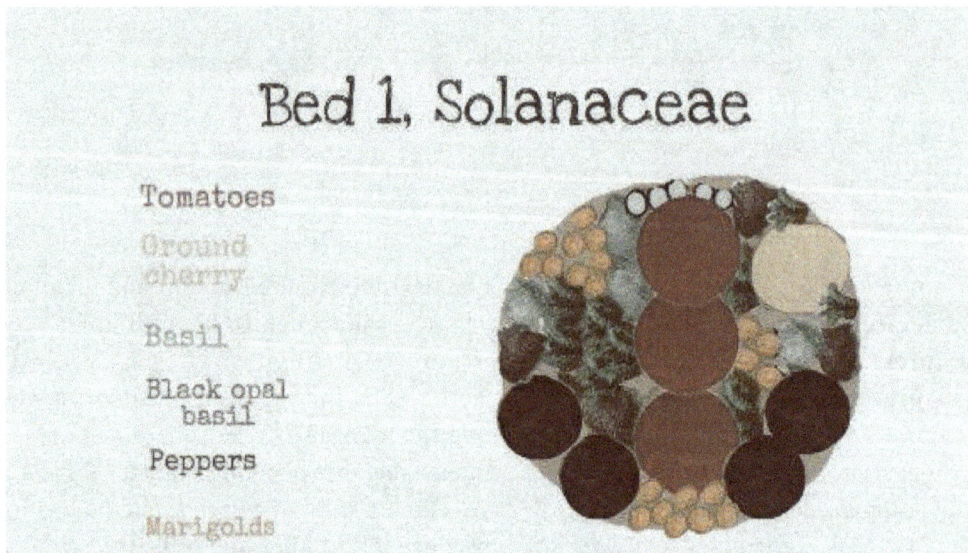

Tomatoes spaced at 1 ½ feet, ideally from north to south.
Peppers at the south of the circle.
Carrots are scatter-sown in around the tomatoes and then thinned as they grow.
Basil is planted in a diamond grid at 1' spacings
Marigolds if desired, are added in any remaining spaces at 6" spacings.

The Beautiful Salad Bowl Polyculture

Bed 2, Salad Bowl Polyculture

Remove mulch

Sow: red mustard, dragon tongue arugula, merlot lettuce, grandpa admires lettuce, walking onion starts, salad cress, fennel, carrots, dill, etc.

For this guild, the mulch is removed and perhaps placed on other guilds. Soil may be lightly cultivated with the goal of reducing soil complexity and setting back succession enough that these pioneer plants will germinate. Pioneers like lettuces and arugula often will have poor germination on good long-term no-till soil! See **Beautiful Scatter Sown Polyculture**.

I aim to sow before a week of good warm, wet weather. This provides a long season of salads, starting with baby greens, which are selectively thinned to allow them good space. Try to always keep a tiny bit of space around plants, and thin when leaves begin touching.

Bed 3, Spring Peas and Winter Brassicas

Spring Peas

Sow at 4 inch spacing in a diamond grid.

Approximately 200 plants

Chop and drop in place when production stalls

Fall bassicas

Purple cabbage

Purple cauliflower, kale, or broccoli

Chamomile

Tansy

The Peas/Brassica Circle
For simplicity, we'll use another "good enough" pattern for a flexible planting design for peas and brassicas.

First the peas:
Poke peas into the mulch in a grid approximately four inches apart. This will be approximately 10 peas per square foot, or 200 peas for our whole bed. If you want fewer peas, just plant a smaller circle, but keep the high density. I use a central trellis or three and plant densely so that the peas hold themselves up. When the peas have nearly finished for the year, chop and drop them in place. We'll plant the brassicas into this mulch.

Brassicas

In early summer I start brassicas indoors. My "good enough" suggestion is to divide brassicas into large and small, with kale, cauliflower, broccoli, collards and brussels sprouts all being large.

Large brassicas: These should be planted in a diamond grid at 1 ½ feet, 18 inches. If you want to plant only these, you can fit 9 in the 5 foot circle.

Small brassicas include most cabbages and fall bok choi (spring bok choi matures fast and can be planted in a scatter-sown bed.)

These can fill in the spaces around the large brassicas in an approximate 1' grid. Any gaps can be filled with virtually any herbs. Chamomile is a good companion for brassicas.

For the holistic natural gardener, I highly recommend purple varieties of brassicas, as these have been found in research to be the most pest-resistant, especially to the cabbage moth. Sometimes I will plant regular varieties as a trap crop, and remove them before maturity to kill a generation of worms, after they flock to these green varieties.

Chapter 14: Nurse Log Patterns for Beauty, Ecology, and Productivity

Nurse Log, Rebecca Stockert

Nurse Logs?

Nurse log picture by Rebecca Stockert

You might not know the term, but most of us have seen this ecological concept in action in the forest or in other ecosystems.

A tree has fallen into a sandy field, and all around it there are new flowers, bushes, and young trees that were not supported before. In the forest, a young pioneer falls, and new species fill in around it, including the most sensitive forest plants, and tree species appropriate to an old growth forest. These are "nurse logs" that help push ecological succession forward by "nursing" vulnerable species and increasing biodiversity.

The effect of nurse logs has long been studied and understood in ecology, but now we have increasingly more evidence that we can use the same tool intentionally in our landscapes for the same purposes: to help establish plants and increase biodiversity. For the gardener, they provide many benefits. They can help us accomplish the same goals as hugelkulturs, but often without the added work of digging the soil. They can help us prepare beds, fertilize plants, build soil, and even help us irrigate the garden. And in the context of this book, they can help us create a finished, architectural look in a landscape.

Because this is the first natural gardening or Permaculture book I'm aware of to include nurse log patterns, I feel a sense of responsibility to do a good and thorough job on one of our most valuable tools. So, let's take a look at just a small sampling of the research that demonstrates the benefits of nurse logs.

The most common designed use of nurse logs is as Log Erosion Barriers, or LEBs, in which logs are laid on contour, especially after logging or forest fires, to increase infiltration and reduce erosion.

There is a large body of research showing that in most cases well-designed and installed LEBs are effective in reducing erosion and increasing water infiltration into the soil. In some studies, they are nearly as effective as established natural vegetation, which does both of these jobs very well (Robichaud et al, 2008.) Manuel Lopez Vicente found that LEBs increased plant establishment by 24% after fire, (Vicente, 2020) The effectiveness did not go down even on steep slopes, making nurse logs appropriate in slopes over 15%, slopes too steep for swales.

In some studies, LEBs increased establishment of target tree species by 10 to 15 times, such as in a study on black pine by Manuel Esteban Lucas-Borja. A study on Andean forests found that nurse logs were an important natural ecological process in regenerating the forests there. Another study found that nurse logs beneficially impacted soil P and N and increased density of plant growth by 15 times and diversity by 5 times, Oreja B et al. In still other cases, they may provide a dry footing to plants that can't stand saturated soils.

In yet another set of studies, nurse logs were found to enhance microbial diversity, and so to help in establishing species prone to disease. D. L. O'Hanlon-Manners, et al.

When comparing potential techniques, nurse logs allow us to use biomass in a way that is powerfully beneficial even without any complicated, expensive or time or energy-consuming processing. For example, in most temperate situations, nurse logs will be far more effective than turning the logs into biochar. Counter-intuitively, multiple studies have found that nurse logs on contour perform better for tree and plant establishment than woodchip mulch, even supporting five times more density in one study. In another mediterranean climate study, trees established with about three times as much success with nurse logs than with processed mulch.

So, now that we see how powerful this simple nature-mimicking technique can be, let's look at some patterns from putting them to use in the garden.

Nurse Log Garden Edges

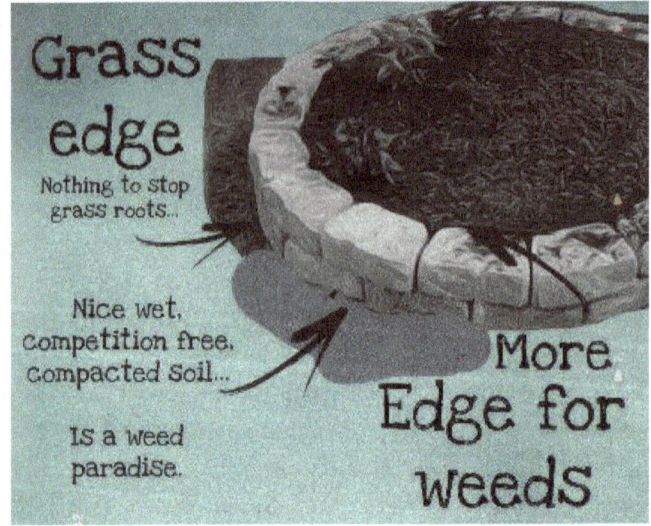

One of the trendy looks in suburban gardens is to edge the gardens with materials to make them look more human-controlled and managed. While I'm not normally a fan of that look, and prefer the look of plants as an ornament to concrete blocks, I understand that it gives a clean edge and look of conformity.

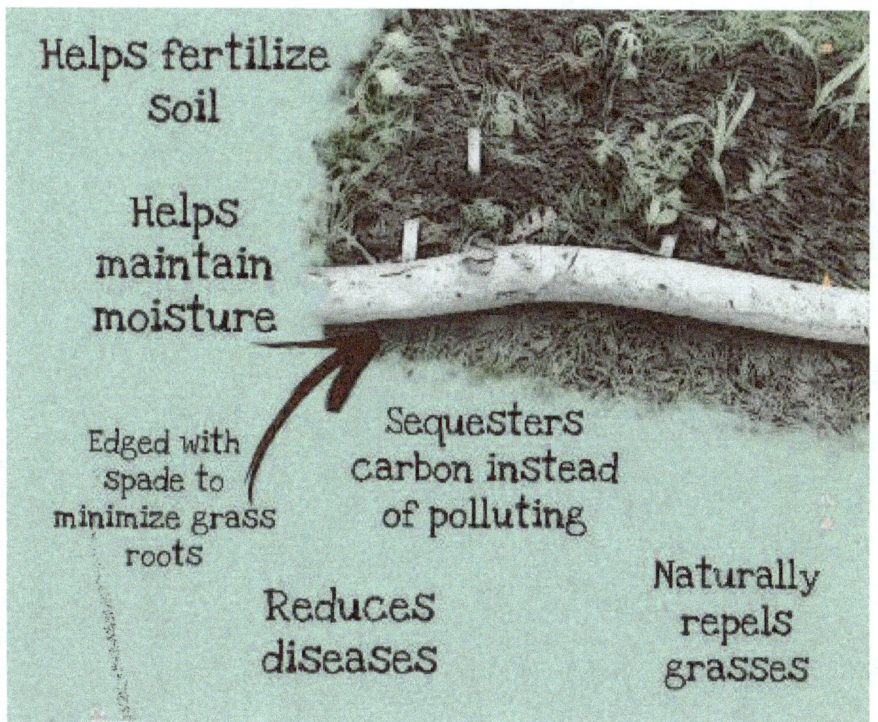

Concrete pavers may look nice, but they can be perfect weed habitat, encourage grass encroachment, may be unsafe for food, and of course, they are a source of carbon pollution. While I still sometimes choose recycled concrete pavers, nurse-log edging doesn't have any of these drawbacks.

The downside is that these edgers are almost always carbon-spewing, potentially dangerous concrete and brick blocks. If they are recycled, it can certainly be sustainable, but I strongly believe we cannot justify the use of carbon-spewing concrete as a decoration in home landscapes. Worse, these create ideal habitat for weeds.

Nurse log bed construction
Taken to the next level, garden beds and retaining walls made out of nurse logs always look aesthetic, natural, and "**terroir**." They are appropriate to rustic gardens or can be done in ways that look very formal. In the pictured example, Stician Samples has created terraces from a mix of techniques, framed with nurse logs. Not only do these provide habitat for plants, but they provide a niche for many other beneficial organisms including lizards and snakes.

We have to take care to construct these so they are stable and sound, and they create ideal growing circumstances for our plants, well-evolved to grow next to large-diameter wood.

I like to drill holes for native pollinators, or even inoculate these logs with mushroom plugs, to add another productive element to these beautiful raised beds.

Transformative Landscape in Vermont with beautiful nurse logs and hugelkultures, by Stician Samples

Nurse logs as tree mulch

Another application taken straight from the research literature is to use nurse logs as mulch around trees. Since in some cases, nurse logs have been found to help tree establishment more than chip mulch, not only is this attractive, it will likely help your planting.

In combination with deep mulch, nurse logs like this can dramatically speed up establishment of trees.

Nurse logs can be laid on the ground horizontally, or large logs can be kept upright. I have used some in my gardens as pedestals for bird baths or for planters where appropriate.

Passive Swales, or "Log Erosion Barriers" (LEBs) The most studied application of nurse logs are as Log Erosion Barriers, which are sometimes called "passive swales" in the Permaculture community.

Passive swales? Yes, indeed, logs on contour will catch erosion and biomass, which accumulates on the uphill side of the log. As long as the LEBs are constructed such that the water indeed slows down, they will start to accumulate biomass behind them, filling in, and—without lifting a shovel—create a mini swale. Many visitors to Lillie House have seen these passive swales built up so much that they have effectively become terraces, and a second round of nurse logs then added to the top to create another swale.

The built up soil behind these logs is rich in organic matter and silty soil particles, excellent growing medium, right exactly where we want it most, next to our crops plants.

Compared to dug swales, in many cases nurse log swales can be more cost-effective, infiltrate nearly as much water, don't require heavy equipment, and have holistic benefits of reducing tilling, reducing erosion or landslide risk, increasing carbon, fertilizing plants, and increasing biodiversity. Since trees will harvest a high percentage of water after three years, dug swales have most of their benefit in those first seasons. But nurse log swales continue to provide fertilizer, water-savings, carbon, and biodiversity benefits for years to come.

Nurse Logs As Swales
Infiltrating water, building soil

Notes:

1. P. R. RobichaudA,C, J. W. WagenbrennerA, R. E. BrownA, P. M. WohlgemuthB and J. L. Beyers, "Evaluating the effectiveness of contour-felled log erosion barriers as a post-fire runoff and erosion mitigation treatment in the western United States," *International Journal of Wildland Fire* 2008
2. Oreja Ba, Goberna Ma, Verdú Mb, Navarro-Cano JAb, "*Constructed pine log piles facilitate plant establishment in mining drylands*" Department of Environment and Agronomy, Instituto Nacional de Investigación y Tecnología Agraria y Alimentaria (INIA), Ctra. de la Coruña, km 7,5 - 28040 Madrid
3. Jorge Castro, Craig D. Allen, Mercedes Molina-Morales, Sara Marañón-Jiménez, Ángela Sánchez-Miranda, Regino Zamora,
Salvage Logging Versus the Use of Burnt Wood as a Nurse Object to Promote Post-Fire Tree Seedling Establishment, Restoration EcologyVolume 19, Issue 4 p. 537-544
as Refuges from fungal pathogens.
4. D. L. O'Hanlon-Manners, P. M. Kotanen, *LOGS AS REFUGES FROM FUNGAL PATHOGENS FOR SEEDS OF EASTERN HEMLOCK (TSUGA CANADENSIS)* Ecology, Volume 85, Issue 1 p. 284-289, January 2004

Beauty in Abundance

Nurse Log Garden Beds at Lillie House.

Chapter 15: Transforming Soil?

Soil comparison. One inch of compost added annually for three years to the soil on the left. No compost ever added to the soil on the right, but it was managed with holistic natural management. The unamended soil now maintains excellent nutrient levels for crops.

Here is another profoundly strange thing about this gardening book. We're well over 100 pages in and we haven't had a chapter on soil yet.

Meanwhile, every sensible modern book on gardening has a dedicated chapter or five saying that soil is the most important factor. This is because for far too long the opposite was true, and soil was not considered at all. Scientific gardening and horticulture considered soil to be nothing but an inert medium to get water and nutrients to the plants. Of course now most us know that soil is a living, important thing and treating it like dirt has dire consequences.

But on balance, I've come to think we spend too much time thinking about and analyzing soil these days. We can even somewhat fetishize it to the point we harm ecosystems!

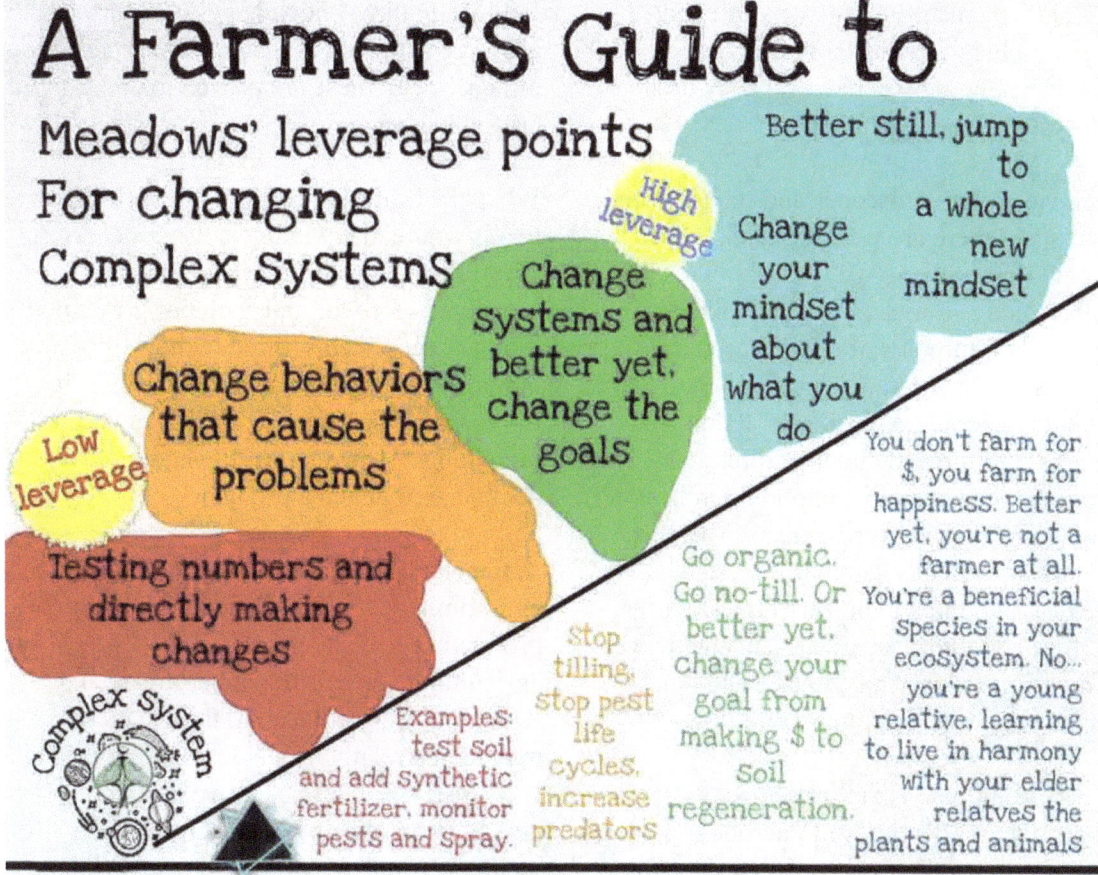

Remember in an earlier chapter we discussed Donella Meadows' places to intervene in complex systems?

You know what else is a complex system?

Soil.

This same Meadows system approach works for soil management, though we have to understand it's a system where humans are only one player. This gives us a nice opportunity to delve a little deeper into some of the ideas of systems theories.

This systems approach is why French Intensive gardeners and Indigenous gardeners were able to create systems we can't beat today, and they did it without microscopes and understanding soil biology. They used high systems leverage points.

When we grab microscopes and start looking at our soil, where are we addressing the system? Brute force—we're tracking numbers and parameters. We're looking at what kind of numbers of organisms and nutrients we have. Tracking actually has very little effect alone. So, then the solution is *usually* brute force methods of directly changing those numbers: import microbes or nutrients with consumer products. These are low-impact methods which is why studies haven't found them to be very effective so far. If we add microbes or nutrients but leave the high-impact systems in place and don't provide habitat, the changes don't last.

The key point here is that if we take a perspective of affecting soil microbe or nutrient levels through management, it is like annual tillage gardening: a job we forever have to do year in and year out.

As we move up in leverage, we can look at harnessing the self-organizing power of behaviors and systems. At this level, we're no longer too concerned about stocks and flows of microbes and nutrients. We're concerned with the behaviors of organisms and their habitats. We know that if we build good habitat, soil organisms will come. We don't need to import them by brute force. This is what our Holistic Natural Management Techniques accomplish. Creating habitat and systems. Now we have transformed the soil to a new state.

This is proven more effective, but yes, this is harder to figure out. We can't just spray somebody's designer microbe spray on the field. It requires knowing systems for growing with habitat, perennials, no till, deep mulch…. But these are proven to be far more profound changes.

When we move to the highest leverage point, this is where many indigenous gardeners put their efforts. It's an entirely different paradigm. Horticultural societies grow food in intact ecosystems, often with fertility systems built right in. That includes fertility flows from flooding or trickling down from mountains. That way fertility and biodiversity decline aren't even a problem.

In the classic "Farmers of Forty Centuries," agronomist F. H. King examines the high productivity, relatively sustainable agricultural systems of China, Japan, and Korea. As King points out, these nations maintained relatively stable populations and food systems on less than two acres per person, more than twice the productivity our "modern" food system can manage while doing rapid environmental damage and rapidly depleting finite resources.

We should also be aware this high-productivity and population density was maintained alongside some of the highest levels of biodiversity and ecosystem health anywhere in the world.

For King, the key to this sustainable high production was that fertility and soil were systematized. Ecosystem fertility flows were widely used to support sustainable food. Fertility generated by forests and wetlands were cycled into productivity. And every scrap of "waste" "was treated as sacred to agriculture." In these nations, human society was at that time truly integrated into ecosystems as one mutually beneficial closed loop.

Add to this one final key point: such a sustainable food system will necessarily require more hands-on food systems and more people behaving in sustainable ways. Combine all these together, and we have arrived at the real thesis of this book. They keys to a sustainable society are for many more people to be involved in food production, for us to live as connected parts of our ecosystems, and grow food off sustainable fertility systems.

This is why in this book I don't tend to focus on soil so much. We're focused on systems. We're focused on transcending the problems of soil decline altogether. That said, microscope work can be useful in showing that these things work and in helping us track our results. But probably not everyone who wants to have a good garden needs to be doing that work tracking numbers and parameters in their gardens. Knowledge of systems that create habitat and keep ecosystems in place is more powerful as a leverage point. But everyone who wants to have a good garden needs to have their fertility systems planned and k ow how to manage things in ways that maintain fertility.

This also helps us follow the Pareto curve or 80/20 principle again. We'll get a far bigger practical impact out of putting 10 hours into establishing good systems than we will out of spending 10,000 hours looking at our soil under a microscope, or applying mineral or microbial concoctions.

That doesn't mean that soil is unimportant. It just means that the HUGE thing to know about soil is that paradigm shift that has already happened from thinking soil is just an inert medium to understanding that it is a living, vital ecosystem, and that it can respond to system-wide holistic investments.

The paradigm shift is is that soil shouldn't be a problem! It's only our careless systems that create the soil problems.

So our focus on *transforming* soil will be to create systems that regenerate soil. Our guiding principle is that each time we undertake a new adventure in the landscape, we build fertility and habitat systems right into it.

In the transformation paradigm we're looking at creating permanent systems and that means knowing where the fertility will come from and building it in, just as in those indigenous horticultural systems, and the systems that have allowed humans to farm for 49 centuries. If we do this, it's not that soil will become unimportant, but that it will not be a problem.

With good systems in place we can let our aesthetic sense of the soil guide us. Most people quickly learn to identify good soil by sight. We can see if it is rich with decomposing organic matter, that it has that beautiful cake-like soil structure, that it is rich and chocolatey in color.

But it is important to not compare our soil to the look of say, somebody farming on the world's best soils in North Dakota or someplace. Look at wild ecosystems on similar soils in your region and get an eye for what they look like compared to depleted soils. Look at wetlands and forest soils to train your eye. For most of us, this approach will be "good enough" that we'll never have a problem.

With well-managed soil, with good carbon, mulch, good crop selection, lots of perennials, and lots of diversity, the soil will soon be good enough to grow a broad range of crops. Soil amendments, conditioners, or tillage will just not be necessary.

Some Common Soil Problems and What to do About Them

Soil Problem	Conventional Solution	Organic Solution	Holistic Solution
Fertility loss	Test the soil,	Add	Use better systems
Poor drainage	Buy fertilizer	compost	Use better systems
Poor texture	Buy gypsum or	Add	Use better systems
Poor microbial biodiversity	other	compost	Use better systems
Erosion	Buy tiller, buy	Add	Use better systems
Soil acidity	additives	compost	Use better systems

Yet another way to understand this is by looking at how to solve common soil problems. In a holistic approach, we'll be looking to solve these problems with systems, not brute force.

Some carbon crops for beauty and abundance

These crops grow abundant material for mulch and compost. If 60% of the land is planted in these it will be easy to create fertility and mulch.

Systems: Forest areas, cut and lay hedgerows, the beautiful abundant mulch strip, edible meadows

Crops: amaranth, corn, phragmites, quinoa, sunchokes, sorghum, heirloom grain crops.

The Beauty of Doodie: Patterns for Poop and Pee

On the way to the outhouse, a young seeker of enlightenment asked Zen Master Unmon, "Master, what is Buddha nature?"
Closing the door behind him, the master answered "dried shit on a stick!"

—*Zen koan*

Well, now here is a zen master after my own heart. If part of our journey in this book is to find the wide open mindset of creativity, which sees the potential for beauty in everything, then certainly there are great riches of beauty to be found in this treasure we typically flush down the toilet and out to sea!

As we mentioned in the last chapter, in "Farmers of Forty Centuries," F.H. King surmised that the sustainable high productivity of historic Asian farming systems was due to their tight cycling of nutrients. And that most certainly included what we will call "humanure." And yet, rather than make a golden resource of it, we make it "waste" and allow it to pollute water and land, and eventually contribute to ocean dead zones. Then we make less-potent synthetic fertilizers in ways that burn fossil fuels and contribute to climate change.

But because the topic is, well, icky, I actually left this out of the earliest draft of this book. I didn't want this one issue to become a barrier to anyone. So, don't let it. If you are not ready for this particular adventure, set it aside until nature calls.

However, we can and should all think about how to better manage our waste, including human waste. This is a huge leverage point for societal change.

In much of the Regenerative movement, we have come to focus on very controversial, small leverage points that we can barely even measure with research, like microbial inoculations and integrated animal systems. Yet, here we have massive amounts of effective fertilizer that everyone agrees could help sustain us, that we are just allowing to become pollution.

Many of the changes we need, especially in the urban areas most humans now live in, will need to occur at a higher level, by changing laws. But the attitude changes we can create by taking charge of our own mess can be key in making those larger changes possible.

And as gardeners and aspiring beneficial animals, nothing can be so powerful as Regenerative Pooping. As we take responsibility for our own waste, we most truly rejoin our ecosystems, becoming part of this cycle of life.

Patterns

Composting Toilets: Composting toilets are toilets which compost human feces into a relatively safe fertilizer that can be cycled back into ecosystems. These are especially important systems in urban areas with inefficient, destructive sewer systems that are frankly impossible to make sustainable, let alone, regenerative. The creation of these toilets is a probably a matter for professionals, and beyond the scope of this brief chapter. The best path is to connect with local expertise on the topic. In many cases, composting toilets remain illegal, despite being safe and more sustainable than the current technologies we rely on. This is an opportunity for local movements to destigmatize regenerative pooping and to get our black gold where it needs to go.

Leach Fields of Fertility: For transformative humans living in suburbs or rural areas, there is one under-appreciated pattern that can be a real boon to fertility cycling, the one which most rural and suburban homes are already outfitted with: septic systems with leach fields. While these often allow nutrients to go unharvested, and so they get a bad rap, well-designed leach field systems can be as regenerative as the best composting toilets, and even provide opportunities for harvesting methane and biogas for energy.

The key to making these work is to ensure that the nutrient is cycled back into productivity.

Many ecological landscapers wonder what is to be done over leach fields. Some experts have recommended removing the sod and building children's playgrounds, others have recommended growing vegetables in the rich soil.

KEY POINT: effluent from leach fields contains pathogens that can make humans sick. The best way to soak up excess nutrients and neutralize pathogens is grasses. Leach fields are designed to work with grasses, which can take up plentiful amounts of nutrients and kill pathogens. This is a terrible spot to grow vegetables and a worse spot for a children's playground. Deep-rooted crops can and will destroy your leach field equipment, reducing its lifespan, leaving it dangerous and polluting.

This is a wonderful opportunity for an ornamental grassland, which can be cut a few times seasonally and harvested for this nitrogen-rich mulch. Leave a five-foot buffer for safety.

This material then should not be used on garden crops. To eliminate any ick factor and any remaining safety concerns, this mulch should be used to feed a fertility or mulch crop, like our **fertility polyculture** in this book. These crops, then, can be added to the compost pile, fed to worms, or used as deep mulch in the garden.

Careful with medical waste and cleaning materials.

If humanure systems are to be used in fertilizing human food, we need to take a little care to not create unhealthy food. Especially with composting toilets, medical residues can pass and persist through the process to be taken up by plants. It is typically recommended that if and when we are on potentially dangerous medications, we take a poop pass, and avoid contributing to the humanure compost pile. (Medications should be safe for septic/leech systems due to low concentrations spread over very large areas.)

For septic systems, we should be very careful to keep things natural and avoid adding bleach or chemicals. These can kill the microflora that help process the fecal matter and can shorten the life of the septic system. A well-maintained, natural, right-sized septic tank may almost never need to be pumped. One treated with chemicals or flushed with chemical cleaners will lack the organisms that can break down waste solids, which accumulate and will need to be pumped. This is a costly and wasteful process.

Fertility Plantings for Avoiding "the Fertilizer Pest Problem." As we've discussed many places in this book, there is a problem with adding fertilizer directly to the garden. It causes spikes in pests, diseases, and weeds. Unfortunately, this also includes urine and fecal matter. Urine especially, when processed in any way, will contribute to problems in crop plants which must then be controlled by the gardener.

Therefore, use these potent fertilizers on fertility crops and strips, such as those featured in this book, instead of on sensitive crop plants. Then, use these plant materials to grow crops.

Alternatively, these can be composted if added to appropriate materials. The compost will not contribute to pest or disease issues.

A Few Systems for Fertility and Soil Health

Holistic Natural Management Techniques are demonstrated to maintain and grow soil fertility: deep mulching, no-tilling, perennial plants, high biodiversity.

Grow BioIntensive System: Plant 60% of the land in carbon crops to make compost. This includes food Forest, hedgerow areas, and three sisters gardens. A landscape with 60% of its area in these systems will grow ample fertility. This material will be more efficient if used as mulch.

Slashmulch: Can theoretically provide 100% fertility if 40-60% of the area is devoted to slash crops, or mulch plants.

Ecosystem Services (Fertility Flows from Nature)

Wetlands: Traditional systems include seasonal flood fertigation.

Greywater wetlands: These can be powerful fertility flows into garden areas and can ensure abundant nutrients.

Forest fertility flows: Forests or hedgerows can be harvested for deep mulch or compost. A forest uphill of a garden will create a reliable flow of fertility.

Nitrogen fixers: Approximately a third of area in N-fixing crops can provide for most of our fertility needs, especially when mixed with other systems.

Animal systems: Integrated manure systems like *victory gardens, which rotate chickens with annual veggies,* can be calculated to provide 100% of our fertility.

Worm Tubes: A system for building vermicomposting, or worm-composting containers right in the garden. These can help provide fertility without unsightly mulch piles in the garden. They can even be designed to be invisible in the planted bed.

Beauty in Abundance

Adventure Step 5: Plan Your Fertility Systems

Just as we did with water, it is important to plan fertility systems for each new garden and bed. We could paraphrase Brad Lancaster and say "plan your fertility before you plant your crops."

- Start by sketching or listing the major crop areas in the landscape. For example, "zone-1 vegetables and fruits, zone 3 nursery beds, zone-3 market garden…."
- What systems will you use to maintain the fertility in each?
- What systems will keep fertility cycling on site?

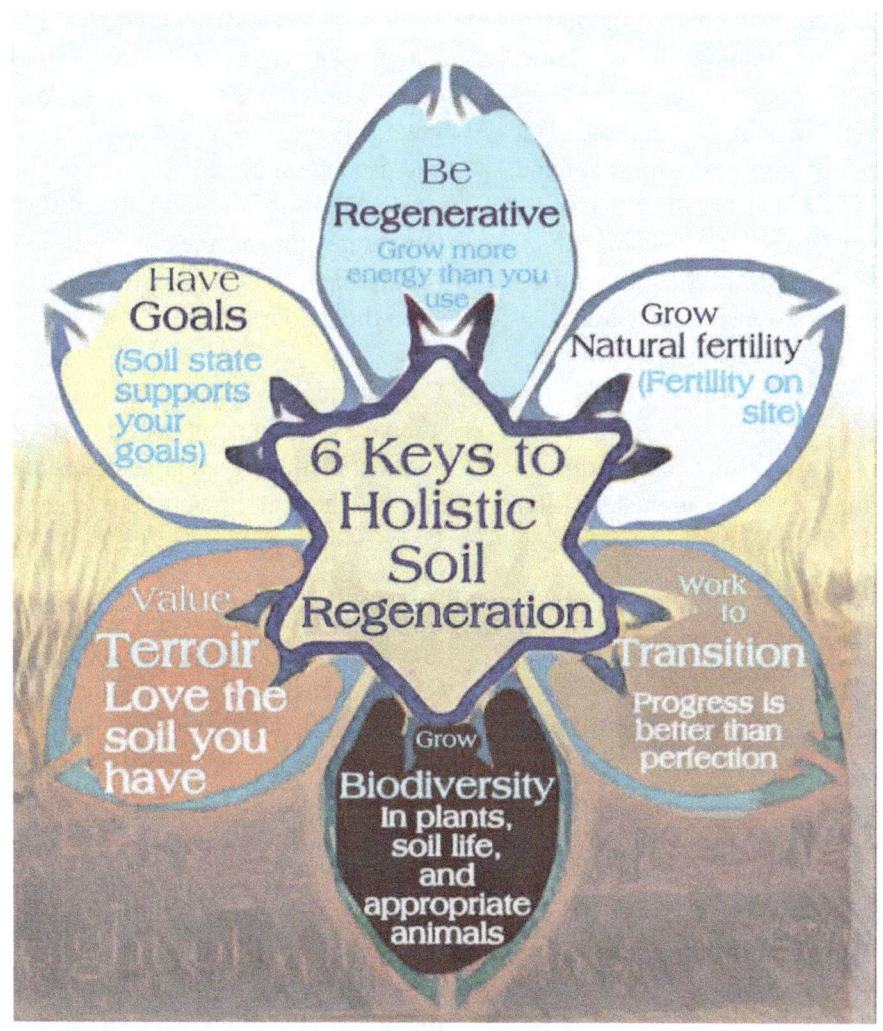

A Zoned Compost System

Compost is of course the classic soil amendment and an important way to reduce waste. While we won't go into detail on compost techniques in this book, we will introduce one pattern to create an easier, simpler composting system: zoned composting.

This is a good example where a little more organization makes something simpler and easier. The basic idea is to compost in place, instead of centralizing compost into one pile. This way, we neither have to cart around compost materials, nor finished compost, yet we always have appropriate compost at hand. The brilliance of a zoned system is that in intensive zones we have intensive, higher-value compost where it is needed. In wilder areas we have more extensive composting systems, where the extra labor is not worth the effort.

Given time, this is always the composting system that naturally evolves anywhere I live. You can Google or find more information on all these systems at Transformative Adventures online.

Zone 0, indoors:
An under-the-sink *worm composting* bin. A worm bin system is handy for food scraps, coffee and tea, and makes the highest quality compost available for intensive gardens. "Vermicomposting" systems use red wiggles worms and a bedding material like coffee grounds. If food scraps are only added slowly, these should never attract pests, flies or bad smells. I usually keep about 5 bins on rotation in the basement and 1 under the sink. When the sink bin is full, I move it and bring up another. Find information at Transformativeadventures.org.

A "bokashi bucket" for liquid compost. This is a way of fermenting food scraps that reduces smells and preps them for the compost pile. Liquid composting is handy for wet kitchen scraps and will not drown the worms. Specialized microbe mixes are not required with organic produce.

Zone 1, intensive gardens: *Worm tubes* in the beds. Worm tubes are easy to make and work great. I like to use an old cracked ceramic pot and use the saucer as a lid. Leave this in the garden bed and fill with garden wastes. In temperate climates they die back in winter, so I bring them in or refresh them each spring from the indoor system.

Composting in place as *sheet-mulch,* also known as "sheet composting." An excellent and fast method to co post scraps and weeds. Just put them at the bottom of a sheet-mulch. Done!

Zone 2, *trench composting* wherever swales are made, or beds are double-dug, we can fill these trenches with waste materials. At Lillie House, Kim and I especially liked to add scraps and plant material directly to our micro-swales, then cover them with a layer of mulch. The composting was rapid and could be dug out each spring. We can even add composting worms to the trench to improve the quality.

In zone 2, I usually keep a single "hot" or thermophillic compost pile using the Berkeley method. This is mainly to have a small amount of high quality compost for plant starts. The downside is this method requires both time and information.

Hugelkulture. Yes, hugelkultur is best when used as a way to compost woody material.

Zones 3-4: Cold compost piles left for a year or so to break down. The most extensive and easiest system. You simply make piles of leaves and add some weeds or grass clippings. If it is taking too long, add grass clippings or urine. We can even add worms to it to speed it up and improve the quality.

Woody debris piles inoculated with mushroom spawn for mushroom compost. Very simple! Get a pile of chip mulch and inoculate it. You can even add whole logs. This unfinished fungal compost is the best material for potting mix, especially for woody plants.

Some Beginner Adventures in Composting

For a beginner, these adventures will get us started with composting in ways that are easy, productive, and move us towards real transformation.

Worm composting: vermicomposting is the easiest way to create the highest quality compost. And the worms can be added to any other compost system during the season. If you only do one type of composting, try this.

In-garden worm tubes. This gets us started with composting in place.

Sheet-mulching. Using waste materials to start new gardens is truly a transformative approach. If you do these adventures, you'll be well on your way to establishing an expert zoned compost system.

Fertility System Design Case Study: Lillie House

In a Landscape Transformation approach, we are always investing in creating regenerative systems with each step. Each time we create a new garden it is good to think about and plan where the fertility will come from, how it will be maintained, so we're not making ourselves dependent on importing fertility or carting around compost.

The gardens at Lillie House were designed this way, with each garden area built on a proven system for maintaining soil fertility. Each area of the yard also has its own composting system.

New garden beds around mulch basins, mushroom composting, and worm composting systems. Each bed was started with built-in water and fertility systems.

Chapter 16: Starting Gardens the Holistic Natural Way

A new garden using multiple techniques: **sheet-mulching, double-digging,** *and* **slash-mulch**.

While many people may discover a no-till system, such as spraying herbicide, or solarizing, or covering with a tarp, and think that is the only way to do no-till, there are actually many ways to do no-till, and each has its own best use, advantages, and disadvantages.

Myself, I regularly use more than a dozen ways of starting no-till beds, usually more than one in each garden. And I never use herbicides or plastic mulches or tarps. Most often, my gardens—imitating the best gardens I have seen—are a mosaic of different techniques, each chosen to be the most appropriate to the soils, crops, and other factors.

While herbicides, plastic mulch and solarizing will all create a clean planting bed, they do not meet my goals of increasing system health, being ecologically sound, and reducing my food risks.

Herbicides are known to cause ecosystem harm, and destroy soil ecology leaving plants more vulnerable to pests and disease problems. Not to mention I don't like the idea of putting poisons on my food. Many of these herbicides are known to be taken up and concentrated by plants, meaning we are ingesting them.

However, we may have no proof these effect humans, but they are known to impact our gut flora, and may create effects of distorting or disturbing our microfloras in ways that are hard to test. For example, it is well known that particular gut organisms may cause us to eat sugar at a higher rate, but the health effect of this increased sugar consumption could not be isolated from some other cause. In other words, diabetes from increased sugar-consumption would not be considered a health effect of the pesticide, even if it was a direct result.

When people talk about the safety of herbicide products, here's why it's difficult to demonstrate they are harmful using modern science.

Think about the "persistent herbicides" we recently discussed. These specifically target a gene that occurs in broadleaf plants, which animals do not have. So ranchers are routinely told their animals can stay right in the fields while they are sprayed with Grazon. The poison, we are told, is inert in the animals, and in humans.

But if a field is treated with herbicide, then a crop planted a year later, that crop will still soak up that poison. Then horses can eat that poison. It remains poisonous. If they excrete that poison it will kill plants. So that poison was active in the horse. If we eat wheat treated with Clopyralid, that poison is active inside us.

Now, we don't have the genes that are affected by the poison, but many important members of a healthy human gut flora do. And there is research showing indeed that these are affected by these biocides.

Our gut microflora balance can cause us to crave sugar, to eat fats, cause irritable bowel syndrome, and have been associated with many autoimmune diseases. Often these can be cured by getting a "poop transplant" from someone with a healthy gut microflora. These are often taken from people living in places where pesticides are not used at such a high rate.

But when researchers go to test the health impacts of Clopyralid exposure, the first thing they do is eliminate any impacts caused by other conditions, such as diabetes, even though diabetes could be caused by increased sugar cravings caused by the harm to the gut flora. Complex effects like this are actually very difficult to document scientifically.

But it makes sense for us to want to limit our exposure, and I wouldn't let industry experts getting paychecks from those companies tell me otherwise.

So it is difficult to demonstrate that these herbicides will cause harm to human health. But it's a fact that these herbicides will affect the composition of your gut microflora (microplastic exposure does, too, btw) so if you don't consent to having your gut microflora dictated by "Big Ag," then you want to try to avoid herbicide exposure.

Technique	Description	Pros	Cons
Dig to No-Till	Start beds, ideally with double digging.	Good results the first year.	Hard Work!
Solarizing or smothering with tarps	Vegetation is smothered for a month or more.	Reliable no-till beds	Poisonous chemicals contaminate food. Soil compaction can be a problem.
Plastic mulch	Plastic mulch is used to cover the soil with holes for crop plants.	No thinking, knowledge or skill required.	High levels of poisonous chemicals and microplastics contaminate food/soil.
Deep mulching	Deep mulch is piled on thick enough to suppress weeds.	Good for starting beds where vegetation is already removed.	Weeds or grasses often grow through creating a problem.
Sheet mulching	Sheet mulching uses deep mulch with a temporary weed barrier which disappears by planting time.	One of the best and most reliable methods.	Slugs love it. If not done enough in advance, the weed barrier can impede water infiltration.
Sheet mulching with Compost	A type of sheet mulching that just uses 4 inches or so of compost each year.	Attractive to those who like a "bare soil" look in the garden. Can germinate seeds easily.	Long term can create soil imbalances. Requires a lot of compost. Rarely economically viable.
Hugelkulture	Stacked wood with soil over the top.	Reliably creates great soil in a few years.	Often fail in the short term without good design.

Plastic is a growing risk as well. Right now, farming is the fastest growing cause of microplastics pollution, and we are discovering this has profound impacts on ecosystems, soil health, plant growth and yes, human health. Research has shown that plants grown in plastic

Animal Tractoring	Get livestock like chickens or pigs to do the dirty work for you.	Effective. Fertile. Often cute.	Animals can often be more destructive to the gardens than valuable, and reduce total food production.
Slash-mulch	Slash resident vegetation and plant into the mulch.	Often called "the most sustainable for of human agriculture."	Requires high skill, good timing and appropriate crops and conditions.
Crimp-Mulch	Like slash-mulch but a cover crop or resident vegetation are crimped instead of slashed.	One of the best methods for organic no-till on scale.	Usually requires periodic tilling. In many systems multiple times per year.
Flooding	Periodic or Spring flooding preps the area for planting.	One of the best human systems. Clears soil while adding fertility.	Requires access to a river or temporary wetland system.
Burning	Controlled burns.	Can create truly sustainable systems.	Requires careful fertility management or "fallow" periods as in swidden systems. May increase particulate emissions.
Chemical defoliation	Spray poisons on the area to kill all the plants.	Can be used on scale for industrial agriculture and requires no skill or knowledge.	Harms biodiversity, harms soil potentially more than tilling. Uses unsustainable materials.
Planting into vegetation	Just plant right into whatever is there.	Easy.	Only works for select crops in select situations.

mulch can have dangerous levels of phthalates contamination as well as microplastics. Even plants grown in solarized soil have that risk. Solarizing also harms soil biodiversity, leaving plants more vulnerable to pests and diseases. For me, it is not worth the risk to the ecosystem or my health.

Luckily, there are easy, efficient ways to start and maintain no till gardens without plastics or poisons.

In this book on aesthetics, we'll just discuss a few, though I've included a chart covering 13 ways to start a no-till garden.

Chapter 17: The Transformative Double Dig

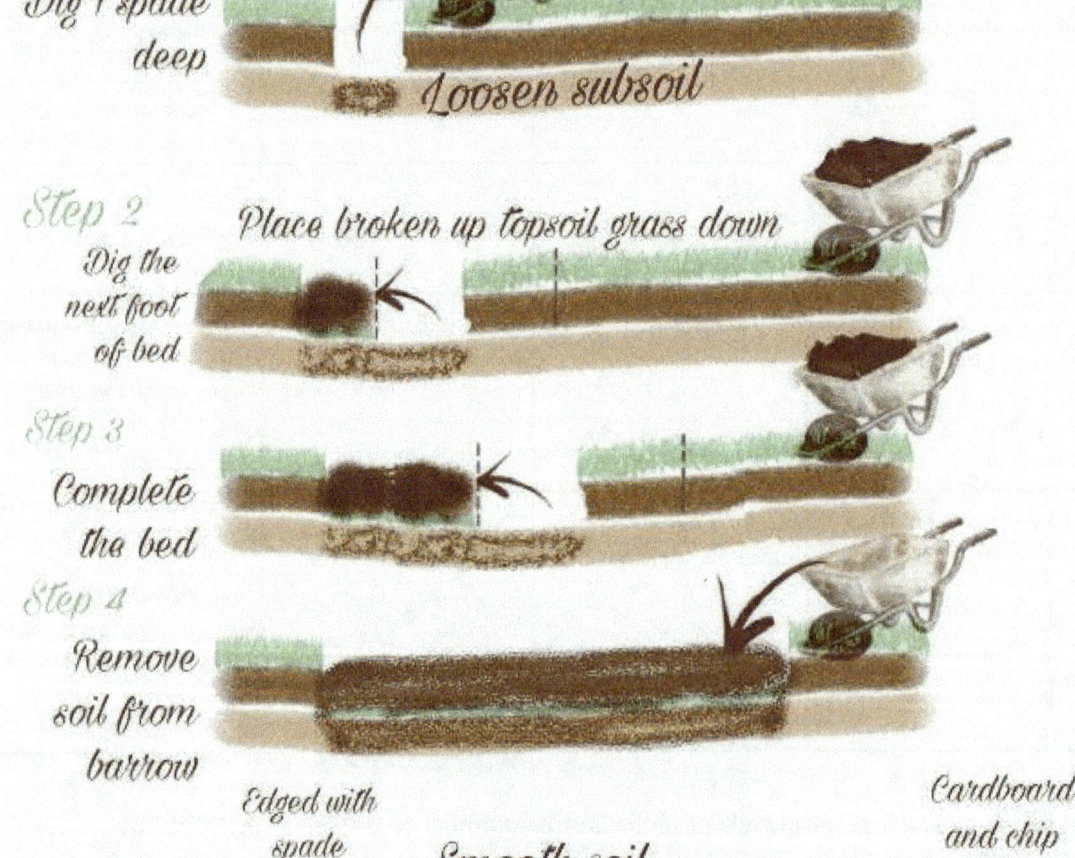

When it comes to establishing a garden with minimal tillage and work, one of the most important techniques for starting no-dig garden beds is: to dig.

That's right, this method can be a highly effective way to start garden beds in the right conditions, which can then easily be converted to no-till management, becoming permanent no-dig beds. This is when double-dug beds become "transformative," when they play a part in creating durable, semi-permanent landscape infrastructure. It can be a very important part of the "tapestry" of different bed types that can help you transform a landscape rapidly.

Usually, this is not stressed enough, and double-digging is used as an annual tillage method, which is far too labor-intensive to be cost-effective.

Overall, double-digging is more work than I generally like to do, especially when no-till is so effective. And there are many cases when it simply is not very effective.

However there may be some good reasons to do a double dig. In this day when no-till has become very fashionable, it's necessary to say a few words in defense of this system, because some people now recoil at any thought of digging the soil.

Not digging is itself not enough to be sustainable. The Grow Bio-Intensive system uses cropping recommendations to grow enough fertility that the system is sustainable, and has indeed been documented to be truly regenerative building soil at a fast rate, even with digging. Meanwhile, it is quite possible to do no-till in ways that require plastics, poisons, or non-renewable resources, which is not even sustainable by definition, let alone regenerative.

Yes, double-digging has the same drawbacks as any tillage in the first year, but it actually takes a no-till system a few years for soil biology to truly build up to the point where they are defending crops well. In the cases where it is called for, double-digging has a lot of benefits while the drawbacks are very short-lived. Most double-dug gardens in the first year will perform remarkably well, while most no-till systems in the first year, like sheet-mulching, are notoriously hit and miss.

So, let's look at which cases merit double-digging and which do not.

First, double-digging can be a great way to casually create effective mini-swales, with beds on contour. I have used this in many gardens for my zone 1 plantings. If you are going to do earth work anyway, on a small scale, you might as well double dig.

Secondly, quite a number of deep rooted crops show remarkable increases in yield and greater drought resistance in research from double-digging.

In one meta-analysis, deep-digging improved drought resistance and crop yields, especially where soil was compacted, (Schneider et al, 2017.) In others studies, crops with deep roots were positively impacted, while in others, shallow-rooting crops had no positive impact. (Holt and Smith, 1998) Yes, this seems pretty intuitive, but shallow-rooted crops like radishes get little benefit from double-digging.

More obviously, double-digging is a heck of a lot of work up front, even if it replaces the annual work of tilling with a rototiller. You don't need a study to show you that.

Therefore, use double-digging for:
- Highly productive beds near to the house.

- For deep-rooted summer vegetable crops like corn, amaranth, and tomatoes.
- When you want to create beds on contour to harvest water.
- When soil is compacted.
- When working for drought resistance.
- When materials to sheet-mulch are not available.

Patterns for Transformative Double Digging

The following patterns will help double-dug beds stay as resilient and long-lasting as possible.

Avoid Grass Edges

One of the ways double-dug beds usually go bad is when grasses encroach. These beds have a lot of surface area, so if they are dug into grass sod, they will necessarily have a short lifespan in which they require heavy weeding.

Therefore mulch paths with cardboard and wood chips if grass is present.

Avoid over-using walkable ground covers between double-dug beds, these can act like grasses and end up eating up the garden beds.

Use Fortress Plants

Later we will discuss fortress plants at greater length, but for now, just know that fortress plants can help us keep grasses and weeds from encroaching on beds. Planting any edges near lawn or weeds with fortress plants will help keep the invaders out.

Use Deep Mulch

While biointensive-type beds are usually kept in bare soil, there is no reason why they can't be grown in deep mulch to conserve water and prevent weeds. Indeed, the French Intensive gardening system was mostly based on using very deep mulch.

Plants can be started from seed in bare soil, and then mulched as soon as they are established and large enough to keep their heads above the mulch.

With the combination of deep digging and deep mulch, beds will be mostly weed-free and any weeds that do manage to germinate will be easy to pull out of this deep, loose soil.

Grow Mulch Makers

Because we will want to use deep mulch, it can be an immense time-saver to grow plants that make mulch whenever we have biointensive beds. Many plants double as fortress plants and mulch makers.

See the Permaculture Garden Bed Makeover Pattern for a guild for annual vegetable garden beds.

How to Do It:

The key concept of double-digging is we are keeping the topsoil layer on the top, minus the sod which can go down by the subsoil to prevent the grass from coming up in the bed.

Step 1. Once your garden bed site is selected, mark off the beds. I like to edge the bed into the lawn before I begin, so that I know I will have a nice aesthetic edge when I am done.

Step 2. Dig one spade deep, and about 1 foot by 1 foot square. Remove this soil to a wheel barrow.

Step 3. Loosen the subsoil with a spade or a pitch fork. If it is badly compacted you may need to break it up by thorough digging.

Step 4: Dig the second square foot of soil, moving width-wise across the bed. Place any sod at the bottom of the first hole, then put the mixed soil on top.

Step 5: Continue in this way until the width of the bed is reached, usually a maximum of 5 feet. Put the soil from the wheelbarrow in the last hole.

Step 6: Move back to the other side and the next trench across. Continue until the length of the bed is finished.

Step 7. If there is grass, use cardboard and wood chip mulch over it, we do not want much grass in the garden near these beds. Edge the bed with an edging spade to make sure grass and weeds will not easily invade. Edging also makes sure that water has someplace to go and soaks into the bed when it rains, instead of rolling off the raised bed.

Step 8: plant crops and mulch deeply as soon as plants are established enough to keep their heads above the mulch.

Notes:

Self teaching resources for double-digging from Ecology Action: http://www.growbiointensive.org/Self_Teaching_2.html

2. Meta analysis, deep digging improved drought resistance and crop yields, especially where soil was compacted. Schneider et al, "The effect of deep tillage on crop yield – What do we really know? Soil and Tillage Research, 2017

3. One example of very positive yields for corn and lab lab. Iddi Mwanyoka, Dosteus Lopa, COMMUNITIES' PERCEPTION ON THE CONTRIBUTION OF SOIL AND WATER CONSERVATION MEASURES IN IMPROVING LAND PRODUCTIVITY IN THE DRY-LAND AREAS OF TANZANIA: THE CASE OF TERRACE, " FANYA JUU " and DOUBLE DIGGING IN SAME DISTRICT *International Journal of Agriculture and Environmental Research*

4. No benefit for two shallow rooted crops, "Small-scale, intensive cultivation methods: The effects of deep hand tillage on the productivity of bush beans and red beets" *American Journal of Alternative Agriculture,* Vol. 13, No. 1 (1998),

Chapter 18: Sheet-Mulching Revisited.

A Guide to Sheet-Mulching, and Lasagna-Gardening, Beautifully

Of the many ways to start no-till gardens the absolute best I know of, and the way I have used the most by far is sheet-mulching. When it comes to transforming landscapes, this is one technique that is probably more useful than just about anything else, since it allows us to "bank" our labor into beds that can be almost permanent. I'm not sure I would know how to garden without sheet-mulching—it would take so much more work. And that's no "bull-sheet."

Sheet-mulching, sheet-composting, or lasagna-gardening is a technique of using a weed barrier and then deep mulch. This technique was made famous by Rodale, and later by Ruth Stout in her book "No Work Gardening," and then revived by Permaculturists like Bill Mollison and Toby Hemenway.

It is hands-down the easiest way to create extremely fertile long-term no-till beds that will be very resistant to weeds. I have had good sheet-mulch beds planted with perennial vegetables and herbs that have lasted nearly a decade with virtually no weeding work, and little maintenance other than harvesting.

And while sheet-mulching is pretty forgiving, I have been consulting long enough to meet folks disappointed in their gardens. Here are the largest problems I've encountered.

The first rule of sheet-mulching is there will be slugs.

Slugs are decomposers that eat organic matter and live without a shell or exoskeleton, which means they are particularly prone to drying out. A sheet-mulch of nice wet organic matter is like a slug paradise. Then you bring them a constant stream of tender young plants to eat… it's like an all you can eat cruise for slugs. And a single slug can reproduce 500 new slugs in a season. And since most sheet mulches are going into new gardens without beneficial insects, there are few predators. So, no predators, and no drying sun, and constant food: slug populations boom.

The best cure for slugs is biodiversity. See, "How I get Nematodally Free Slug Control." Research has found that slug damage can be minimal in wild gardens with high biodiversity. My own experience is that after a few years, with good diversity, slug damage becomes almost impossible to notice.

But in the first year, slugs in sheet-mulch can be brutal. Beer traps, using cheap beer in a shallow buried dish will kill many slugs quickly. Make sure they can climb into the beer easily. Leaving a lip over the beer will keep slugs from joining the pool party. Hand picking slugs at night can quickly reduce slug populations, too.

Often, we hear that ducks or chickens are the cure to slugs, but in almost every case I know of, the ducks and chickens end up doing more damage to the garden than the slugs. It's very difficult to have a high-productivity intensive garden with fowl, though ducks may be appropriate to more extensive food forests with a lot of aggressive plants like mints dominating. Of course, that sort of planting usually requires no slug control.

Planted sheet-mulch

The Second Rule of Sheet Mulch Is: Don't Let Grasses Win!

A sheet-mulch overrun with grasses is one you usually just have to start all over again. So it is imperative to keep grasses out or keep them minimized to the point where we can "spot mulch."

Spot Mulching is a mini sheet-mulch, usually just a pizza box and a small pile of mulch. This can take care of a small grass or weed encroachment without having to redo a whole mulch.

But the edges of the sheet-mulch need to be watched after.

We have a rule for guilds, pick one: 1, plan to keep grasses out, 2, plan guilds that can thrive with grasses, or 3, plan to do a lot of garden work weeding grasses out.

Anyplace sheet mulches contact lawn I recommend edging it with an edging spade a few times a year. In addition to edging, I recommend planting edges with fortress plants, especially for perennial beds. Together these should make sure grasses aren't a large problem.

A third problem is possibly caused by using the wrong material for the crop. When to Go "Back to Eden" with Wood Chip Mulch? The "Back to Eden" approach is a sheet-mulching variation using only wood chips, and sometimes fertilizers like urine to provide nitrogen. As we've talked about before, mulches, including wood chips should provide ample fertilizer as they break down. The drawback with wood chips is that they can break down so slowly.

A common problem I have seen observing chip-mulched gardens is that the break-down can be too slow. In early seasons the grass beneath the cardboard feeds plants, but once this is used up, it is replaced too slowly. The chips break down too slowly, so if more are added they begin to accumulate. Often, as they break down, weed seeds start to germinate on top of the mulch, and what do we do to suppress them? Add yet more mulch? At this point very deep layers of chips can cause real nitrogen binding. Usually gardeners at this point either till in the chips or begin the arduous task of raking off the chips into a pile somewhere to compost.

And so, I recommend using wood chips for establishing perennial guilds like tree guilds or perennial vegetable gardens, where the plants and diversity will supply both future fertility and weed suppression and the mulch can be allowed to break down slowly. The fungal dominated soil from chip mulches will also be better suited to these crops than to annual vegetables, which will grow better in a different mulch like straw mulch.

When and Where to Lasagna?

Less of a problem than a consideration, we need to understand that no tool is good for every job. Sheet mulching properly requires a lot of material. One pickup truckload of mulch will do a 50-100 Square Foot bed, at best. So, this is not a technique for broad acre or farm scale. But it is a technique for high production, low work beds right next to the door or for special fruit tree guilds, established one or two at a time per year. The fact that sheet-mulching "doesn't scale" doesn't mean it isn't worthwhile. Sheet-mulched beds on a 100 acre site will still be the lowest work, highest productivity beds and they may last for a decade without much new labor. That is worth the investment!

Lasagna beds are particularly worthwhile on very poor soils in droughty regions. In places where gardens always look a bit rough and scraggely, a sheet-mulch can make all the difference in the world.

Therefore, plan sheet mulches in the right place, with the right crops, and with a mind towards slugs and grass encroachment. Fortress plants, beer traps, edging, and using the right materials can all help you achieve success with this method.

Patterns for Making Sheet-mulching Work Wonders

Look at the Lawn

When we're assessing a sheet-mulch, we need to look at what we are replacing, usually lawn. A good, rich, luxurious patch of lawn will make for an amazing sheet-mulch. That lawn will contain enough nitrogen for any crop and as it breaks down it will feed soil life, ensuring that under the mulch there will be a rich chocolate cake like structure with great drainage and nutrient holding capacity. This is when it might pay to invest extra in a sheet-mulch designed to be bomb proof.

Toby Hemenway's Bomb-Proof Sheet Mulch

When you need a sheet-mulch to do great in year one, I recommend Toby Hemenway's Bomb-Proof Sheet Mulch, as presented in Gaia's Garden, also available online at Toby's website. You can find this with an internet search and enjoy the thorough instructions. The basics of the recipe are a cardboard or newspaper layer, an inch of good compost or other N-rich materials like manure, then eight inches of bulk organic matter, and finally a layer of finish mulch.

How-To: Mike's "Good Enough!" Sheet Mulch for Aesthetic Landscapes

My "good enough" recipe aims to use a few rules of thumb to get carbon/nitrogen ratios close to what we see in a compost pile, so that we end up with great fertility quickly.

Ingredients:
1. Cardboard or newspaper. No tape, glossy surfaces or colored inks, these could be toxic. Large appliance boxes are the best and easiest to use.
2. 6-8 inches of bulk organic matter, especially fall leaves. Unless it is a tree guild or perennial vegetable garden, try not to use more than 2" of wood chips. Do not use saw dust.
3. An inch of "finish mulch," ideally grass clippings or cocoa hulls. Shredded straw is not ideal, but works.

Steps:

1. Choose the location, make sure it's a good spot for the chosen crops and that it is ideally close to the house.
2. Choose the materials, see the considerations below, but usually, whatever is available is the best material to use! Store the materials close to your mulch area, or even on cardboard on top of your chosen spot.
3. The best time to start is fall. The sheet-mulch will be ready for spring planting. But you can start any time. A sheet-mulch ideally takes 1 month to break down for planting. In some cases, we can plant into a new sheet-mulch, but just expect it won't be ideal.
4. Mow down the existing vegetation and leave it.
5. Wet the ground thoroughly, or better yet, aim to sheet-mulch right after a good soaking rain.
6. Lay out cardboard, making sure there are no holes for weeds to get through.
7. Layer on your bulk materials. Wet each later as you go if you can.
8. Layer on your finish mulch. Do a little celebration dance.

How to plant a sheet-mulch

When planting time comes, there are a few ways to plant a sheet-mulch.

1. **Plant starts**. Sheet mulch is most ideal for plant starts. If a heavily mulched bed is planted out with starts, there is no room for weeds so the work for the season is pretty much done. I use a hori hori or knife to pierce the sheet-mulch and make a small "pot" to put the plant in. I like to keep a little compost to add to the hole to make sure there is root contact. The plant doesn't have to be planted down entirely into the soil, but the cardboard layer needs to be pierced well so the roots will be able to penetrate. Place the plant into the hole, making sure to spread the roots out some. Add compost, water the plant in well, and cozy the mulch back in around the plant. Don't worry if it doesn't seem perfect, plants are resilient and will figure things out.
2. **Planting seeds**. Seeds can also be planted directly into a sheet-mulch, but ideally it requires a bit of compost. Again, make an opening like a "pot" in the mulch, making sure to pierce the cardboard layer. I've had high germination rates with large seeds like corn, Turkish rocket and squash, dropped right into the hole without compost. For smaller seeds you will want to fill this hole with compost then plant seeds into it following the directions on the package. A sheet-mulch can be planted out at grow biointensive spacings, for those familiar with that system.
3. **Scatter p-sowing**. While often recommended, I do not consider this an ideal approach. It is possible to scatter sow onto a sheet-mulch, but it requires spreading an inch of compost over top as a seed bed. The problem is that it loses all the benefits of the sheet-mulch. The bed will be especially dry until roots penetrate the soil and weeds may land and germinate in the mulch as well as our crops.

Some Variations and Considerations

Look at the lawn, the basic recipe is for places with a decent thick healthy lawn or grassy vegetation. If the lawn is patchy, or scraggley, add extra fertility.

Layered Fertility

A general rule is, when fertility counts, for every three-four inches of bulk organic matter (leaves, chips, branches) add one inch of "greens," compost, grass clippings or coffee grounds. Start with one inch of compost on the cardboard, then alternate three inches of leaves and then 1 inch of grass clippings, then three inches of leaves, then finish mulch.

Shredded Leaves

Fall leaves are probably the best sheet-mulch material. But they are even better when shredded. Lay them out on the lawn when you're mowing—the mix of grass and shredded leaves is sheet-mulch gold.

Bulk It with Branches

Fallen branches, sticks and logs can be added on the cardboard to bulk it up. If fertility is a consideration make sure you're using compost and grass clippings.

Rough Mulching with Mulch-Makers

Weeds (and especially weed trees) are incredibly helpful for sheet mulches. I do not shred these; I just use them whole, which is what we mean by "rough mulch." If sheet-mulching in season, catalpa or mulberry leaves and branches can be especially useful, cut and added in bulk to the sheet-mulch. Rough mulch with young tree leaves and branches will be close to the perfect C/N ratio, even with other mulch matter added. Cover with a finish mulch. Don't use toxic trees like black walnut.

A final word is necessary on the topic of the temporary weed barrier, because a few authors have disparaged sheet-mulching on the grounds that the weed barrier impedes water infiltration. There was a study testing cardboard as a weed barrier (without the total sheet-mulch system) and found it impeded water and air circulation.

But as all the resources on sheet-mulching explain, this barrier is temporary. It is used to remove the resident vegetation, and then it disappears, converted into worm food. So, in a planted sheet-mulch, there should be no existing weed barrier to impede water. To discount sheet-mulching based on this error in understanding is quite illogical and extraordinary.

I can say from experience on dozens of sites that the claim that mulching alone works as well as sheet-mulching is false. My experience on probably more than two dozen iterations of both sheet-mulching and mulching without the barrier is that 100% of the time, mulching without the barrier has weed re-emergence as a major problem. I have actually not had an experience where this did not occur. In fact, I count on it and this is usually why I've omitted the paper layer, Meanwhile, sheet-mulching has been 100% successful at removing resident vegetation without failure. Mulch without the temporary barrier has its place, but if bed creation for a garden in a place where there is vegetation is the application, I strongly recommend sheet-mulching with a temporary barrier over just a deep mulch.

Notes:

Toby Hemenway's Bomb Proof Sheet Mulch from Gaia's Garden:
https://tobyhemenway.com/resources/how-to-the-ultimate-bomb-proof-sheet-mulch/

Gardening Without Work, by Ruth Stout, Norton Creek Press, 2011

Hugel-terraced hillside, Rebecca Stockert

Chapter 19: Hugelkulturs that Are Cool and Cultured

An Updated Guide to Wood Mound Beds

Hugelkulturs (pronounced something like "hoogel culture," or better with the German umlaut over the hugel, if you know what that is) are a gardening technique popularized by Permaculturist Sepp Holzer and promoted by "Permies" forum founder Paul Wheaton. In short, it's using piles of wood to create raised beds, usually covered in a layer of soil.

The wood breaking down attracts fungi, which can help create a drought-proof garden, fertilize the plants long term, and you end with a pile of amazing soil. It's basically a way of composting tree debris by gardening on it. This "stacks functions" in the most literal sense. It's especially useful if you happen to have a pile of logs you want to compost.

When I made my first hugelkulturs in 2011, I had never actually seen one, even after a decade of visiting Permaculture sites. Now, I have seen probably hundreds of hoogels. The "Hoogs" have become one of Permaculture's most prominent and well-known patterns. In fact they have become so popular that it isn't uncommon for people to think that

Permaculture is just a form of gardening by using hugelkulturs.

As a soil-building technique they're hugely fool-proof. In 3-5 years they usually yield a big pile of amazing wood compost you can then turn into gardens.

And yet, having seen hundreds of hoogs, and read about hundreds more, I must admit as a gardening technique they're hit or miss. I'd estimate that as a gardening technique, people are unhappy with them a third of the time or more. Sometimes, they will be the best garden people ever had in year one. Other times they can be a total flop. Many people don't mind too much because they didn't have high expectations for them and mostly wanted a cool landscape feature.

Let's talk about that third of people who are unhappy with them. The life-cycle of the hugelfailure is that it starts with great promise and high expectations. It gets prepped and planted and then things start to go wrong.

I said above they *can* be drought-proof, but sometimes they become water-proof instead. If the big pile of woody material dries out before becoming saturated, the soil can easily become "hydrophobic," meaning it repels water. Fungi in particular can create waxy residues that coat the soil so that water can't soak in. The tend to do this as a competitive advantage in situations with low water.

Then, because they are big mounds , it's nigh impossible to get water into the dang things until winter comes and they get a good soaking under constant snow. Only then will organisms start to build up that can break down the waxy compounds. So, the first year's veggies die of drought. Building hugels in fall so they have a chance to break down and soak up ample water is a good start. If it's dry, making sure they are well watered at creation can help.

A second common problem is critters. This is a matter of perspective. hugelkulturs are one positive pattern for inviting wildlife into the garden. Moles and voles love burrowing in hoogels. In many cases, this does not seem to hinder the performance of the hoog to much. But in some cases, they can do a lot of damage, especially if a woodchuck decides to call the hoogel home. Which brings to mind the familiar old nursery rhyme:

How many hugels could a hugelchuck chuck if a hugelchuck couldelchuck hugels?

What's the point? If a woodchuck moves into your hugel, you'll want to chuck the whole thing because you ain't getting any vegetables out of it.

So, if there's a risk of creatures you may want to plan for that accordingly. Some hoogs I've seen built over a hardware cloth mesh, though I suspect that may become a problem as the hoogel comes to the end of its life and it's time to harvest the compost. Another option is to plan for wildlife tolerant crops.

A third major issue with hugels is grass encroachment. Sometimes this starts even in the first season, but more often the grass encroachment starts right when the wood has broken down enough to finally become truly fertile. Once grass gets into a hugel, it's game over. It's impossible to get grass roots out of the log pile and in a year it's the richest mound of lawn you've ever seen. Many illustrations of hugels show them covered in grass with a few vegetables. This is probably accurate.

The final issue that tends to make unhappy hugels is unhappy neighbors. If hugels have become one of Permaculture's most prominent features, it has also sometimes become one of the most prominent complaints. Let's face it, big piles of wood and yard waste aren't exactly pretty and big piles of dirt with branches sticking out of it isn't much more inspiring.

Therefore, the best, contemporary hugelkulturs will plan for these potentials up front. Design hugelkulturs to be grass-proof, wildlife conscious, and aesthetic.

Grass-Proof Hugels

Remember our rule for successful guilds: design to keep grasses out, or design to grow grass-tolerant crops, or plan to be intensively hand-weeding grasses.

Ideally, hugelkulturs should not contact grasses or lawns. A hugelkultur built on top of a lawn is inviting disaster. Hugels near lawns should be edged with an edging spade twice a year or whenever grasses are starting to spread. Fortress plants along grassy edges may provide an extra layer of barrier when combined with edging. Grass landscaping barriers are rarely effective.

Hugelkultures that Won't Make Your Neighbors Hate You

Since I do Permaculture programs that specifically incorporate designing for beauty, I have now seen plenty of my students make hugels that are very attractive landscape features. These include hugels outfitted with nurse-log frames, those in interesting shapes, and those with artistic elements.

Hugelsculptures,

is what E James Spielmaker calls his designed mounds of organic matter, turning wood compost mounds into yard ornament.

Hugels can be walkable, can outline spaces, can be integrated with nurse-log steps to form beautiful, designed terraces. These hugels designed by Mikel McCormick look like an ornamental element. Large, 6' hugels can be partially buried to not create quite such an interruption in the landscape, or alternatively, they can be used to hide ugly views or create a feeling of enclosure.

Mycohugelkulturs

While we're speaking of "cultured" hugels, why not culture them with mushrooms? Winecaps can be grown by purchasing a bag of mushroom spawn and crumbling it into the edges of the mound. Winecaps love hugels and soon the mound will be growing pie-plate sized beautiful, edible mushrooms in addition to plants.

Hill-gelkultur, or Hugelkultur Terraces.

(See illustration at top of chapter)

Hugelkultur variations can be especially useful on hillsides or mountain slopes, such as the ones Sepp Holzer is growing on. In such situations, erosion can be high and soils bare to begin with. In these cases well-designed terraces with woody debris and soil can provide a growing medium for plants that otherwise would have to contend with few resources. We've already seen how nurse logs can regenerate hillsides and infiltrate water, and hugelkultur terraces take that to the next level, speeding up the process with deep wood and soil.

Use of nurse logs as ornaments and stabilizing features can create truly beautiful landscapes, such as the one created by Stician Samples, here. It's important to make sure these are well-secured into the hillside with deep stakes and that logs are placed in a stable configuration.

Walkable hugelkulture with nurse log edging and log-round stepping stones, by Mikel McCormick.

Hugel sculptures in an urban front yard Permaculture garden. A transformative landscape will be good for wildlife, including one very special type of magical beast, kid-critters. Here, a hugelsculpture provides play habitat and a garden trellis, which will feed soil as it breaks down.

Children seem to love these landscapes, filled with places to hide, insects to observe, magical plants to discover, berries to pick, and seasonal fruits. Kid-critters would often run around the paths at Lillie House, often calling it a "maze" or "labyrinth."

Image concept E. James Spielmaker, art, Andrina Learmonth

Part 4: Patterns and Projects for Abundant Beauty

Chapter 20: Herb Spiraling out of Control

Spiral by Donna Petrocella

If you're fairly new to gardening, the herb spiral is one of the best beginning gardening adventures. Creating one of these beautiful, functional pieces of landscape art will naturally teach you some of the most important gardening wisdom, wisdom that's often left out of other "beginning gardening" classes and programs. If you've been gardening a few years growing ornamentals or annual vegetables, making an herb spiral might just transform the way you garden forever!

This is a spiral garden bed design promoted by Bill Mollison. It is a great model of what we're aiming for with "Transformative Adventures:" a learning experience that transforms the way we live and do things. Of course, they're not necessarily the best tool for every site and every gardener, but for most homes, homesteads, and newer gardeners, they're transformative!

So, what are the important natural gardening lessons imparted by this simple spiral? Here are a few:

1. That growing food can be fun and easy! Buying these hardy perennial plants and planting them is a much more reliable way to learn and be successful with gardening than trying to create an full annual vegetable or flower garden as a beginner. In this way it helps build confidence.

2. Using nature patterns, such as the spiral to create beauty and function.

3. It teaches us to prioritize the most "transformative" efforts first. An herb spiral will last years and allow us to include the most highly nutritious and health enhancing plants on a daily basis. The herb spiral might be the most health-enhancing bed in the whole garden!

4. It teaches that a garden is more than just a place to grow food. We can create a beautiful natural environment for ourselves.

5. To "catch and store" our energy into permanent well-organized projects with perennial plants, instead of just creating annual garden beds that require lots of maintenance.

6. Microclimates, the herb spiral has different climates for different plants, everything from cool wet areas, to hot dry ones are created by the spiral's design. That observation in turn teaches us to value the microclimates

we have at home, instead of thinking that everything has to be a sunny garden bed with "perfect soil," we can make use of the special characteristics of the soils and aspects we have to work with!

7. "Right plant, right place." The herb spiral teaches us that each plant has its "niche" and encourages us to learn a little about these dozen plants and where they thrive.
8. It teaches us to look for "design problems," such as for grass encroachment, and to solve the problem with design instead of hand weeding labor.
9. Efficiency, it teaches us to grow plants we use often right near the back or front door. And in a good herb spiral the only maintenance work can be harvesting, which is an amazing lesson in efficiency.
10. Balance with nature. It teaches us that we can grow "intensive" gardens with food in permanent beds near the home. On the other end of the spectrum a semi-wild "guild" teaches us when and where to let nature do more work.
11. It ideally uses recycled materials like "urbanite" (recycled concrete or bricks) to teach us that we can make beautiful functional things from what society would consider "garbage."

And finally, as visitors stop to admire the garden, the herb spiral teaches us how we can create learning adventures for others. An herb spiral workshop is one of the first workshops community transformation activists around the world have used for decades to start community organizing and teaching others about more sustainable life ways. Sharing such a highly transformative garden design with family and community members builds consensus and vision, as well as our own social capital, head and heart authority.

Steps on the Herb Spiral Adventure:

1. Explore images and resources for instructions, plans and aesthetics. Consider making your own pinboard with your favorites, and posting a link to it in the comments!

2. Find a place to put it. Ideally, a traditional herb spiral will do best at the sunniest corner of the house, as close to the kitchen as possible! It should pass the fuzzy slipper test, allowing you to get to your herbs without getting the slippers wet after a rain. If it's shady, how will you adapt the design? Maybe you don't have the perfect spot! That's okay, that teaches us to adapt and work with what we do have.

3. Start making a list of herbs you want. I recommend growing a good variety such as you'll find in traditional designs, even if you aren't familiar with them. Over time, you may decide to pull a few things and or perhaps use some herb spiral space for peppers or lettuce. This teaches us to allow our designs to evolve.

4. Collect materials. My favorite herb spirals use recycled materials. Try looking on Craigslist or on FB Marketplace. Stopping by construction sites can also be a good way to get materials. Perhaps your town has a municipal drop-off for construction materials. If you can get extra materials, and have a

place to store them, do! You could use them to set up an herb spiral workshop!

5. Look for design problems. With herb spirals, the biggest problem is encroachment of grasses. If it's by the lawn and there's no barrier, how will you keep grass out? Some ideas: weed barriers, digging around it with an edging spade, using fortress plants, creating other garden beds around it, mulching around it…. Have a design solution in mind!

6. Lay down cardboard to kill sod, if the spiral's going on lawn. Then lay out the materials to form the spiral. Alternately you can dig up the sod, compost it and use this magical material to fill your spiral.

7. Choose a "filling." Composted sod, especially composted clay sod may be the best growing medium on the planet! Just make sure the grass is dead. You can use regular garden soil, or a potting mix like "Mel's Mix." Or use "Mike's Mix," Which is 1/3rd pearlite or vermiculite, 1/3rd compost and 1/3rd rough unfinished mushroom compost or bark fines. Then top with cocoa mulch or chip mulch.

Put it all together!

Chapter 21: Herb-Inspiraled Spin-offs

The Lillie House unwound herb "spiral."

Of course, complaints and criticisms of herb spirals are more common than actual herb spirals. This can teach us a great deal about design, transformation and gardening.

Some of the common complaints are: "I have a big garden, an herb spiral is too small to be practical." "It's too much work for such a little space." "A spiral doesn't fit my space."

Square spiral, by Dennis Hamilton

"I grow herbs out in my field so I don't need a spiral."

None of these should be seen as problems with the herb spiral. They are design constraints that can make the design better. We can look at the underlying micro-patterns that make up the herb spiral concept and see that we can mix and match them to suit our specific cases.

What is the herb spiral? Here are the patterns that make up the herb spiral:

1. Invest your time into permanent landscape transformation instead of annual gardens. This alone makes the herb spiral highly valuable. Rather than growing herbs in annual tillage fields, we can create a spiral garden that - if designed as above - can last for decades using very little extra maintenance.
2. Use permanent double reach beds to minimize work.
3. Put herbs and high-value greens close to the kitchen to maximize the use of these highly beneficial plants. The herb spiral should pass the "wet slipper test," allowing you to run outside in the morning for some herbs and greens without getting your slippers wet.
4. Use microclimates and niches to give plants better growing conditions to minimize labor.
5. Use natural patterns to communicate natural beauty and appreciation for nature.

Add these patterns together and - voila - we have the herb spiral.

But that doesn't mean we can't mix and match them. We could add other patterns, or thoughtfully cast off patterns that don't meet our needs.

To give an example, I will present the blown up herb spiral at Lillie House.

This herb spiral uses all of the patterns above, but throws in a few extras, and does so on a much larger scale. The main reason is that as a chef for perennial vegetable and herbal cooking classes on site, I just wanted to have way more herbs than could fit in a small spiral, but I still didn't want to lose all the benefits of the herb spiral design mentioned above.

And so this large, 2 part herb spiral guild still uses permanent paths and permanent beds, it is still an example of long-term landscape transformation, it uses microclimates well, and creates a large, visual spiral element in the landscape.

This design uses urbanite to create raised, well draining areas for sunny climate herbs like oregano, thyme, tarragon, rosemary and lavender. The apricot tree is happy in such a microclimate as well. Then there are shady areas to the north of the trees, with deep woody debris and swales for ample water. These ares grow herbs like sweet cicely, ramps, marshmallow and valerian. Sunny, rich soil beds are found between these two microclimates, ideal for herbs like chives, walking onions, leeks, basil, mints, and tulsi, as well as some greens. And the plan included a small pond for watercress and other aquatic herbs. Even the stepping stone paths create a microclimate for creeping thymes, lemon thyme, clover and chamomile. And yet, everything is still in a double-reach, easily accessible bed, close to the door to keep the slippers dry. The spiraling oregano edging on a raised bed of urbanite and woody debris acts as a fortress planting to keep grasses out.

The Lillie House "spiral"

Chapter 22: Planting into Resident Vegetation

(*small cardboard and newspaper collars can be used with deep piles of mulch to plant starts into mown vegetation.*)

A final method essential for starting beautiful gardens without tilling is simply planting right into resident vegetation without preparing a bed.

Yes, it is indeed possible to plant right into lawns or weedy overgrown fields, but it does take some consideration, proper design and planning. This can especially be a good way of converting large acreage areas to productivity, that would not be possible with sheet-mulching, or digging.

The first major consideration is that most of our garden plants, vegetables, ornamentals, and even fruit trees and cane fruit will be overrun and choked out, if not badly impeded by the competition of grasses. So, we'll need to select only crops that were evolved to be up to that particular challenging of succeeding into grassland.

A second issue is that even with the right plants, we may need some special preparation, such as mowing frequency or shashmulch planting to give our chosen crops a heads up over the competition.

Another thing to think about is whether the spot is appropriate to the technique and the plant choices.

And finally, the major consideration in the context of this book is that without special design, the whole thing can look quite awful and lead to grumpy neighbors.

Therefore, use good guild design such as presented in our patterns on edible meadows and "terminator guilds," design to be attractive, and plan an implementation strategy that will appropriately prep the area for planting.

While the section on **edible meadow design and installation** will go into more detail, one technique that can really help with beauty is to invest heavily into planting grass-tolerant spring bulbs. When neighbors see flowers first thing in spring, they will say "oh, it's a flower garden" instead of "my what ugly blight!" Keeping things flowering through the season can have a huge positive impact.

Cut timing of mowing can also have a major positive impact on establishing more flowers into such a system, and the more species there are, the more niches there are for us to establish other valuable species, vegetables, or tree crops. A tight fall mowing when plants go dormant, will help kick off seed germination. Delaying any mowing until after daffodils begin to die back will have a big impact in increasing species. Limiting our landscape work to just one or two real mows per year will also have a positive impact on our ability to grow vegetables and flowers.

Crops appropriate to being planted into resident vegetation, with the right technique and preparation include: Corn, beans, and squash (see the section on an **aesthetic three sisters planting**,) amaranth, potatoes, Jerusalem artichokes, cup plant, teasel, monardas, mints, blood-veined sorrel, vigorous bush and tree crops, possibly wild strawberries, and a whole variety of bulbs (covered in the later chapters mentioned above.)

Notes:

Jean-Eves Humbert Does delaying the first mowing date benefit biodiversity in meadowland? Environmental Evidence, 2012 https://environmentalevidencejournal.biomedcentral.com/articles/10.1186/2047-2382-1-9

Corn germinating in a pile of slashed vegetation with some compost on top to help out.

Chapter 23: Overview Patterns for Beautiful Transformative Landscapes

"When you put beauty in a place that has none, that's a game changer."
— Ron Finley

To live in a beautiful, uplifted situation with dignity should not be seen as an exclusive privilege for the elite. It is the birthright of every being born to this earth. The idea that the rest of us should content ourselves with stodgy utilitarianism is oppressive. We all deserve beauty.

To surround oneself with beauty is old, powerful magic.

Look! The elk in the wood carries itself with pride, dignity, its chest uplifted, its head and shoulders like a king, its every movement, its very presence is an incantation:

"I AM HERE!"

When we dare to create a beautiful, uplifted environment for ourselves, it allows us to be in the moment boldly, rather than shrink to some sense of shame or inferiority, or worse, an aspirational fantasy of salvation – some savior, the lottery, that promotion, the dream-job, the business breakthrough, the revolution – which may never come.

Instead, we can each make our worlds beautiful and our lives rich. This does not require money. It's the ultimate taboo of our age, the truth almost everyone wants to hide from you, but you have more power over the way you will experience your daily life than anyone else on this planet does.

The world we want to see starts with us.

As we've said before, looking for "patterns" and models to replicate and adapt is both a good tool for improving function as well as creating beauty. This pattern language paradigm makes beauty and good design available to anyone, not just those who can afford designers. So, with a firm grounding in the techniques we'll use to establish beautiful, healthy ecosystems, let's turn our attention to finding patterns for beauty.

Whenever I approach the teamwork of design, and design is ALWAYS teamwork, I begin by finding patterns that fit the landscape, the architecture, the history, ecology and community character of the space. Each space will be different. What is the "genius of the place" trying to say? How can I help it be heard, rather than stifle its voice under my own song?

As we search for patterns, the most important to begin with are the overarching, or "meta" patterns for a site. If you wish to develop a farm, then you might find a Regenerative farm or Permaculture project that is a good model for meeting your own goals and start there. If your goal is more on the homestead scale, you might look to some truly regenerative and successful homesteads, such as Bealtaine Cottage, Lillie House or Silk Farm. Usually, when we have this overall idea of what the site can be like, the design can flow from there.

We can do the same for aesthetics. We can find a place that is inspiring to us, matches our neighborhood character, makes us feel good, and works with the kind of feeling we would like to recreate. Then we can design from that.

However, when it comes to aesthetics for Transformative Landscapes, not all patterns are a good fit. The modern HGTV suburban landscaping look, for example, is based so deeply on unsustainable materials and high maintenance cost with - at best - very low utility and productivity. It would be difficult to adapt that patternto meet the goals of a Transformative Landscape. As much as I love touring Versailles, unless your home comes with a fleet of workers to maintain the garden, it's not going to be a pattern that meets your needs. The same could be true for some famous Permaculture sites that require a constant flow of interns and volunteers to keep them functioning. If you want to have a beautiful home landscape, the best models will be home-scale projects like Lillie House, which worked with no volunteer or intern labor.

So the patterns in this chapter are particularly useful to those of us creating Transformative Landscapes, gardens and farms that don't require a fleet of servants or volunteers. These start with ecology, productivity, and ease of maintenance at their heart. Instead of handing work over to volunteers, we're finding ways to hand work over to nature, while still looking safe and beautiful.

Let's look at how this pattern design process played out at Lillie House.

One of the first patterns that inspired Lillie House was the famous "Italian House" at Kuskova. I wanted to invoke this "allee" of trees and hedges framing the entry like a visual mandala setting the mind to a state of focus and attention. And the model was appropriate to the Italianate design of the home, the climate, the neighborhood, and my goal of promoting sustainable living through beauty. And of course, it was appropriate to having a tree-filled landscape of productive plants.

The next is the model of the Jardin de curé, a historic form of European food forest garden. These were beautiful spaces for growing fruit, vegetables, cut flowers, and medicinal herbs while creating a healing, rejuvenating space. We'll cover those in this unit.

Of course, I did not copy that design completely. I just used it as an idea of what kind of space and feeling I wanted to invoke.

Here we see those jardin de curé colors in our gardens, with an edible meadow inspired by gardens at, Highgrove, the residence of Prince Charles.

And finally, the most important, inspiring and beautiful pattern we looked to was the "home garden" pattern found throughout the world in indigenous societies. It is the pattern of a just and sustainable society, and one known for its beautiful, life-enhancing environment. The home garden is the first pattern we'll cover in this unit.

Jardin de curé, Photo, Andreas Praefcke, wikimedia

Chapter 24: Thematic Patterns for Beautiful Gardens

The Theme of a garden is what gives the garden its overall feel and character, the major impression and idea it will communicate. Will it make people think of a golf course and suburban sprawl? Or will it invoke an ecological civilization of abundance and harmony?

This overall character or 'style" of a garden will be the number one thing people take away. It will guide each aesthetic decision we make.

Therefore, as we start thinking about designing or transforming a landscape, the very first step is to explore patterns for garden theme and style. The following patterns will be a good starting point for exploration, but you may find inspiration beyond this book! You may find inspiration in your favorite natural places, or perhaps in a movie. I have often wished to recreate the pastoral feel invoked by The Shire in the Lord of the Rings movies.

Meta Pattern 1: The Traditional Forest Garden: The Home Garden Pattern

Images: Opposite, an early design sketch of Lillie House, based on the Home Garden pattern, adapted from Home Gardens in Nepal. Above, the finished implementation after further design revision and inspiration from the Jardin de curé, itself a form of European Home Garden.

There it is, the basic "Home Garden," an ancient, evolved forest garden system and found, in one form or another, almost everywhere trees will grow.

At its most basic level, it's a home or small village unit in a protective forest clearing or surrounded by a forest garden planting or enclosure, a mixed function productive hedgerow membrane designed to exclude harmful energies while welcoming and containing beneficial energy flows.

Outside, there are typically more gradients of wild and human-managed uses, like pasture land, wood lots, and traditional commons food forests like the Kyrgiez apple walnut forests. These systems go beyond the scope of this book, but some are mentioned in the worksheet on agriforest systems to explore. And of course, there are fully wild areas, too.

Inside, there are areas (could we call them organs?) devoted to all the functions necessary for the unit to survive and prosper, areas for vital exchange of information (social functions,) areas for production, areas for growing food, sustaining internal health, and so on. It has units for photosynthesis, converting sun energy into useful energy for the homestead. It has various appendages for affecting the outside world, for reaching out and bringing in vital resources, sharing information with outside units, and exporting its wastes. It has a sort of central nervous system which regulates the distribution of resources within the unit and responds to external stimuli. It even has a nucleus home structure to protect its most vital information processing, organizing, and storage units, us humans.

The very word "garden" comes from the word enclosure, "garde." This is true in many languages.

As these units grow, they eventually reach out with their genetic information, sharing it with other units and reproducing itself, splitting into separate units to reproduce the pattern. Each new cell adapts to its specific environment, its place, its resources and landforms. It fulfills its role in maximizing life enhancing energies and performing particular duties as part of a larger fabric of human communities. Each naturally grows, shrinks, exchanges resources and acts according to the needs of the whole.

It is likely that this is the fashion in which humans, and our companion species, generally spread around the globe. It's a pattern that follows the dictum "life makes good for life," enriching its ecosystem and ensuring biodiversity and resilience for all species.

And so this pattern is found throughout the tropics, but also in temperate climates. They are found in Nepalese home gardens, and in Japan. In Europe they were common, making up the "toft and croft" medieval landscape, of communal fields for grain and cottage gardens of vegetables, fruits, medicines and flowers. Buffalo Bird Woman, a Native American traditional gardener described a similar gradient of personal and communal gardens and commons for forest gardens in the book "Buffalo Bird Woman's Garden." Images of early Eastern Woodland native villages showed similar patterns of enclosure, usually on a small village scale.

Algonquin village, John White, 1590

Is this self-organizing cell-like basic home garden structure a "poly species" super organism? (super-organisms are those made up of multiple individuals, such as a bee hive.)

Is this some basic, hard-wired pattern of biology that we evolved in order to adapt our environment in a way that made us beneficial keystone species?

Do we instinctively utilize the rigid, lignin-rich properties of woody perennials to create our "cell walls" and define our "tissues" (communities) just as plants do?

Is this pattern the ideal result of simple, logical ways that biological organisms thrive in interaction with each other? Is it shaped in some fundamental way by the laws of nature on our world, regardless of scale: cell, tissue, organ, organism, community, ecosystem?

Now, I do not want to make the error of giving the impression this pattern was universal. It certainly was not. For example, images of traditional fishing villages in Europe, North America, and Asia seem to take a different form. If most of the food comes from the ocean, enclosure is not necessary to ward off pests from the garden. Instead, access to water is the priority, so yards are often small and lack significant gardens. Even then, small enclosures may hide a few vegetables. But if the goal is to live off horticulture, this pattern comes with many advantages.

Architects such as Christopher Alexander have pointed out that we seem to innately feel "right" inside these enclosed, cell-like spaces. We love plants and rich, lush gardens filled with fruits and vegetables. We thrive on the loving companionship of non-human animal companions. We do indeed seem programmed to order our environments to create this home-garden pattern.

So, how did we get stuck into the self-destructive, eco-cidal patterns of agrarian civilization, where "cells" stop acting beneficially, and instead maximize "consumption," the conversion of raw energy into ever more humans, ever more cells, without regard to role or the health of the whole? ...into "expansionist" units that infiltrate, and convert healthy cells into similarly single-minded, anthropocentric, tissues with poorly defined borders? Until inevitably, the growth causes the breakdown in the healthy function of the whole super-organism "ecosystem," and with it, the humans that depended upon it.

Of course, it's probably a stretch (to say the least) to compare the self-destructive nature of agrarian civilization units to cancer cells. But the underlying biological mechanism and behavior characteristics look very similar, and unless we're careful, they'll have the same end result. This civilization is killing us.

Whether or not our homesteads are surrounded by the literal "cell wall" enclosures of traditional home gardens, it's interesting to me that many sustainability and Permaculture folks seek to solve our society's problems by recreating this basic "closed loop," cellular pattern of humanity, in various ways.

Take a walk through this Polyculture of fruit trees, herbs and vegetables, to pick a salad of red roach, miner's lettuce, lettuces and herbs. Rhubarb, potatoes, garlic, leeks and other vegetables await the fall harvest. Christina Perma Dumbrava

As these new transformative landscapes emerge, are we witnessing the death of a destructive, old system, the agrarian civilization?

Are ideas like Transformative Landscapes, Permaculture and biomimicry an instinctive reclaiming of our biological role as beneficial organisms? Is this the basic "genetic" motivating information that will organize a new, healthy, diverse ecological civilization "super organism" to take the place of destructive society?

Of course, I'm being metaphorical more than metaphysical. Ultimately, these questions are unanswerable, unscientific, and verge on being meaningless.

But I do believe that when we turn away from the destructive global system to build our own personal Permaculture systems and communities, based on caring, mutually beneficial interactions between species, we create a powerful, beautiful way of life that will spread, endure, and provide comfort and security for generations to come.

Lillie House, an urban food forest in Michigan, is abundantly loaded with food, and it's a beautiful place to be.

What is a Traditional Forest Garden?

Defining Characteristics of Traditional Forest Gardens

In *chapter 24*, we'll look at overall patterns that we can use to inspire better, healthier landscapes. But for now, there's one over-all pattern so important for creating transformative landscapes, that we'll start to introduce it now: The traditional forest gardens (also called food forests or home gardens) of the world. This basic pattern is so useful to the creation of truly sustainable landscapes and societies that nearly every garden inspired by this book is likely to at least resemble a traditional forest garden.

So what is a traditional forest garden? First, let's be clear about what it isn't. Every book on the topic going back to the first publications to use the term have clarified that a forest garden is not growing food in a forest, it is growing food like a forest. Which means in most climates and circumstances, forest gardens are recommended to have much lower densities of trees than an actual wild forest does, so that sunlight can infiltrate enough to grow the intended crops.

So, if a forest garden is not an actual forest, what is it? Here are a collection of defining characteristics taken from the Permaculture and academic literature on the topic. Not only can these help us understand what a forest garden is, but help us grow good, functional forest gardens. For example, if one thinks it must have a true closed canopy like a forest, one is unlikely to create very functional forest gardens.

1. **Layers**. A forest garden has layers like a forest, including tree crops, bushes, annual and perennial plants, climbers, vines and so on. Different authors cite different numbers of layers, but the important thing is to think of crops in layers.
2. **Climate appropriate**. Like a forest, these gardens are adapted to the climate. For example, in temperate areas, it is usually recommended to use sun traps and open canopies so that light can infiltrate. In areas where too much sun is a problem we can use a closed canopy.
3. **Work with natural succession**. Like a natural forest, these gardens follow the arch of *"natural succession."* This is the process of how ecosystems mature, typically gaining in diversity and mass over time. For example, a grassland left untended will eventually become a forest. Nature starts with annuals first, followed by perennials and ultimately shrubs and trees. A good forest garden will also start with heavy annual crops and have more perennials as time goes on.
4. **They use Heterogeneous textures**, the somewhat randomized uneven spacings of a forest, rather than homogenous ones like with rows and even spacings like an orchard. Research on forest health has found this increases both diversity and health, reducing pest and disease issues.
5. **Polyculture and guilds**. Like a forest, these gardens use many crops interplanted together, "polycultures," instead of monocultures of one crop. Ideally, these crops are grown in ways that mean they function well on their own, which is called a guild. We will go into depth on creating beautiful guilds in a later chapter.
6. **High biodiversity.** These systems are biodiversity hot spots, which is good for stewarding wildlife, and the biodiversity helps to control pest and disease issues. That includes a diversity of yields, so if one crop has a problem, other things will still work.
7. **Flexible adaptability of inputs and outputs**. This is perhaps the most important defining characteristic of a forest garden. The goal of a traditional forest garden isn't to have the highest possible output. In most cases, this is just a poor goal for real people. We don't want to waste time to have the highest possible output, we want the highest possible value for our time! The goal of these landscapes is that they have something like a dimmer switch built in. When we are busy, they can be set to low management mode, and with very little maintenance they can still produce a lot of luxuries like fruits and vegetables. But if there is a need, they can quickly be scaled up with more inputs, and have a much higher output.
8. **Integrate living space and social functions**. The traditional forest garden isn't valuable simply because it produces calories, like a corn field. They are sources of wealth and security, they are living and recreational spaces, they are places to raise children, have celebrations, keep traditions, or simply experience beauty.
9. **Favor holistic natural management techniques**. Because spraying poisons harms biodiversity, it also ironically harms our ability to grow clean produce without more poisons. One spray kills biodiversity, requiring more spraying in the future, a phenomenon Rachel Carlson called the pesticide treadmill in Silent spring. We'll talk about holistic natural management in a later chapter.
10. **Balance between human management and wild.** Including appropriate kinds of enclosure in most **cases.**

Resources on Home Gardens and Forest Gardens

1. Edible Forest Gardens, David Jacke and Eric Toensmier, Chelsea Green, 2005. An encyclopedic two-volume set that was a great leap in the art of forest gardening. The "characteristics of a forest garden" presented here are largely adapted from this work with some additions from the academic literature.
2. Creating a Forest Garden: Working with Nature to Grow Edible Crops, by Martin Crawford and Joanna Brown, Green Books, 2010. A classic from one of the major practitioners in the field.
3. Integrated Forest Gardening, by Wayne Weiseman, Daniel Halsey, Bryce Ruddock. Chelsea Green, 2014. This book provides us with theory and models for guilds.
4. Forest Gardening, by Robert Hart. Green Books, 1956. The original classic on the topic is more of an inspiring read than an up-to-date how-to, but provides a beautiful vision.

Beyond these how-to texts, there are a number of academic studies on home gardens which have provided a theoretical basis for this chapter, especially the characteristics and goals of forest gardens. An almost constant characteristic in this literature is the integration of social function as one of the most important elements in the forest garden. A second is the flexibility of inputs/outputs, another almost universal feature in the literature. Western forest gardeners looking to create good systems would do well to appreciate why these two patterns are universal in the literature. A few of these are particularly accessible as reading:

5. Home Gardens in Nepal, Resham Gautam, Bhuwon Sthapit and Pratap Shrestha, Editors. Biodiversity International, 2004.
6. The Forest Farms of Kandy andOther Gardens of Complete Design By D.J. McConnell, K.A.E. Dharmapala, S.R. Attanayake, Routledge, 2003
7. Home gardens of Kerala: Structural Configuration and Biodiversity, Allan Thomas and Sajan Kurien. Indian Journal of Research, 2012.

Meta Pattern 2: The Cottage Garden

Thatched Cottage in Bernaval, Hippolyte Camille Delpy, public domain.

The cottage garden is an informal, romantic style that mixes flowers, herbs, vegetables and fruit trees in a way that fits many homes and neighborhoods.

These traditional gardens of Europe are informal in style and integrate ornamental plants and edible plants, usually with gradients of enclosure around the home.

Looking for easier, more useful ways of managing landscapes, and better connection with nature, gardeners like Gertrude Jekyll and William Robinson repopularized this ancient form of garden. In some ways, this was the beginning of Permaculture and Transformative Landscaping!

These gardeners were influenced by the pre-raphaelites and again by William Morris and the mystical view of nature they shared.

The cottage garden doesn't rely on rules. It's more a set of principles, an informal feel, room for wild spaces, beds with semi-wild plantings of tightly packed ornamentals, natives, and edible plants. Use of fruit trees, vertical space with arbors and fences. Materials are rustic, recycled, and antique feeling, rather than new and modern. In this way it is similar to wabi-sabi.

Plant choices should include anything that makes the gardener happy, but some plants are specifically chosen to reflect this old-fashioned utilitarian feel. Roses, especially multi-use varieties like rugosas with their big, delicious hips. Climbing plants on arbors. Edible hedges are typical. Useful flowers like calendula, violets, hollyhocks, and campanulas. Foxglove is a classic, but lupines give a similar feel. Herbs like sage, lavender, thyme, wormwood, catmint, feverfew, soapwort, hyssop, sweet woodruff, and germander are welcome. Usually all these are in soft, comforting colors, whites, pastels, blues, and purples. Guild suggestions and color ideas will be found in a later chapter of the book. Native plants from wherever you are add to the "wild" and informal feel, updating the cottage garden look to your own region.

All of this makes the look ideal for most Permaculture sites, and even adds function to most landscapes and farms by providing hedges, zones, and enclosure without having to "design" them in. A traditional farm that starts moving towards a cottage garden feel will necessarily be doing good Permaculture design!

For many European-style houses, the form works well, and in most neighborhoods. Most Victorian and mid-century homes were actually built with something like this garden in mind. Larger suburban homes with the look of estates might benefit from a more formal design, like the Jardin de curé, our next pattern.

Patterns to Get the Cottage Garden Look

—Keep things informal. Informal, natural and recycled materials, things that look antique.
—Emphasize the usefulness of the landscape. Seating areas, work areas, clotheslines, etc.
—Same goes for plantings: Integrate herbs, vegetables, fruits, and flowers together.
—Use color schemes and thoughtful planting design to create beauty instead of expensive materials and formality.
—Preferably use comfortable, soft color schemes with pops of complementary colors.
—Include water in the landscape (see section on ponds.)
—Signature plants include culinary and medicinal herbs, foxgloves, borage, multi-use roses, primroses, bellflowers, hollyhocks, lupines, delphiniums, and legumes like sweet peas and runner beans.
—Curved paths and beds are less formal than straight lines, but if formal bed layout is used, go with wild polyculture plantings.

Meta Pattern 3: The Jardin de Curé

The French "Jardin de Curé" might just be the original "Permaculture garden" of temperate Europe: diverse, beautiful, abundant, easy, and custom designed for the circumstances at hand.

If the cottage garden is an ideal pattern for cottages, informal homes, country houses and mid-century neighborhoods, the Jardin de curé maybe better where a more formal look is desired, or where the architecture is larger or more stately.

Again, I believe we have a lot to learn from these old, evolved gardening systems of traditional cultures, which often have better worked out maintenance and plantings than modern "forest garden" versions that lack long histories of practice. The front yard at Lillie House was modeled off this style of garden.

The "Curate" or "Curé" was the head parishioner in the French Presbyterian Church and his garden had to be multi-functional and easy to care for. The curate himself would have planned and maintained his garden, with the help of some volunteers from his congregation. So, not only did it provide an important source of fresh fruits and vegetables for the curate and his family, it also had to help provide for the needs of his congregation, and be easy to maintain in the busy curate's free time.

As the most learned man and woman in the community, the curate and his wife were often called upon to be healers of physical ailments as well as spiritual ones. The curate would often act as physician while his wife would be midwife and herbalist. So this garden needed to be a true "physic garden," one of the town's most important "gardens of herbal medicines."

And, of course, the curate had to attend to the spiritual needs of his flock in times of crisis, when they were in need of comfort. Soft colors dominated by shades of blue, lavender, and white painted a soothing spiritual backdrop. These colors shine best at dawn and dusk, the times of day when the busy family would be home to appreciate them, and to walk with those in need.

And when it was time for celebration, the curate's garden provided the church with a source of beautiful cut flowers, again in soft, spiritual colors. Even today, many french flower and rose cultivars in these soft colors bear the names of saints, a testament to their history in the Jardin de curé´. In fact, curates and their gardens were important in the history of French plant breeding. The "dames rocket" would have figured as both a vegetable and a flower in the curate's garden.

And, since this multi-purpose garden had to be both beautiful and easy to care for in the spare time of the curate and his congregation, a polyculture system of formal beds with informal plantings was used to cut down work and keep things tidy.

Hidden inside these geometric beds called "parterres," vegetables, medicinal herbs, fruit trees, flowers, and what many today would call "weeds," grew together in a semi wild profusion similar to the English Cottage Garden, only surrounded with edging of box, or with useful herbs such as thyme or lavender. At Lillie House, the oregano and lavender hedge imitates the low box edging in the exotic formal gardens that the well-educated curate would have seen at gardens like Versailles. Unlike a boxwood hedge, when we trim our edging, it's time for oregano pesto, or Greek potatoes.

And finally, at the end of the day, this edible, medicinal, flower garden had to be a spiritual retreat and meditation sanctuary for the curate.

It had to be a meditative space of natural beauty, but also of spiritual importance. Formal beds were typically laid out in the form of a cross, with other symbols and spiritual reminders woven throughout. Our secular Permaculture-inspired version takes

the shape of an ankh, an ancient symbol of permanence and a fitting symbol of the goals of Permaculture which are spiritually important to us.

Beauty in Abundance

The garden was enclosed by walls, mixed hedges or espalier trees–or a mix of these–to create the feeling of sanctuary, in the manor of Christendom's oldest spiritual gardens the Hortus Conclusus, which is so often seen depicted in medieval art.

This was thought to symbolize the garden of eden or perhaps to even invoke heaven.

And–just as in Permaculture–water was a mandatory element in the garden, usually located at the center of the cross, in the form of a simple pond or well.

Today, there's renewed interest in this very old style of multi-functional garden, but it's not for the first time! Back in the Victorian era this form of garden became a brief fad, especially for victorian entrance gardens in the "beautiful" style of architecture, making this style of garden appropriate to many historic homes and neighborhoods.

Related Patterns for Transformative Landscapes:

1. **Formal beds with informal plantings**. Oregano or Thyme "edging" can help keep things looking neat, while using "messy" looking polycultures that require less maintenance than "tidy" plantings would. Mix up fruits, veggies and flowers together, as in a "forest garden."

2. Include **water.** Humans and wildlife are naturally drawn to water.

3. Create comfortable **places to sit** for contemplation, with **beautiful views**.

4. Make nice **paths** for a walk, with interesting things to see along the way.

5. Use **spiritually meaningful symbols**, these could be symbols of nature, secular humanism, philosophy or religious meaning–Anything you can connect with personally. This will add a layer of depth to your garden and what it has to offer you.

6. Rely on **cool, calming colors** like the blues and whites of the Jardin de curé´. These provide an overall theme for the garden that can still harmonize well with pops of other colors, such as the reds of roses or the yellows of brassicas gone to flower.

7. Include gifts (plants, statues, pots…) from friends and family, and your garden will speak to you on a personal level.

8. Create a feeling of **enclosure and privacy with mixed hedges** of fruit and flowering plants.

9. Plant many **aromatic plants.** The multi-functional Permaculture garden should appeal to all the senses.

Meta Pattern 4: The Post Wild Garden

For those looking for a more modern aesthetic, the "post-wild" gardening movement, as covered in beautiful coffee-table books like "Planting for a Post-wild World" and "The Rambunctious Garden," has gained popularity as both a critique and outgrowth of the "wild" gardening movement of the last century that advocated gardens of native plants. Often these "wild" gardens were planted more in the disciplined monoculture battle formations of botanical gardens than anything you'd ever find in the true "wilds."

Such gardens of site-appropriate native plants were proposed as being more environmentally friendly, since the native plants would be well-adapted to the conditions, and not need as much watering or fertilizer. And gardens populated by the local flora and fauna would be more "terroir," more appropriate to the local character.

However, the downside is that in reality, in a world with non-native soils, climates, and exotic species, these gardens often take more maintenance time than traditional flower gardens, and sometimes require heavy spraying and tilling. Because the idea was to plant "well adapted plants" little care was given to things like the soil and caring for water, or wildlife.

Post-wild carries the same thoughts further down the stream by emphasizing naturally tight plantings of ecologically modeled but well-designed plant communities. Post wild plantings copy what you see in lots maintained by mother nature herself. We can of course easily take this to the next level of Transformative Landscapes, by also planning for wildlife, catching and storing water, and avoiding pollution and energy intensive materials.

Permaculturists call these wild assemblages "recombinant ecosystems." You'll notice, this isn't the idealized "untouched nature" of pure native plants that the wild gardening movement dreamt of. Post-wild gardens prioritize the resource and energy efficiency of truly "wild" plant communities as they pop up in the *real* natural world, "weeds" and all, over the exclusive use of natives when the

This edible polyculture by Christina Colis Perma Dumbrava has subtle color echoes, sprays of grasses that blow in the wind, and of course, plenty to eat as well as look at.

cost to people or ecosystems is too high. It recognizes that non-natives can play an important role in bringing stability to ecosystems dominated by non-native soil organisms and human use patterns that are inhospitable to native plants.

The idea is that unlike many native gardens that often require continuous work weeding and spraying to keep out "invasives" and watering and soil amendment to keep plants healthy, nobody does ANYTHING to these wild recombinant ecologies and they just "keep on keepin' on," as they say. Due to the diversity, it's extremely resilient. If it's a dry year, something in that tangle will thrive. In a wet year, something else will step up to keep things looking lush and healthy.

Post-wild plantings also give up the mulch-obsession of modern landscaping. Because, spreading those orange-dyed wood chips everywhere? WTF? Stop it. Just stop it now.

Mulch once and you'll mulch, mulch, mulch… it never ends. You'll be back mulching year after year. Which isn't always a bad thing, but it takes time, work, and resources. But, ecosystems are SELF-MULCHING and clever gardeners design gardens that do the same! In well-functioning plant communities, the plants look after each other, and do all the gardeners' work: watering, weeding, fertilizing, pollinating and mulching.

And because these plant communities resemble true wild nature, there's a better chance that they will capture that underlying magic, the beauty of nature that stills our breath and quiets the mind. Such plantings have the opportunity to transcend being "gardens." They may become ecosystems. Which isn't just good for the garden, it's something we can see and feel. Healthy ecosystems – that's something we've spent our whole EVOLUTION learning how to spot and everything in our being tells us that's the kind of place where we want to be.

And there's no reason that post-wild gardens have to be messy postmodern art. Hardscaping can add structure and tidy formality while reducing maintenance. No mowing or weed whacking is necessary to keep these lines looking crisp and formal.

Careful use of color can bring amazing beauty, character and variety. Good designers have been painting post-wild landscapes worthy of museum walls.

And post-wild designers often tie wild plantings together with formal beds and garden layouts, such as we have done in our EDIBLE post-wild front-yard Jardin de Cure at Lillie House:

In my gardens, I have used the ideas of guilds and ecological modelling to create edible ornamental plantings that are filled with both flowers and food. You'll find many examples later in this book. That's right! There's no reason why the post-wild aesthetic that's becoming so popular can't be applied to create beautiful EDIBLE landscaping.:

For those who are interested, the books listed above are loaded with "garden porn" (as it's known on the interwebs) and practical advice are a good place to start. But in my opinion, some of the best tools for designing great plant communities, whether edible or ornamental, come from the field of Permaculture. The Permaculture concept of "guilds" can be a great tool for analyzing the naturally occurring ecosystems around your home and designing gardens that work as well.

The guilds in this book are a departure from some Permaculture books in that these are designed to be stable ecologies, rather than mere companion plantings that require human maintenance.

Some patterns for Post Wild Edible Transformative Landscapes:
1. Materials: Modernist materials work well with clean edges. To be Transformative, choose recycled or durable materials.
2. Formal beds, informal plantings.
3. Use of subtle color is especially important for the style.
4. Patterns for sound and aroma.
5. The section on guilds and polycultures are particularly relevant to this style.

(Another post-wild edible garden designed by Lillie House. This one is an ecologically modelled polyculture.)

Meta Pattern 5: The Japanese Garden

The Japanese Garden is an overarching theme I must mention, because it is ideally suited to Permaculture. And yet, a whole book could be written on the regenerative Japanese Garden. And I am not the right person to do it. But I will say that for someone who is drawn to this aesthetic, there are many wonderful books on the tradition, its rules, its symbols, and techniques. These may be combined with the patterns in this book on edible plants, especially the patterns on **coniferous plants** for edible landscapes.

Meta Pattern 6: Novel Themes for Novel Times

I am convinced the very best themes for transformative landscapes have yet to be named. They are the truest representations of person and place, and the genii loci. And of course, these strange times we find ourselves in. I want to see gardens that make me envision the survival of humans into the distant future, the gardens our sustainable successors will make when our way of life ends. I want to see gardens in the rubble of this cruel civilization. I want to see gardens of hope and transformation. What are the elements of such gardens?

Terroir

A romantic path of alternative ground-covers leads past an abundant polyculture to a terroir structure that seems so natural that it could have grown from the mountains itself. Does this garden by Christiana Perma Dumbrava contain the blueprints for a sustainable, rich future?

Terroir is the French word for "earth," and most people know it—if they know it—from wine. It means that the wine has the character of its place.

The idea is that a flawed wine will have character of place that makes it preferable, more special, more unique and charming than a wine that has been perfectly balanced with additives after the pressing process. I like my people to have some terroir to them, too.

Beyond wine, this concept has been used with fruits and vegetables. Some of the famous produce of the world has this distinct taste of place.

And we can use the same word for the aesthetics of place.

A building or garden that is not keeping with the local aesthetic is often considered an eyesore, and will have the neighbors wanting to tear it down for the greater good.

For me, the best gardens appear as though they grew out of the landscape on their own, making use of local materials.

It is also important that we reflect local tastes. Some wine makers even consider this part of their terroir. A "sophisticated" wine might go over the heads of folks in rural places, while a simpler, sweeter, more accessible wine might be more honest to the area. What it lacks in overt sophistication it will make up for in local charm. The same can be said of gardens. We don't have to have Versailles in our backyards! We just need to have a nice place where our friends and family want to spend time.

So the best gardens will always be a collaboration, a co-evolution, both with nature, and with the neighbors.

A terroir approach to beauty will also help head off gentrification cycles. Gentrification is driven by the silent code of conspicuous consumption and "property development." An HGTV version of landscaping, with expensive industrial materials is not only bad for the climate and earth, but it can trigger gentrification cycles, and also alienate us from our neighbors. This is of course very important if we are new to a neighborhood. If we are not careful, we can unintentionally impose an outsiders aesthetic on the neighborhood, trigger gentrification, and destroy the very reasons why we moved to the neighborhood in the first place. A beauty that seems to grow from the place and culture of the neighborhood may do the exact opposite. Signaling that there is internal investment that will keep prices stable and prevent gentrification opportunities for investors.

So landscaping with a sense of "the earth" brings both social benefits and greater beauty.

Tips for "terroir" in the garden
—Use some local, common plants. For example, at Lillie House we used rose of Sharon because it was common in the neighborhood, and edible!
—Use locally available materials. If you are in field stone country, quarried shale will look out of place. If you're in the mountains, rounded field-stone may look odd. Concrete blocks are only terroir on Mars.
—Nurse logs of local timber are usually terroir.
—Garden to the local ecology. If it's pine forest, think of that. If it's prairie, build on that.
—Use native plants.
—In cities we can be bolder, have more fun with gardens. In the suburbs we may want a more formal look. In the country, we can have a relaxed casual garden with more wild areas.
—Garden for the neighbors and folks in your town. Gardening is performance art, and who will be the people who will see the performance?

Garden style recommendations for different settings:

—The Home Garden. Anywhere around the world where a beautiful sustainable life is desired.

—Jardin de curé is recommended for stately historic homes, Italianate and Eastlake homes, and large ornate Victorian houses. It is also a good fit in modern neighborhoods with homes in a contemporary style.

—The Cottage Garden is good for gothic homes, cottages, woodland retreats, and mid century modern homes. Small urban Victorians look char ing with a front-yard cottage garden.

—Japanese gardens are good for homes built in a Japanese style, or where culturally appropriate.

—Post-wild edible landscapes are excellent for prairie style homes, and modernistic architecture.

—Mosaic landscapes are good for larger farm properties look beautiful when treated as a tapestry of ecosystem types. And they are most sustainable that way.

Meta Pattern 7: The Mosaic Landscape

From the point of view of a beautiful sustainable human habitat, our goal is almost always a "mosaic landscape," one that weaves different ecotypes and uses together into a form with abundant habitat, lots of edges between elements, and lots of variety.

A typical Mosaic Landscape following the adventures and patterns in this book might include forests and hedges, fields and gardens, herb and greens gardens outside the door, some annual vegetable gardens on the French Intensive or BioIntensive model, perhaps some integrated animal systems like a victory garden with chickens, it may have areas of forest garden, more natural food forests or native forests, edible meadow areas, edible hedges and hedgerows, all integrated in together into garden rooms or areas, and with food, herbs, native plants, and habitat all throughout.

You will notice that is exactly what you will naturally arrive at if you complete several of the adventures in this book.

Such landscapes have been found in a multitude of research to provide great habitat to wildlife and native species, while sequestering carbon, caring for water, and meeting human needs well.

Chapter 24: The Edible, Aesthetic, and Mighty Hedgerow, and/or Tapestry Hedge

You want to LEARN JUST ONE WEIRD TRICK that will have a guaranteed positive effect on a huge range of features including:

—Farm and garden productivity and profitability
—Reduced irrigation requirements
—Reduce cost of fencing and livestock management and feeding
—Home and garden security
—Improved livability and reduced home heating and cooling costs
—Reduced pest and disease pressures
—Increased pollinators, native flora and fauna,
—Increased soil health and fertility,
—Increased water health,
—Increased biodiversity and overall ecosystem health
—Fight climate change by sequestering a whole lot of carbon
—And provide more food for less work than just about any thing else you can do?

Then here's your word for the day: "hedgerows."

Hedgerows, or living fences and boundaries have been planted by humans for at least 6,000 years (Mueller, European Field Boundaries) to simplify landscape management, and cheaply and easily provide food, medicine, fuel, craft materials, and a long, long list of other services.

Across climates, including in the tropics, most researchers are stressing the importance of these systems to the health of ecosystems, with research demonstrating their value to wildlife, their ability to increase biodiversity, bolster native bird and invertebrate populations, act as a buffer from agricultural pollution, clean water, recharge aquifers, mitigate soil loss and erosion, sequester carbon, and so on.

Throughout Europe and Asia, in cultures where hedgerows were a common landscape

feature, they are now being recognized for their value, and cherished for their associated cultural traditions and character within the landscape. In many countries, such as the UK or Japan, they are associated with culinary traditions, are considered an important part of the sense of place, such as the "bocage" landscapes of France, or have long-standing spiritual symbolism or religious and cultural traditions associated with them.

The word "garden" itself comes from "*garde*," the enclosure created with hedges. They have been seen as a metaphor for the boundary between worlds: the human world and the wild, the civilized world and the outside, and of course, our world and the spirit world. The hedgerow is where one may find the entrance to the land of the fae, or fairies.

In North America, we culturally lack the deep appreciation for hedgerows, "bocage", and "mosaic landscapes." Worse, environmentalism is more influenced by the racist mythology and assumptions about a wide-open American "wilderness" untouched by humans (despite the large First Nations population.) And of course, modern environmentalism is largely driven by corporate products and services like poisons, and hedgerows can't be maintained with poisons.

And yet still, American ecology departments widely reinforce the high-value of hedgerows and windbreaks for ecosystem health and agricultural productivity, and decry the loss of these features due to poor economic choices and poor understanding of food safety practices, as a tragedy.

And while some agricultural authorities working under Food Safety and Modernization Act (FSMA) regulations superstitiously eye hedgerows with suspicion (or suggest that removing some hedgerows might be a necessary "balanced approach") research continues to show that hedgerows do NOT pose a risk for contaminating food, and actually reduce risk while removal and "mitigation" efforts such as tilling around hedgerows may actually increase food safety risk.

But many studies have shown that the improvements to ecosystem health from restoring hedgerows do carry benefits to humans in farming, gardening, home or other productive landscapes, including reduction in pests, and increase in pollination.

In addition to a positive effect on ecosystems, hedgerows can have many direct benefits and yields to humans (depending upon their design) including firewood, building materials, reduced irrigation, increase soil fertility, free fertilizer, improved soil carbon, reduced soil loss, garden stakes and trellises, tool handles, better plant growth from shelter, and of course, foods and medicines. In most places where hedgerows exist, they have long been seen as an important source of all of these.

Designing an Edible Hedgerow or Tapestry Hedge

While one can use similar techniques for broad-scale windbreaks or livestock enclosures by making appropriate adaptations, this article will focus on the small-scale edible hedgerow or tapestry hedge, based on European designs and traditions.

This design was inspired by a couple of my favorite foraging spots. One in particular, was a naturally occurring hedgerow that produced a large variety and quantity of fruits, nuts, and vegetables throughout the season—all with little to no annual maintenance from humans.

Compared to my hard work as a guest farm laborer, the high rewards and low maintenance of my favorite hedgerow seemed like a great idea, and I decided to "take it home" with me. I will probably never live without one again.

But when I began looking into expert recommendations on how to grow an edible hedge in the U.S., I was surprised to see everyone advocating against the species and spacings that were common to our favorite foraging spots.

Was Mother Nature growing these natural food forests all wrong, as the experts suggested?

Traditional Spacings: Tight!

However, when I looked into resources on traditional hedge culture, species, and techniques in other countries, I found systems that looked very much like what was occurring naturally at my favorite high-productivity, low-maintenance spots and European models I had visited. (Mueller, European Field Boundaries Volumes 1 &2, etc.)

While modern U.S. recommendations look very different in terms of species selection and management, the biggest difference between traditional hedge, hedgerow, and windbreak technologies, and modern recommendations (at least in the U.S.) is in spacings. Most common recommendations I can find from U.S. sources recommend planting at spacings where crowns just intermix or touch at maturity (perhaps 7-10' centers for many shrub species, with many suggesting 5' was too tight) whereas traditional forms typically space plants at 1- 2 1/2'.

While these seem very close to a gardener, these spacings approximate what one commonly sees occurring in natural thickets in many biomes, including our favorite foraging locations.

But, despite the expert recommendations, these tight naturalistic plantings found in traditional hedgery have proven their effectiveness in experiments over millennia. And now they have become the cutting edge of scientific forestry (as well as many Permaculture circles) throughout the world, since the 1980s when a Japanese forester named Akira Miyawaki took notice of the way Mother Nature grows forests.

It's no coincidence that this evolved technology of hedgerows closely resembles the cutting edge reforestation program referred to as the **Miyawaki Technique**. Miyawaki was a forester interested in regrowing healthy forests in Japan for purpose of maintaining habitat, controlling erosion, sequestering carbon, and solving other problems through the ecosystem services. He noticed that the plant spacings and communities used in "conventional" forestry practice looked nothing at all like the types of spacings and communities that would occur during natural reforestation. He also noticed that the conventional approach often performed poorly in comparison to the way nature solved this same problem.

He hypothesized that the conventional recommendations were tested and developed to optimize commercial yield in various ways, not to optimize fast, easy, cost-effective establishment of healthy forest. Basing his approach off of the observation of naturally occurring rapid reforestation, he developed a system of using seed from locally adapted, free specimens, with intermixed stages of succession, at very tight plantings, and he found that nature was solving the problem the right way. Across many climates and ecosystem types, Miyawaki's technique has been replicated and found to far out-perform conventional practice, with far less cost and fewer destructive chemical inputs.

Indeed, some U.S. sources appear to criticize Miyawaki's methods as inappropriate to North America's idealized "natural landscapes," because they do not leave enough room for our most important North American "nature area" keystone species: tax dollars, corporate petrochemicals, herbicides, and heavy machinery.

For some, it's counter-intuitive, but research has found that even on spare soils in dry climates with a risk of desertification, the tight plantings of the Miyawaki technique lead to rapid establishment with little after-planting care, even where conventional techniques failed with continued intervention!

This goes well with current ecological theory and research, which has found that in such tight, naturalistic plantings, cooperation outweighs competition and gives the individuals an advantage compared to situations with unnaturally wide plant spacings.

Moreover, with rapid, dense growth, we humans can begin to reap the rewards of ecosystem services in short order. In just years, we can have windbreaks, enhanced microclimates, erosion-prevention, water harvest, herbicide buffering, wildlife habitat, biodiversity and food (under tight plantings most species will bear "precociously" or at an early age.) This is certainly better than struggling with establishment for 15 years for a wind-break to finally serve its purpose. With all this, it's no wonder that cutting-edge Permaculture designers like Geoff Lawton have begun to implement Miyawki's methods for such applications as forest gardens, alley cropping and, of course, hedgerows.

As a very rough guideline, my recommendation for a hedge where production is the main goal is to plant into a 10' wide strip of at least 30' of length (to have much room for diversity) with larger woody perennials at 2'-3', and with smaller woody shrubs and herbaceous perennials filling in the gaps to create approximately 1' spacings. I typically don't plant anything directly between woody perennials, except perhaps for short-term crops like Jerusalem artichokes, which can provide a big yield in early years, ensure a complete hedge effect in the first season, provide ample biomass for mulch, and then die back as woody perennials establish.

For best results, vary plants by height and species, planting main species such as hazelnuts at standard spacings, and filling in the gaps with smaller species. For larger hedgerows and windbreaks, I recommend multiple rows of woody perennials, but still at tight plantings. For hedges where security or animal-enclosure is the main goal, I recommend 1' spacings, with a high percentage of thorny species such as hawthorn, blackthorn, or sea buckthorn.

Permaculture Design Parameters

In addition to siting the hedgerow to provide as many benefits as possible, one should consider mechanisms to ensure adequate water and appropriate drainage. Hedgerows are excellent features on swales or micro-swales. I have used a system of net and pan micro-swales along with trench-composing swales to make sure that water would flow down the hedgerow, providing adequate water to young trees during establishment while also keeping water from completely filling planting holes on our very compacted soils with extremely poor drainage.

Integration with keyhole gardens and "edible border gardens" are an excellent idea, and common in the English and French gardening traditions.

Species Selection for Temperate Regions

For our main woody perennials, we'll need plants that share a few major characteristics:

1. They can thrive in tight, wild plantings.

2. They are disease-and-pest-resistant.

3. They take well to coppicing, or hard pruning techniques where they are periodically cut back to the ground.

Traditional hedge management simplifies pruning and maintains health and productivity.

Major species will make up 50-60% of the multi-purpose edible hedge, and include hazelnut, hawthorn, blackthorn, or bullaces, sea buckthorn or autumn olive if it is not opportunistic in your region.

Note that European wild plums take to coppicing well, but according to USDA research American species of wild plums do not. Also note that there are some American species of hawthorns that take well to coppicing and are high-quality edible fruits, which far surpass the European species. To maximize productivity, these major species should be spaced at essentially orchard spacings, approximately 12-15' depending.

Support species should fill in the gaps between, still alternating heights when possible. My recommendations for these include: Asian pears, wild pears, rugosa roses, goumi, elderberry, medlar, currants, and brambles.

Note that I do not include apples, European pears, or other common culinary fruits, as they are unlikely to be healthy and productive in such conditions. Occasionally I encounter the exception that proves the rule, but in most cases, I would not expect them to do well in an edible hedge, and are more appropriate to other growing systems.

Finally, many perennial vegetables can add to the productivity and biodiversity of the edible hedge, including: asparagus, good king Henry, Turkish rocket, sunchokes, tuberous sweet pea, sweet woodruff, Caucasian spinach, perennial arugula, fennel, pokeweed, endive, garlic, walking onions, ramps, monarda, mints, groundnut, passion fruit, yams, and many more.

Resources

NOTE: when using the following list, care should be taken to ensure that species are not problematic in your region, are appropriate to your specific sites and soils, and appropriate to "coppicing."

1. Suitable species for Australia, from a Permaculture perspective: https://permaculture.com.au/venerating-and-regenerating-hedgerows/

2. Suitable species for humid tropical climates: https://steemit.com/permaculture/@reville/living-fences-in-indonesia-50-species-to-use-in-the-tropics

3. Hedgerow medicines: http://www.hedgerowmedicine.com/

4. Hedgerow foraging in UK, applicable to many temperate climates, http://www.wildfooduk.com/hedgerow-food-guide/**Selected Citations (more research sources appear in text above)**

1. Contour Hedgerows increase soil and water health in the tropics: https://www.sciencedirect.com/science/article/pii/0167880995010122

2. Increase wildlife and biodiversity, UK: http://www.hedgelink.org.uk/index.php?page=21

2. Hedgerows reduce risk of food contamination: http://ucanr.edu/blogs/blogcore/postdetail.cfm?postnum=26370

3. Hedges reduce pests and increase pollination.http://calag.ucanr.edu/archive/?article=ca.2017a0020

4. Hedgerows and bocage in France: http://www.polebocage.fr/-Bocage-and-hedgerows-in-France,136-.html

5. Miyawaki technique effective in dry Mediterranean climates: https://www.researchgate.net/publication/226157594_Effectiveness_of_the_Miyawaki_method_in_Mediterranean_forest_restoration_programs

reduction in pests, and increase in pollination. 1- 2 1/2'. 1- 2 1/2'.

This chicken apothecary guild by Michael Wardle of Soil Savour Permaculture, is functional and beautiful. Just add chickens.

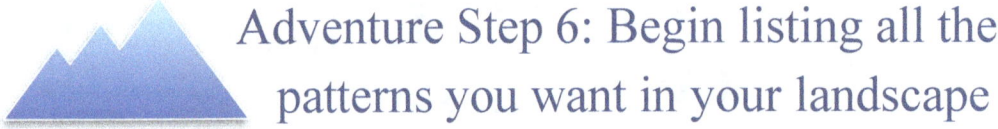 Adventure Step 6: Begin listing all the patterns you want in your landscape

I suggest listing patterns by Permaculture zone, as well as by areas you recognize, like "front yard, back yard, behind the garage, etc." Start with major themes like "jardin de curé." Then add your social functions, as these are a top priority. As you go through the rest of the book, and continue to explore, add all your ideas to this list. The act of making the list is more important than actually having the list. Hopefully, by the time the list is on its way, you'll have a good idea of what the garden will be like.

If you have a "uses" map, then add these patterns to each use. For example, if the use is "veggie garden," you might add "veggie circle beds," and "worm tubes."

Chapter 25: All-Star Ornamental Edible Plant Profiles

While the standard approach to beautifying the landscape begins and ends at buying industrial materials like pavers and garden boxes from big box stores, we will take the position that a truly beautiful landscape begins with plants, and can be achieved with plants alone. This has been the way many gardens throughout history have achieved their beautiful effects, not to mention the beautiful places we may find in nature. Pre-tumbled pavers are not necessary to give a place beauty. Putting beautiful plants and flowers in an ugly concrete "box" does not add to their beauty. We can create beautiful gardens with zero cost.

In the right place, every plant can have its beauty and charm. People often remark at how beautiful potato flowers are in context in my gardens.

Yet, there are some plants I find so useful to creating edible ornamental landscapes that I'd like to give some space to a few recommendations. We'll see these plants again and again in this book.

Many of these plants I've chosen to spotlight because there is little about them in the rest of the Permaculture and sustainable gardening literature. So there are plants you will see in this book and in the recommended guilds that are not featured in this section. For these, I recommend the online resource Plants for a Future, PFAF.org. You can look up any of the plants in this book on that site.

Most of these plants are in this particular book because they fulfill a number of special roles: they are beautiful, easy, perform ecosystem functions for us, and are edible for humans. Some will only fill a few of those roles, but they fill them so well that they are invaluable in the beautiful, bountiful garden.

Edible, Useful Trees and Bushes

Trees are the keystones of most ecosystems, and if we want to create edible, ornamental ecosystems, we need trees. Trees are also major stewards of biodiversity and powerhouses of sequestering carbon, helping us transform our landscapes in ways that meet our conflicting goals of securing human needs, fighting climate change, and preserving biodiversity.

For the beautiful landscape, trees are a must. In addition to giving dimension to the garden and framing views, spring blossoms fill the landscape with beauty and aroma. Plums, cherries and apple blossoms can bring absolute, senseless beauty.

Yet, with the number of beautiful, productive trees available, I'd need multiple volumes to give profiles on each one. Instead, I'll summarize a few main suggestions, both for their importance and a few because they are often neglected in other texts.

Fruit trees in the home landscape? Yes! No! Yes! Maybe?

A few generations ago, people were eager to plant fruit trees at home. It was just what you did. Everyone had an apple or pear, and maybe a few fruit bushes.

Then things changed. For a few generations, planting fruit trees was just not something you did. And if you did, the trees usually went unharvested and you were generally unhappy with them. Now, as things shift back, it's common to hear that people just sort of became stupid and stopped planting useful trees. Why *wouldn't* people want to plant fruit trees?

But the truth is people didn't get stupid—they had very good reasons for the choices they made. The landscaping industry was motivated by making customers happy, and largely, that is what they did by planting low-maintenance ornamentals. And understanding the reasons will help us make better choices.

Over the last 20 years, as I've watched a new generation attempting to correct the "mistakes" of the previous generation by planting out loads of their favorite fruit trees and guerrilla-grafting fruit trees to ornamentals in public spaces, more often than not they are generally unhappy with them, and the trees go unharvested, often becoming sources of mean yellow jackets and mess. Their children will be another generation that just doesn't plant fruit trees.

Remember, the opposite of a bad idea usually isn't a good idea, it's another bad idea.

Going back those few generations, there were evolved cultural traditions for planting fruit trees. If something worked well, and cared for people, and helped to create a beautiful life, it caught on. Most families had one or two locallyadapted easy fruit trees like apples and pears, and access to lots of extensive fruit in foraging systems like hedgerows, food forests and wild thickets. When it came to the intensive fruits, we all know certain regions were famous for having THEIR apple or pear, because that is what they found grew best in their region and everyone stuck with what worked.

A few years ago I discovered an old McSweeny apple in Kalamazoo, which seems like one of the worst, most pest-plagued regions for growing apples in the world. The McSweeny was once Kalamazoo's apple, and it was said that nearly every house had one. In the middle of dozens of apple trees completely obliterated by pests, these McSweeny apples had very little pest damage and lots of beautiful clean fruit.

Most of those traditions, like planting local varieties, have been lost or dismantled to promote a consumer lifestyle and booming markets, including globalized nursery and fruit markets. National nurseries marketed "the best" fruit trees and pesticide companies mass marketed the poisons to allow us to grow whatever we wanted, wherever we wanted. Hippies with dreams of simple lives going "back to the land" bought them up (along with other corporate gardening products and services.) Even today, these companies sell these same burdensome trees with ads reading "you can never have too many fruit trees!"

But reality set in, and most families discovered that they didn't want to keep up with an expensive, time-consuming spray schedule, and that they didn't want to waste time harvesting bug-ridden, half-eaten fruit that wouldn't keep when they could just buy nice clean fruit inexpensively at the store. The trees became disappointments at best, or outright problems of rotting fruit mess and hornet food. And the back-to-the land hippies largely came to believe that growing food and community sufficient lifestyles are actually much harder work than they really have to be. And these hippies turned into the cynical consumer culture yuppies who wanted lawns and low-maintenance forsythia bushes.

People discovered that fruit trees are like egg-salad sandwiches, they follow the law of diminishing returns. If someone gives you one sandwich it will alleviate your hunger. A second might fill you up for the rest of the day. But they don't keep well, so if you have a third, you'll need a friend. If you have five, now you've got to do the work to organize a little egg salad sandwich party. If somebody gives you 27, now, either all those extras are going to go to waste, or you have to do even more work to get them to somebody who can eat them. A whole truckload is going to take a lot of work, probably won't earn any money, and most will probably go to waste. But it's still probably not enough to start a full-time egg-salad sandwich business. Once you have more than a few big intensive fruit trees, all the extra will just take more time and not deliver any extra useful fruit. Until you have enough for a business.

An informed view of fruit in the landscape for happy fruit-growers.

So, we can learn from the wisdom of both sides. Yes, fruit can add to our lives tremendously. And yes, if we aren't smart about which ones we plant, they can become a huge waste of time and a problem in most landscapes. In my experience, the happiest fruit-growers follow the old wisdom of just growing a couple high-value intensive fruit trees, and then have hedgerows and forest gardens with easy, no-work fruits that won't become a menace if they don't get harvested.

Of course, moving up in scale, there is a point where we may want to have an orchard. At that point, my recommendation is to not just plant a bunch of fruit trees willy nilly, but to do a good permaculture design process, with thorough goal-setting, realistic economic viability analysis, and to calculate exactly how many fruit trees you need.

One major consideration is if you are in an easy-fruit region or not.

Some regions have the "difficulty level" of the gardening game set to easy, while others have it turned up on moderate, difficult, or nigh-impossible mode. This is a long-known fact of life that has rarely been discussed in natural gardening or Permaculture texts.

For example, the dry season of mediterranean climates often disrupts pest and disease lifecycles, while old abandoned fruit belt regions may have countless orchards of untended, highly susceptible commercial fruit selections that keep pest and disease populations turned up on high, with enough still being sprayed to keep natural predator populations from getting established.

Plum Curculio Hell Realm
Plum curculio effects all stone fruit, and most apples and pears. Depending on variety, they can cause fruit damage or a complete crop loss.

Spotted Wing Drosophila Pergatory
SWD effects soft fruits like raspberries, blackberries and blueberries. Healthy biodiversity and consistent picking are required to get fruit without heavy spraying

Maps of two major fruit pests. We could add maps for fireblight, fruit moth, apple maggot, rusts, borers, spotted lantern fly risk, and others and we'd find a great deal of overlap.

While writing this book I visited an established old food forest filled with plums, gages, figs, apples, pears, and berries. It was no longer being tended but was still highly productive. Most of these were commercial varieties, and they were thriving, with hardly any pest or disease damage in sight anywhere.

The same exact garden in the "pest hell realm" of Southwest Michigan's old fruit belt wouldn't have produced a single clean fruit.

The moral of the story is you need to know what the level of difficulty is in your region.

If you are in an easy region, you hardly need to think about which varieties to plant. Simply follow your intuition and plant what you like to eat. That said skimming over the following recommendations (especially on how many to plant) will give you some "food for thought." If you're in a moderate region, you need to plan a bit more, have better systems, more biodiversity, and better selections. If you're in a pest and disease hell realm, then we can and should still grow fruits at home (this may be the only way to get clean, healthy, poison free fruit!) But it will absolutely require smart planting choices and good systems, or usually it will lead to disappointment.

Even in a difficult region, a home grower can grow virtually any climate appropriate fruit on a small scale with good yields without spraying. The key is to keep trees small and bag the fruit with silk or cotton bags. These can be as decorative as they are functional.

But on a larger, landscape, or commercial scale, it takes special crop selection for fruit trees to be more than trouble.

Basic considerations:
For Transformative Landscapes, we'll take the position that the highest use of human habitats is meeting human needs. That frees up natural areas to be better habitat for wildlife. But keep in mind, human uses extend beyond food. Trees can provide shade, privacy, materials, or simply beauty. But I do believe we have an ethical obligation to grow some food plants, most especially where it is easy. And in many cases, food plants for humans make the very best habitat plants for wildlife.

We'll also take the position that antique, heirloom, and naturally selected varieties tend to be of more value to both wildlife and humans. And, they tend to be in more danger of going extinct or being lost, so we can double up our impact by planting these and maintaining this biodiversity.

Fruit Trees: How many to plant?

1 tree each, max 3-4 per family: Apples and pears; peaches, apricots, plums, cherries; citrus, figs. Apples and pears are high-yielding trees. Traditionally, families planted only one of these, which will realistically produce all the fruit a family needs and some for selling or giving to friends. Peaches, plums, and cherries will produce less, but they do not store well. If you plant more than 3-4 of these, the additional trees will require you to do additional work to get that fruit to people who can use it. If you want more varieties, then try grafting more varieties to one tree. At this point, the utility of more fruit trees goes down until we get up to feasible commercial orchard size, which might be dozens or hundreds of trees.

1-2 each, no urgent maximum: Paw paws, mangoes, and quinces, fruits that require very little care and the mess usually isn't too burdensome. Still, 1-2 will provide all the fruit a family needs.

2-3 each, no maximum, plant as many as you like: These are lower-yielding, don't create messes and are of very high wildlife value. And they tend to be beautiful. Serviceberries, crabapples, hawthorns, dogwoods, and mulberries.

The best time to plant a tree was 20 years ago. **The second best time is now.**
—*Chinese proverb*

Beneath these fruit-tree boughs that shed
Their snow-white blossoms on my head,
With brightest sunshine round me spread
Of spring's unclouded weather,
In this sequestered nook how sweet
To sit upon my orchard-seat!
And birds and flowers once more to greet,
My last year's friends together.

—*William Wordsworth*

Beauty in Abundance

Best Trees and Shrubs for Difficult Regions

Nut Trees: Most nut trees will produce quality nuts even in poor regions. The biggest difficulty is competing with the squirrels for them.

Intensive Fruit Trees:
First we'll start with trees that produce large grocery store fruits.

Asian Pear

Aka, "not Bradford pear." This best of the ornamental edible fruits for temperate regions. It has amazing dessert fruit, and has all the good looks of Bradford pear, without being a public menace. In regions where fireblight is a big problem, Chojuro is the only variety rated as highly resistant.

Over large swaths of difficult pest regions, Asian pears can grow a high % of beautiful, clean fruit without spraying or even bagging. Stink bugs may still cause damage, which can then be exploited by wasps and birds. Still, Asian pears may be some of the easiest fruits for home growers.

American Paw Paw, Asimina triloba

The paw paw is an exceptional fruit for the home grower. Trees are highly resistant to pests and wildlife and it's rare to see fruit eaten by bugs or critters. The trees are beautiful in the landscape as well, and a stand will dominate the ground, creating a weed-free understory in most cases.

The biggest design constraint is that they are very slow growing. A few varieties have been noted in the literature to be fast growing and precocious, and my own experience backs that up. These include the varieties "mango" and "prolific." I've had these varieties fruit heavily in the second or third year, while seedlings have taken 10 years or more. I now recommend starting with a fast variety and then grafting scions if desired.

Figs

Wherever they thrive figs are one of the best edible ornamental trees available, producing a luxury fruit with pest damage being exceedingly rare. Vigorous trees are very easy to grow.

In colder climates, there are varieties that can be grown outdoors with no protection (other than good microclimate selection) down to zone 5. I have documented two specimens in Michigan, both planted on the south sides of houses, which grow each year to be larger than lilac bushes, 10-12' tall and bearing hundreds of fruits. While they could also be grown in pots, the care and small harvests do not seem worth it to me, especially when they can be so productive outdoors with no help.

Chinese mormon apricot

Apricots are not easy fruit trees to grow. They blossom very early and these blossoms are very susceptible to frost damage. But in regions where they do well, the Chinese Mormon apricot is my favorite, produces excellent fruit, and probably the easiest to grow. It also has the side benefit of having a sweet pit, rather than a poisonous one, which can be eaten like a sweet almond. This half-wild vigorous variety is highly resistant to borers, diseases, and other insect pests. It is a fast-growing and beautiful tree.

Best ornamental mulberries

There are mulberries that grow almost everywhere; and almost everywhere they grow, they are easy, highly productive fruit with excellent berries. In addition, many have edible leaves which can be high quality cooked greens.

Contorted mulberry is a favorite of mine, which I call the "Harry Potter Wand Tree." On top of its fruits and edible leaves, it produces contorted branches that are put to excellent use in flower arrangements and crafts. Many children have used sticks from my tree as magic wands. Note, that some trees are listed as non-fruiting in the trade.

Weeping mulberry is another beautiful, edible ornamental.

Beautiful Bountiful Elderberries

Highly medicinal, great for wildlife, and second only to grapes for wine-making, elderberries are another beauty in the landscape.

Black Lace Elderberry is a personal favorite, with high quality edible berries, buds, and flowers. One of the top ten complaints I hear about landscaping projects is that red Japanese maples are out of control, were planted too close to a house or foundation, and have become menaces. Black Lace Elder is edible, is a high-value pollinator plant, and fills this same role as the Japanese maple without creating the problems of that tree.

Variegated elderberries. There are now a few named variegated elderberries on the market. All are attractive, edible, and good-quality habitat plants.

Hawthorns

When it comes to ornamental trees, the hawthorn has a long history as a preferred selection for landscapers, because it is easy and low-maintenance. Many species are also noted for being good for wildlife.

If you've ever dared to nibble a few of these berries while passing a stand of street hawthorns, you'll probably have a very negative impression of the taste. Most are bland at best, and a few are downright spitters.

Nevertheless, there are quite a few very good eating haws (the fruit) out there in the world, many native to the same regions where the exotic inedible hawthorns are planted as street trees. Take my personal tasting notes with a large "grain of salt," as my sample sizes were small, and I can't be really certain I correctly identified all the species.

I cross-referenced a list from Plants for a Future with my own experience and the literature on their value to wildlife to come up with this list of my top recommended hawthorns. Take my tasting notes with a big grain of salt as my tasting samples were small and it's possible I mis-identified the species. Fruit is listed on a scale of 1-5, following Plants for a Future.

C. holmsiana (4/5)
C. douglasii (4/5. I've tried this fruit and considered it an acceptable trail snack.)
C. macrosperma (3/5, but often noted for large fruit and historic food use. I've tried it and thought it too mealy for fresh eating.)
C. mollis/submollis (4/5, but receiving special mention for commercial production. I believe I've tried these and found them a nice trail snack.)
C. pedicelata (5/5) Sweet-tart fruit that is great out of hand.
C. Pennsylvanica (5/5) Good quality fruit.
C. champlainensis (4/5) Good flavor, but mealy.
C. illinoisensis (4/5, with special mention for commercial production)
C. ellwangeriana (5/5. I've tried these from a few individuals and found the color very appetizing but the flavor less attractive than other haws. Often the most recommended for commercial fruit production, but most specimens have sharp thorns. Some specimens available without thorns but fruit quality is unknown. Some are now reporting this as another synonym for mollis, which seems likely)

Serviceberry Selections

If the goal is to balance beauty, edibility, and wildlife benefit, serviceberries get a true gold star.

In difficult regions plagued by pests, serviceberries can often produce an abundance of beautiful, clean fruit many people will consider better-tasting than blueberries. And they are very high-quality pollinator and bird-feeding plants, perhaps especially where they are native.

Serviceberries will grow in full sun or even in fairly dense shade, though productivity will suffer. Most species will deliver an amazing spring flower show, a beautiful form throughout summer, and then a spectacular show of fall color.

However, not all are created equal in the taste department. So here are my recommendations for eating:

A. alnifolia. Zones 1-9, 4 feet to 40 feet, depending on the selection. These are among the best tasting serviceberries, though there is still a lot of variety in the fruit. **"Regent"** is a favorite selection, often grown as a small landscape bush, usually 4-6 feet tall. Very productive, my favorite serviceberry fruit. **"Northline and Smokey"** are great-tasting varieties, but less predictable in terms of height in my experience.

A grandiflora, Zones 4-9, possibly down to 2. 10-40 feet height. These are very good tasting berries, in some cases better than alnifolia. The biggest drawback is they tend to grow taller. However, many will have an umbrella shape that allows easy harvest from underneath. **"Autumn brilliance"** is a favorite selection.

The Best Edible Crab Apples

While, may Permaculturists are recommending we get out and guerrilla-graft ornamental crab apples with large, cultivated apples, I am always recommending we get out and graft all the local desert apples with nice crab apples.

Of course, in regions where apples are easily grown, there are a high number of exceptionally beautiful, and delicious varieties to choose from. This again, could fill a whole book, and there are very many on the topic.

But in many regions of the world where apples are grown commercially, the presence of high-pesticide zones guarantees we will face nearly insurmountable pest pressures growing apples in home or public gardens. As the saying goes, "you can grow 100 acres of apples and not get a single clean fruit without spraying weekly." If you're in one of the magic good apple regions (they're out there!) then plant all the apples you like. Otherwise, we need to be thoughtful.

On the home scale we can grow espaliers and bag some fruit. We can get all the apples we want that way. But unbagged apples will USUALLY be pretty well pre-eaten.

Some folks respond to my crankery by saying "well, buggy fruit can still be used!" Sure. It can. But nobody does. Even in areas with lots of foragers and Permies, I know a dozen apple trees planted in a mile of my house and nobody is harvesting all that buggy fruit, not even me. You can make home cider from buggy fruit, but it can't be used commercially. And buggy fruit won't store, so unless you can eat 100 gallons of apple sauce, it's probably not going to be worth harvesting. I see new people get into foraging yard apples every year, and they usually do it once, then move on to spend their time on all the great non-buggy fruit out there.

Meanwhile, there are quite a few crab apple varieties that are gorgeous, have absolutely amazing fruit, and produce a huge amount of nice clean fruit. Some of the very best apples I have ever eaten have been large, clean crab apples, with all the complexity, juiciness, and crispness of the famous apples of the world. The trees are easy to care for, usually naturally smaller, highly ornamental, with easy fruit that's also excellent for cider.

Here are a few of my favorites:
Chestnut - One of my favorites for fresh eating due to its nice crispy texture. A nice sweet flavor, too.
Dolgo - A white flowering variety with beautiful fruit. The fruit is somewhat mealy to me, but the flavor is very distinctive, almost reminiscent of cranberries, and it adds a special flavor and color to ciders, and juices. Certainly worth snacking on fresh.

Red Vein - A red fleshed variety with a crisp texture and a nice tart flavor that I find quite addictive fresh. You have to like tart fruit to appreciate it!

Whitney - A pink flowering highly ornamental tree with fairly large sweet fruit great for snacking.

Wickson - A cross between a crab and and pippin, this small apple seems to have the pest resistance of the best crabs, but a flavor comparable to some of the best apples I've eaten.

Dogwoods

Cornus Kousa is right up at the top of this list for me. There are few trees so beautiful in the landscape or so easy to grow. What kept it from getting its own profile is the variability of the fruit quality between varieties and that I have no idea what to do with it other than eat it fresh.

Cornelian Cherry, Cornus Mas

If you like to pucker, then you'll want to get acquainted with the sweet-tart flavor Cornelian Cherries.

The Cornelian Cherry Dogwood, or Cornus Mas, is a native of west Asia and carries an ancient culinary pedigree, being pickled as "olives," used as gourmet, floral-scented preserves, and even as the original sweet-tart "sorbet" in Persia. "Floral, complex, intriguing, distinctive, rich, unequaled" are often found in the long strings of adjectives writers use in describing the flavor of the cooked fruit, when sweetened or added to alcohol.

Or to be quite honest, other seedling varieties may be mostly astringent and sour, then turning sickly sweet and mealy. There seems to be a large difference between seedlings in this species! So this is a place it may pay to get named grafted varieties from a reputable source.

It grows to a large, showy shrub (12' x 12'.) The spring flowers are small, yellow and attractive and the summer foliage is a nearly waxy beautiful green.

There are already some excellent resources for Cornus mas online, including profiles at Plants For a Future and Mother Earth News, but after recently finding this beauty in "the wild," I'd like to add my own recommendation that this would be an excellent addition to the Permaculture landscape or forest garden.

This was a shrub we passed over when planning our landscape at Lillie House, based on its reputation for hogging space and the rather lack-luster recommendations we had read elsewhere. But one taste of the fruit grown in S.W. Michigan was enough to instantly correct that mistake.

What's more impressive is that in S.W. Michigan it's tough to get real "cherries" without spraying and fighting off birds, but the Cornelian fruit was large, overwhelmingly "clean," (no bugs or

diseases,) very abundant, and apparently productive over a fairly long harvest period, making it ideal for a home landscape. This is a true example of a "no-spray," no-maintenance fruit bush which will produce a high-quality fruit. With its edible fruit, spring flowers, and dense habit, it's also recommended as a wildlife planting, making it a multi-functional plant in the Permaculture garden.

The fruit we found, grown on several bushes, was quite variable, but tasted similar to the cherries it's named after. By most accounts the fruit is too sour to eat fresh, but I enjoy the "sour patch kids" tartness and the ripest berries are actually quite sweet, similar to a plum. I ate dozens out of hand while picking.

The drawbacks are primarily related to picking, packaging and processing. While ideal for the home grower who wants daily fruit in their freetime, for commercial picking, the long ripening period would require many days in the field hand-picking. The wood is reportedly very brittle and easily damaged during picking. Ripe fruit seems to easily fall from the bush. Underripe fruit is astringent and very sour. Once picked, the fruit does not keep and must be used immediately, so commercial shipping isn't an option. The fruit has a large pit and from most accounts is usually "clingstone" making it difficult to remove without destroying the fruit.

So, overall, the bush takes up a fair bit of room, and processing takes some time and energy, but since the Cornelian Cherry will fruit with no spraying, fertilizing, or other maintenance, it seems that a little extra work on the processing side would be well worthwhile, at least on the home scale. Difficult processing would seem a barrier to its value in commercial production, so I wouldn't recommend it for that purpose without further study.

Author harvesting cornelian cherries in a hedgerow.

Cherries

Virtually any cherry tree will make a fine ornamental, and if you can manage to get a few before the birds do, then they're also a quality edible. Standard cherry trees will grow to 35' in height, making them tall trees in the landscape. Many homeowners are taken unawares by the size of their cherry trees, expecting smaller trees. Without commercial harvest equipment, these tall home cherries often become trees for snacking from the low-hanging fruit. But their beauty is still unharmed by the birds.

Bush cherries

For most home growers, I have taken to recommending bush cherries instead of cherry trees. I find my clients will have less to complain about this way. It is easier to harvest and easier to keep birds away from bush cherries.

In regions with the dreaded plum curculio, an evil weevil that eats prunus fruits, cherries are the hardest fruit to grow without spraying. I can't recommend bagging the fruit, because I can't imagine bagging enough fruit, then individually unwrapping them to eat. But with bush varieties, I can throw a net over the whole bush in the short month or two it takes for them to fruit and ripen, without it being too obnoxious in the landscape. And these are reliably gorgeous plants. **Nanking cherries** will be the first fruit to produce heavily in most forest gardens, and then die back after a few years. It's not the greatest tasting, but the early yields are encouraging. The U of S **"Romance Cherries"** are a marvel. Beautiful plants with small but great-flavored fruit certainly worth snacking on.

Plums

In regions where they are easy to grow, there are many plums which double as landscape stars and great eating. Among the most beautiful for the landscape are:

Purple-Leaved Plums

In warm climates, purple-leaved plums will bear nice crops, and the fruit is quite delicious. Unfortunately, in most climates these early bloomers rarely set fruit. For fruit, zones 9-10.

Weeping Santa Rosa

This beauty is among the tastiest plums, it's self-fertile, one of the easiest to grow in difficult regions, and it's absolutely covered with blossoms. The weeping form makes it easy to keep small and to harvest. Grows to 12-16 feet if left to its own devices. Zones 5-10.

Quinces

This is a rare beauty with a rugged form, fascinating bark, candy-striped blossoms, and in fall, large, bright yellow ornaments that fill the home with a spiced-citrus aroma.

If you want to grow much of your own food on trees–and there's a lot of good reasons to do so–then you traditionally would have an easier time of it in the tropics, where there are a lot of "tree fruits" that can be used as staple vegetables, like plantain, avocado, jackfruit, etc. With the powerful productivity of trees, this means that tropical Permaculturists with established food forests report feeding 30 families from a small fraction of an acre year-round from their "vegetable" CSA. Of course, most of the "vegetables" are actually tree fruits!

In cold climates, our ancestors didn't cultivate many fruits to fill this role. While we do have quite a few tree vegetables, like toon tree and linden, we don't have many traditional calorie rich, starchy, staple crop trees. The only ones that come to mind are Chestnut, possibly the mulberry, and, of course, the quince.

Most people these days aren't too familiar with Quinces, but our great grandparents were. There was a time when this was the desert-island tree of choice for American settlers–if they only had room for one fruit tree, it would be a quince, for its versatility and many uses. Not only was the quince preferred over other "pome fruit" like apples and pears for desserts, but its high pectin content, acidity, and spicy aroma made it the perfect sidekick for any fruit preserves, jam, or jelly. The same pectin content also made it great for cosmetic purposes, including the original "hair jelly," or "pomade." As a matter of fact, I've been using a home-made pomade for the last month to keep my coiffure in order and I'll be sad to go back to salon products when it's gone. It works great, doesn't get greasy or stiff, and as Ulysses Everett McGill said of his "pomade" in the film Oh Brother Where Art Thou? "I like the smell of my hair treatment–the pleasing aroma is half the point."

But more importantly for northern growers, the quince has a tradition of being used as a staple in savory dishes going back to antiquity. It has been featured alongside olives in tapas dishes, added to stews in hearty chunks, paired with meat and poultry, used in stuffings, and even stuffed and roasted in its own right, with cheeses, bread crumbs, and herbs.

Growing Quinces
Although it is generally renowned as the easiest-to-grow of all pome fruit—being all but indestructible—growing quinces in regions with fireblight requires some special consideration. In my region for example, we have occasional outbreaks of blight, and the

Quince can become problematic. Many orchardists have complained that the species can act as a carrier, and that normally "blight immune" varieties of apples or pears will be killed by blight if they're too close to an infected quince. (I know of one S.W. Michigan grower who reported severe blight outbreaks every year in her "blight tolerant" Seckel pears, which went away completely after removing a nearby quince.) Indeed, that's the old saying. "If the orchard is plagued with fire blight, find the quince and cut it down." So, in my food forests, I've

planted quince far from any other pome fruit.

Unfortunately, there are no varieties currently marketed as "blight-tolerant" in the United States, as the fruit is so uncommon that it hasn't been cultivated or researched much here. However, in the Caspian region, where quince is still a staple and blight a major problem, this has been worked on, so varieties from this area are often considered by knowledgeable growers to have better blight tolerance.

According to my sources, the variety "Aromatnaya" is considered to be the most blight-tolerant cultivar. I have tested it and found it to be true. It is also commonly recommended for ripening well in northern climates and cool, cloudy summers. Luckily, this has the reputation of being one of the best-tasting quinces and it has been my favorite in my tastings.

In regions with fireblight, **Aromatnaya** is the only quince I can currently recommend.

Medlar

The poor medlar is the neglected stepchild of the pome fruit world. Perhaps this is because it requires aging, or "bletting," for the fruit to become edible, and modern people just generally don't have time for that. Well they would if they would taste some good medlars. The usual description of the ripe taste is "spiced apple sauce," which is about perfect.

For the beautiful Transformative landscape, medlar has other benefits. It is an ideal "extensive" fruit, which produces lots of fruit for little work, even in difficult climates, where few pests seem to bother it. And it is another quite beautiful fruit, with rose-like flowers that are quite larger than apple blossoms. Its bark, somewhat twisted habit, and rich leaf color give it a romantic or medieval charm.

Other Edible Ornamental Trees and Shrubs:

Besides fruit trees, there are some very nice, edible, ornamental trees, and for me, the inclusion of these is one feature that distinguishes the best well-designed and professional food forests and forest gardens from amateur models that never live up to their potential. This is what makes a modern "food forest" more like the highly practical, evolved systems around the world, like tropical spice gardens. As we saw earlier, there are certain diminishing returns on growing fruit in a home landscape. At that point, why add more fruit? Why not add more diverse uses and yields instead to get even more high-value from the landscape? In this regard, models like Martin Crawford's forest garden and books can show the benefit of diversifying yields.

Here are just a few of the best for the beautiful abundant landscape:

Toon tree, Toona sinensis

The toon tree is a great ornamental in the right position, with beautiful fall foliage and leaves and shoots with a flavor of onions and garlic. The selection **"pink flamingo"** is very flamboyant, with a pink-reddish coloring to the leaves that would make a statement in a bold-color garden. Zones 6-10, though in my experience it has generally failed to thrive in most zone 6 gardens I know of.

Redbud,

Redbud capers are quite a treat, the flowers dress up a salad beautifully, and children love these pink-purple buds raw. I heard one young lady squeal "garden candy!"

Redbuds are a personal favorite of mine in any landscape. I just adore this beautiful tree for its form and the bold spring colors.

A very adaptable tree native to North America, with some evidence of N-fixation, though the effect is considered weak. It can grow in full sun or tolerate some shade. Zones 4-9.

Sumacs

This beautiful, useful genus of trees has species around the world. In North America, the many native species are much maligned, often being sprayed by overzealous native plant enthusiasts and governmental agencies for being "invasive" in the regions where they are native. A look at the beautiful fruit spikes in summer or fall will confirm that this is a super-powered insectary plant, as a huge variety and number of insects are drawn to them. Many of these are natives, too.

Most sumacs can have the fruit steeped in water to produce a lemony-flavored beverage rich in antioxidants. Most also produce edible shoots in spring. The "bobs" (berry spikes) can be used for smoking, barbecue, incense, or beekeeping.

And the plant has both a beautiful form, creates lovely dappled shade, and the flowers and fruiting bobs add a bold red color to the landscape. These bobs mimic the shape and color of amaranths for a beautiful color and form echo.

Where over-abundant, they make a first-class mulch-maker plant, providing a free resource many go to buy from Home Despot for after spraying these natives with poison.

Sichuan Pepper Trees, Zanthoxylum species

Sichuan pepper is a sort of addiction to me, which is lucky because these plants are generally easy to grow and beautiful in the landscape. This is the pepper that is less "hot" than somewhat numbing, but it has a bold flavor when put to use in the kitchen.

The actual sichuan pepper tree Zanthoxylum simulans is a little more picky and grows in zones 7-10, while the Japanese relative Z. peperitum is hardy down to zone 6.

Redbud at Lillie House

Espaliered Fruit Trees

A particularly beautiful pattern for fruit trees for the home scale is espalier. This technique of growing fruit trees as two-dimensional fences has particular value for those looking to grow fruit at home on a small scale without spraying. Although, it is worth mentioning that in some regions of the world, espalier has been used for commercial fruit production, and is considered highly profitable once established, compared to standard fruit tree spacings. Still, despite it being a well-established research-demonstrated superior technique for profitability, many will complain that such a fussy system "won't scale." These critics have a point that the labor on a large landscape seems prohibitive.

Espalier has several important benefits for productivity. Because sap runs vertically through trees and promotes green growth, turning branches horizontal slows sap and puts more energy into fruit production and less into growth we'll have to prune. Greater air access means less disease. There is more sun access to all the leaves, which means maximum photosynthesis and maximum immune response. Prunings can be used for mulch or scionwood. And it is easy to grow many varieties on one tree using this technique. Finally, it is easy to bag fruit so we can get good quality fruit without spraying. Espalier is the highest-productivity way to grow fruit per square foot there is.

Aesthetically, there's a huge set of benefits. Espalier can define spaces with a wall that can still be seen through, allowing a feeling of privacy, yet still allowing discussions with neighbors. They can create a huge visual impact when flowering, when covered with fruit, and even in winter when the architecture of the trees can be a show-stopper in the landscape. Espalier is sculpture made of fruit.

And in some municipalities front-yard fences are not allowed, but they may not have restrictions on fruit trees. This was the case at Lillie House, where rules wouldn't permit enclosure in the front yard, but the "apple trees" were not a fence.

So, consider espalier in designs where the extra human labor will be worth the added beauty, formality, and fruit production.

How to do it:
1. **Select appropriate varieties of fruit.** Pome fruits are the most common to espalier, though it is possible to espalier many fruits including stone fruits, mangoes, persimmons, paw paws, citrus, and etc. Cherries make decent espalier and can be covered with a net to keep birds and bugs off after flowering. The biggest problem is that fruits must bear fruit on "spurs," rather than the tips of new growth. Spur-bearing fruits are most common, but there are some, mostly apples, which only bear fruit on tips. One of my clients planted an espalier of Granny Smith apples, then wondered why he never got any fruit. Since Granny Smiths are tip-bearing, all the fruit buds would get trimmed off with pruning. Luckily, the trees had lots of energy stored in the roots and could be regrafted with spur-bearing varieties.
2. **Choose a root stock.** Many modern sources recommend dwarf and super dwarf roots for espalier, with the thinking that these will naturally stay small. But the historic espaliers of the world were all grown on semi-standard root stocks. Dwarf stocks may take too long to grow, and may never grow to full size easily. And the standards will grow deep roots, which is associated with greater flavor. Yes, there will be more work, but the benefits of the increased vigor are probably worth the extra labor, especially since espalier will require yearly pruning anyway.
3. **Plant the trees and start the training.** Here's the scary part. In year one, we must cut the tree down to the level where we want the lowest rung in the "chassis," which is what we call the structure of the espalier. Learning to call it a chassis increased my enjoyment of owning an espalier by 73%. Usually this height is about 1 ½' to 2' feet. Make sure there are two buds left under the cut, and position these two highest buds to run in the directions you want your espalier to go.
4. **Invest in the first rung.** The first rung is the most important investment, and in the historic literature on espalier it is always clarified that we need to spend two if not three years on developing this lowest rung. The time is worth it. Once we add higher levels, the sap will go to them and growth in the lower rung will be slow or stop. If you look at images of espaliers online, you will see many where the lower rungs are thin branches while higher branches, or even just the top levels end up very thick. On such trees, the lower branches can actually die back or easily break.
5. While you're waiting for the lower branches to fill in, you could **consider running the wires** for the higher branches, and ornamenting these with fabric or solar lights so it looks intentional and "done" while you wait patiently the five years it will take to get to the highest level. This may also help the espalier act like a fence, the decoration can be an additional repellent to deer.

Summary of Recommended Tree Crops for Home Gardens
Excerpted from
http://lilliehouse.blogspot.com/2016/04/selecting-trees-for-forest-garden-what.html

Moderate Maintenance:
These options provide big, sweet, valuable fruit without such a problem with bugs. It's possible to get good, clean fruit without spraying or bagging. The biggest work problem with these after establishment is harvesting, and storage, which can become a burden.
Asian pears
Jujube (zone 7 and up)
Nakita's gift persimmon (the only persimmon I can currently recommend for homeowners.)
Cider and cooking apples and perry pears (if you don't mind buggy fruit.)
Quince (A known fireblight host, Aromatnaya's the only variety I recommend.)

High-Value/Low-Maintenance Extensive (Feel Free to Plant Lots of These)
You almost can't plant too many of these. Where it seemed appropriate, I made recommendations to feed a family of 4. Having extra could provide the opportunity for trade, sales, or value-added products without creating a burden. These have almost no maintenance requirements, and if they're not harvested they won't create a huge mess.
Blackberry (Plant in an "island" where it can be maintained by mowing.)
Raspberries (same as above.)
Goumi (4 bushes)
Elderberry (edible flowers and fruit.)
Serviceberry
Hardy kiwi (Issai is the only variety I recommend, 2 vines)
Paw paws (3 - 4 trees)
Honeyberry (5-10 bushes)
Strawberries (25 plants)
Aronia berries
Nanking cherry (4-5 bushes)
Medlar (1; Beautiful tree!)
Mulberry (Illinois Everbearing. 1 tree.)
Cornelian Cherry (2 bushes)
Hazelnut bushes (look for selected varieties that are blight resistant and have large nuts.

Nut Trees for the Home-Grower
Black walnut. Valuable tree, difficult nuts to crack. Select easy-to-crack varieties.
Carpathian walnuts.
Pecans.
Butternut.
Heartnut.
Hickory nuts. My favorites.
Chestnuts. Staple carb crop highly recommended for larger gardens or forest systems.
Korean pine.

Chapter 26: Selected Plant Profiles

Alliums, Onions, Garlic, Chives, and Other Stinky Delights

There are probably no plants as useful to us humans in the garden as the allium genus, which refers to onions, garlic, chives, and all the ornamentals that simply get called "alliums" in nursery catalogs.

It's rare even in the boring Standard American Diet (SAD) to go a day without a meal that doesn't have an allium in it, at least in the spice blend. In fact many people are proud that when a recipe calls for garlic, they triple it.

And that's probably a good thing, since these plants are also quite miraculous for our health, often enhancing immune response, fighting off bacteria and viruses, and curing us of the affliction of finicky lovers.

That's right—as my Lithuanian friend Julius told me—they had a traditional saying "if your lover doesn't like the smell of garlic on your breath, get another lover."

What's more, these easy-to-grow crops tend to be high in calories, improving the productive utility of any landscape.

And glory that they are, the alliums are a true beauty in any landscape.

This is why when I started the garden at Lillie House, my very first goal was not to plant trees, but to make sure I had a collection of a good dozen vigorous, attractive, and useful alliums.

Garlic is highly underutilized and misunderstood. It's often grown as an annual for the bulb which is like buying a car just to use the radio. It can be grown as a multi-purpose perennial, and when it is, it is hands-down THE most important plant to grow in the home garden. There is no food that is research-proven to be more health-enhancing than garlic. And, in addition to the bulb, it produces high-quality, flavorful greens all year long (even in winter) the scapes (when harvested while tender) are an amazing vegetable, and the bulbils are amazing and incredibly useful, too.

Ecologically, it can repel pests and can be used to make a pest spray. It can fit into many polycultures easily without reducing yields. And it can thrive in a wide range of circumstances. And I've documented several stands persisting and spreading for long periods of time in the wild with no human care.

Virtually all of that can also be said for Egyptian walking onions, which behaves similarly, but has an onion flavor. Also greatly misunderstood, these produce several HIGH QUALITY vegetables on one plant. Top sets can be encouraged to grow quite large, and when forming, can be eaten without peeling. Pickled, these make fantastic cocktail onions. They produce green onions almost all year long. And if treated right, they can size up to be nice quality storage onions, too, about the size of cipollinis. Once the peel sets on top sets, these can be used to flavor broths and stocks, and the peel gives a nice color to them (this secret is actually how most commercial broths get their rich color.) This is the most used plant in our garden. There are practically only around 2 dozen days a year when Egyptian onions don't get used in our kitchen, yet I've spoken to many gardeners who simply had no idea what to do with these, some thinking they can only be used as scallions, others that only the top sets are usable, and others growing them only for the bulbs.

On top of these two powerhouses, there are perennial leeks, and a whole host of species that can be used as "chives." We have "chives" flowering in all different colors, all season long. My personal favorite, and the one that often wins our taste tests is *allium unifolium*.

Let's look at a few of the most useful for the beautiful, productive landscape.

Egyptian Walking Onions, *Allium* x *proliferum*

This strange perennial vegetable comes first, right at the top of the list, because it is always the most used and useful plant in any of my gardens, and one of the most striking visually adding a character that exceeds the pop of color of a flower.

Visitors to the garden are often startled by this "onion that grows onions on top of onions." Rather than flowering, walking onions grow "top-set onions" on top of their green onion stalk. These topsets will start off white, then get a red skin which can be nearly as striking as a flower. Then, these onions will start to grow new green onions, which themselves can form new topsets, with new green onions with new topsets! I have had walking onions that were as tall as I am, with four sets of onions stacked on top of each other.

Finally, this is TOO MUCH ONIONS and the plants fall over, with each set of bulbs rooting, giving them their name as they "walk" around the garden.

But that is, of course, if you can resist eating them.

Resistance is futile, because every part of this plant is edible and of high quality. The green onions can be nearly evergreen, and with a bit of snow cover are more often than not available through the worst of a Michigan winter, right underneath the snow. If not, the bulb can still be dug, and this onion bulb can have a strong, red onion sort of flavor with lots of bite, or lots of sugar when caramelized. The topsets are a special vegetable, and when they first form, they have no skins and can be taken and thrown right into a dish as pearl onions without peeling. And with some luck and folding over the greens as with annual onions, the bulb may swell up to be the size of a small storage onion.

Soil and Aspect: Egyptian walking onions are happy in most soils and in most conditions, thriving even in poor soil and drought. They can live in full sun or even in a considerable amount of shade, so long as they do not have better adapted competition. They will perform best in fertilize soil with mulch.

Propagation: You'll hardly have to worry about it, as they will propagate from both bulb division in the soil and from the plentiful topsets. Thinning when clumps form is desirable, but these I usually use as onions.

Kitchen Uses:
Leaves: cut as needed and use as green onions. An excellent flavor, especially in spring.
Bulb and leaves: Can be used as scallions, and caramelizes quickly for a very sweet spring caramelized onion. I will use as many as 20 spring onions in one batch when I caramelize.
Bulb, harvested when the greens go dormant, usually in a late summer drought. Bend over the plants to promote bulb swell, then harvest before rains come. Keep in cool, dark, dry storage.
Topsets: harvest before skins form to avoid peeling, though they can be used after. I use then whole in soups and stews and pickle them for one of my favorite pickles.

Garden Uses: These can make an excellent pest repellent spray, can be planted near trees to repel animals, and used as a nitrogen rich mulch if they become over-abundant.

Guild roles: Guild-matrix candidate, pest-repellent, can be part of a fortress-planting, may have mild grass-repelling properties, nutrient-accumulator, mulch-maker.

Chives, *Allium schoenoprasum*

This is a common perennial bulbing plant in herb gardens that deserves to be even more widely grown and used in ornamental gardens. New varieties mean there are chives in shades of pink, lavender, white, and purple, and all can be used in the kitchen. In the garden, they can be placed to create spring color echoes.

Soil and aspect: highly adaptable to both, but thriving in full sun and garden soil.

Kitchen uses: edible flowers, edible leaves added to almost anything.

Garden uses: part of a fortress planting, insectary plant. Color echoes

Guild roles: Mild pest repellent, possible fortress plant, insectary plant,

Alternative "Chives"

There are a variety of other alliums that can be used in a similar way as chives, providing similar splashes of color, edible flowers, leaves, and "spears," and some are better than allium schoenoprasum. For a full season of "chive" spears and flowers to use in the kitchen, plant, chives, Garlic chives, A. tuberosum, A. Carninatum, A Cowanii, and A. unifolium. I like to have these around in quantities where I feel I can use the spears, leaves and flowers abundantly and not feel like I'm robbing my eyes to feed my stomach.

Allium Unifolium, Single Leaf Onion, "Pink Chives"

I'm giving this lovely perennial allium its own profile because it has been underused in Permaculture gardens and cuisine, and I find it one of the most useful of the alliums.

A North American native plant of the mediterranean Pacific, this gem thrives off spring water and is ideal for Permaculture gardens on depleted soils or gardens with a goal of drought tolerance.

Its gorgeous and delicate pink flowers make a real show in the garden, and are also the usual winners in "chive" taste tests I've done. The "spears" hold up better than most chives making it a fine vegetable, and the flowers and leaves have an especially zingy allium flavor.

Soil and aspect: Full sun, well-drained soils. Can do well in poor soil to rich garden soil, provided it is well draining. Can compete well with grasses and persist in grassland.

Kitchen Uses: spears, a very fine vegetable added to salads, stir fries and sautees at the end of cooking a dish. Leaves, used like chive leaves. Flowers used like chive flowers. Bulbs. These are unusual for onions, and form a very crisp tuber that can sometimes be harvested right from the soil surface in a dry season without digging and used without peeling.

Garden uses: grass tolerant guilds and edible meadows, insectary, possibly pest repellent, use in drifts or scattered into meadow plantings, or plant in clumps like chives.

Varieties and variations: some specimens may flower then produce bulbils afterwords. On some, these bulbils can have a beautiful bright red color looking like flowers. These bulbils make an interesting vegetable or garnish, used or pickled before they grow a skin. They have the same tangy taste as the flowers, and add an additional aspect of propagation for rapidly spreading them. This seems to be a limited genetic trait as some populations will almost never produce bulbils, while others produce them reliably year after year.

Leeks, *A. ampeloprasum*

Among the most ornamental of alliums, leeks have long been used as a flower in potager gardens. They can flower in purple, lavender, pink, and white, and can easily perennialize, though it's usually grown from seed as an annual. Dividing specimens can form large clumps that can be thinned and spread around.

Culinary leeks are grown for their swollen, succulent stems, which require a trick in perennial culture. Clumps of perennialized leeks will have thin stems similar to garlic. To size them up, we must thin them out to single plants and give them a generous topdressing of compost.

Soil and aspect: Full sun, rich, or well-drained soils. High fertility is required to grow the thick stalked vegetable used in gourmet cuisine.

Kitchen Uses: The swollen bulb and stem, often thin-sliced as a true gourmet vegetable. Flowers used like chive flowers.

Garden uses: A beautiful flowering allium, possibly pest-repellent, add a round bright flower to the garden.

Varieties and variations: There are some leeks sold specifically as perennials and as ornamentals and these have reliable color characteristics for the ornamental gardener. Ornamental leeks are edible, though they may not size up as reliably.

Anise hyssop, *Agastache foeniculum*

Anise hyssop is another beautiful, perennial North American native plant that deserves to be grown in the edible garden more often. Leaves of some specimens make ideal, sweet, licorice-flavored salad greens (while some are slightly tough or bitter), and flowers which can make a root-beer flavored tea.

In the landscape the flower plumes are highly charismatic, echo liatris and monarda, and attract a wide profusion of pollinators and beneficial insects. In some cases it appears to tolerate grasses.

Ken Fern of PFAF.org reports young plants may be very susceptible to slug damage.

Zones 4-8, though related species may have similar uses and be available for warmer zones. *Agastache Mexicana* fits zones 8-10. *A ruguosa* is perennial down to zone 8 and has a strong licorice flavor.

Soil and aspect: Full sun, rich, or well-drained soils.

Kitchen Uses: leaves and flowers used in salads and teas.

Garden uses: A beautiful flowering plant and amazing pollinator plant.

Propagation: Easily grown from seed, or transplanted after self sowing. The genus has often become commonly self-sowing in my gardens, providing many free plants, or free salads (as young seedlings seem to make the best additions to salads) though Ken Fern (pfaf.org) reports young plants may be very susceptible to slug damage, so that may be a problem in some locations.

Varieties and variations: There are some hybrids and named varieties with wonderful colors and characteristics within this genus and I could imagine a garden using these as a major element. Some have reliably tender and mild leaves.

Amaranths

Underrated amaranths are a true gift to farmers, home gardeners, and those looking to achieve greater self-sufficiency. Grain amaranths are second only to maize as an easy calorie crop.

Grain amaranths are exceptionally easy to harvest, store, and use (with the right culture) and very flexible in the kitchen.

In the landscape, the large plumes of color and often beautiful leaves can be exceptionally charismatic. Nothing adds such a bold statement to a landscape than the bright-red, large, feathery grain heads of "Hopi Red Dye" amaranth, and even the Chinese "orange giant" grown for grain can make a bold statement. There are a large number of amaranths grown as ornamentals, including cockscomb, and celosia, and all have edible leaves.

A skillful gardener can create echoes of color and structure by repeating amaranth forms, with cockscomb in layers around the garden.

Soil and aspect: Full sun, well-drained soils. Can do well in poor soil to rich garden soil, provided it is well draining. Can compete well with grasses and persist in grassland.

Kitchen uses: spears, a very fine vegetable added to salads, stir fries and sautés at the end of cooking a dish. Leaves, used like chive leaves. Flowers used like chive flowers. Bulbs. These are unusual for onions, and form a very crisp tuber that can sometimes be harvested right from the soil surface in a dry season without digging and used without peeling.

Garden uses: grass tolerant guilds and edible meadows, insectary, possibly pest repellent, use in drifts or scattered into meadow plantings, or plant in clumps like chives.

Varieties and variations: some specimen

Artichokes, Cardoons, *Cynara species*

If you wish to make a bold statement, the Cynara genus of perennial edible plants including Artichoke (**zones 7-10**) and Cardoon (**zones 5-10, possibly 4 with help**) will do it for you.

These loud, bright purple, iridescent flowers can catch your eye a whole block away in the city, so it's no wonder pollinators seem to swarm to it. And I take special pride when lawn-loving suburbanites eye it with horror, being the the largest monster thistle they've ever seen. One man, fresh off his lawn mower spat "it's. A. THISTLE!" in utter disgust after I told him this was the not-so-secret ingredient in artichoke dip. And so, in American gardens it may have special significance as the antithesis of the lawn.

These plants—with their silvery foliage, architectural form and beautiful flowers—will usually be the center of attention wherever they are found.

For gardening and kitchen purposes, I'll treat them about the same, other than Cardoon is reportedly more hardy, and individuals have overwintered 4-5 years unprotected in my Michigan garden, even through the 2 coldest winters on record (zone 5, dipping into zone 4 temps.)

Personally, I'll say I'd rather grow cardoon. I know most people consider this the wrong answer. The stalks and meaty leaf petioles (leaf stems) have a flavor very much like (probably) from artichokes when grown on rich soil, and some varieties will produce globes as large as artichokes, though covered with harder spines. But I'd rather let most of these flower anyway, and harvest an occasional side stalk or large leaf. I've seen these harvested after flower, leaving giant flower stems in the garden that look like something from a psychedelic rock video. Besides, artichokes are easily found in the stores in most places, whereas cardoons are still hard to come by, especially in the U.S..

Soil and aspect: They deserve full sun and your best, rich, garden soil. Despite being a thistle, they do not thrive in competition.

Kitchen uses: The swollen bulb (especially the heart,) stems, and leaf stems are used as a vegetable. Both species can actually be used interchangeably, but artichokes are better grown for the flower heads and cardoon for the plant parts.

Garden uses: A focal point, a dramatic gesture. I've seen them advised as a mulch-maker plant, but it would seem a shame to me to chop and drop them. In the winter they will reliably reduce to a big pile of mulch for themselves, anyway, and like most broad-leaved plants that conserve soil moisture and make habitat for soil life, they appear to grow very rich soil.

Image by Jeffrey Barnette, Landscape Designer

Propagation: Easily grown from seed, or transplanted after self-sowing. In colder climates it's probably better to start them indoors and transplant them out.

Varieties and variations: There are some hybrids and named varieties with wonderful colors and characteristics. The best of both used to be traditionally grown from clones, keeping their fine characteristics, but these are never available in the U.S.. There is a red cardoon that is particularly stunning in the garden, and can still be used in the kitchen.

Beans

The common culinary vegetable and garden plant can also be a winner as an ornamental with the right varieties. Especially gorgeous when trellised or growing up ornamental corn stalks.

Soil and aspect: Full sun into moderate shade. Can do well in poor soil to rich garden soil, heavy or light soil. Cannot compete well with grasses.

Kitchen uses: Flowers, green pea pods, and shelled peas, and dried as a protein-rich crop.

Varieties and variations: The most common ornamental is the scarlet runner bean, a vigorous growing plant with beautiful flowers and edible beans. There are now some varieties of this species with delicate pink or other colored flowers to look for.

Hyacinth beans **Dolichos Lablab,** have beautiful, colorful foliage, flowers, and beans, though the raw beans are poisonous. When cooked they are edible.

The common bean, **Phaseolus vulgaris**, also comes in some very beautiful edible champions in different forms. Purple dragon dwarf bush bean makes an incredible ornamental in borders, and has beans that are great for homesteaders, being both excellent as a green bean and as a dry bean

Bellflowers, *Campanula species*

Distinctive bellflowers, usually in blues and whites, are all edible and all have edible leaves, though some are more "edible" than something you'd actually want to eat. Meanwhile a few have very good eating qualities. Here, I'll discuss a few of the most useful for the edible ornamental garden.

Campanula percisifolia is probably the best of them in my opinion, which is probably influenced by it being the one I can most easily grow in the Great Lakes region of the U.S.. The blue or white flowers have a sweet pea-like flavor and are nice in salads. One of my favorite edible flowers. The evergreen leaves can be a bit tough and are best in the middle of winter when you kick the snow off them.

Zones: 3-8

Soil and aspect: Full sun to moderate shade. Can do well in poor soil to rich garden soil, heavy or light soil. Cannot compete well with grasses in my experience, though I heard of one gardener having this plant colonize her meadow, probably just to be stubborn since she was hoping for a pure native flower meadow mixed with the Eurasian grasses.

Kitchen Uses: Flowers, raw, used in salads or as a garnish. Leaves, though I don't find enough of them to usually bother cooking, and they are a bit tough in a salad. They seem best to me in winter, when I'll harvest them and add them to eggs for some green.

Garden uses: In a happy spot this plant will finally spread enough to harvest some greens in quantity, the blues and whites are perfect for a Jardin de curé color pattern.

Varieties and variations: The variation crystalocalyx is probably best for eating, with reportedly larger leaves (it is the only version I have grown,) but this may be hard to obtain seeds from in this day of careless internet seed companies.

Campanula rapunculus: a dreaded plant in some regions, this runner-spreading bellflower has edible roots, leaves, shoots and flowers. I've avoided growing it because I don't want the harassment of my local native plant activists, but with careful garden culture and harvesting it could be welcome plant in stable guilds and the tiny seeds are not a large risk for dispersal by other species.

Zones: 4-10

Soil and aspect: Full sun into moderate shade. Can do well in poor soil to rich garden soil, heavy or light soil. May stabilize in meadow guilds.

Kitchen Uses: Roots, cooked. Flowers, leaves, shoots.

C. versicolor. One of the best for edible leaves of the bellflowers I've tasted. This is another¡ bellflower option that is not a risk for spreading, though I've had some "invasives" extremists eye it with suspicion anyway. It does not spread by runner, which is kind of a shame since the leaves make pretty darn good eating, and it gives a great show of flowers, too.

Zones: 7-10

Soil and aspect: Full sun into moderate shade. Can do well in poor soil to rich garden soil, heavy or light soil. Cannot compete well with grasses.

Kitchen Uses: Leaves used in salads, even in winter where it can survive. Flowers, raw, used in salads or as a garnish.

Borage, *Borago officinalis*

The annual blue star flower adds subtle beauty to the garden, cool cucumber flavors in the kitchen, and a wealth of resonance of story, and mythology.

Reportedly repellent to tomato hornworms and other garden foes (some say this is due to its attractiveness to predatory wasps) it is often advised as a universal companion plant, and I follow BioIntensive Gardening proponent Alan Chadwick's rule of allowing it to grow wherever it comes up.

Some people have told me they find the plant to look shaggy and homely, but I think the silvery foliage and blue flowers are very attractive, so here it is in this book. To each their own. It has a long history of culinary use going back at least to the Romans who put it in wine and the Celts who made a soup of it. I do both.

Soil and aspect: Full sun. May self-sow into lawns.

Kitchen uses: Flowers, raw, used in salads or as a garnish, or frozen into iced cubes. Leaves, crushed and added to teas or alcoholic drinks for a cucumber flavor. I make a "cucumber water" with borage leaves that most people think is just cucumber water. Young stalks, peeled and filled with a cream cheese mix for an unusual and refreshing appetizer. You have to get them before they become fibrous. Young stalks, peeled and blended into a soup. I like to make a cold gazpacho of borage stalks, sorrel leaves, and green garlic, blended into a cold-steeped tea made from the borage leaves. Then thicken with bread and emulsify with oil like a traditional gazpacho. A favorite cool dish for hot weather.

Garden uses: Self-sowing annual that fills in gaps in guilds.

Varieties and variations: There are blue and white-flowering varieties.

Brassicas, kales, broccoli, cauliflower, etc.

There are quite a few beautiful, perennial brassicas available for ornamental edible landscapes. Of perennial kales, Daubenton and Cosmic have beautiful variegated foliage and are reliably perennial.

I have grown a black perennial kale (rare on the market) which has been the most cold-tolerant perennial kale in my collection, surviving the two coldest winters on record in Michigan. Some specimens have lived for five years or longer. These cannot be propagated by seed and must be created by rooting side shoots.

Tree collards and purple broccolis may also perennialize in zone 7 and above.

In general, I highly recommend these purple and black varieties, which research shows have a much higher resistance to pests like the cabbage moth. In my gardens, green brassicas will be loaded with caterpillars while the purples go almost entirely untouched. The anthocyanin pigment is also delicious, reducing the cabbage flavor, and highly nutritious.

Camas, *Camasia quamash*, etc.

It's often reported that the roots of this species were so valued by certain Native American tribes that they were used as currency. I get it. I value them both for their edibility and for their beauty in the edible meadow, where they seem to be quite at home, providing beautiful, structural purple flowers to the mix. The roots are sweet, and people say they (like everything else) taste like chestnuts, though most modern people don't know what chestnuts taste like. This is the forager equivalent of "tastes like chicken." Anyway, I like them, and the texture cooked, though there's a danger that if the preparation is off they'll get rubbery. In an edible meadow, they may spread by seed and root division, slowly growing in numbers until it's possible to harvest them in quantity for meals.

Zones: 3-9 (possibly some variant species up to 10.)

Soil and aspect: A mediterranean-climate plant native to North America, it thrives in rich seasonal wet followed by a dry season in midsummer where it goes dormant, though it can adapt easily to wet summer climates. It can grow into moderate shade and competes well even with runnering Eurasian grasses.

Kitchen uses: Bulbs, roasted, or sliced and caramelized. It's rumored you can caramelize these into a molasses-like substance, though I haven't tried that myself. I usually add them to stews or in roasted root medleys.

Garden uses: Can be used in garden polyculture beds or in edible meadows.

Varieties and variations: Over-eager visitors have voluntarily "weeded" these from the garden on multiple occasions when they were in beds, looking a bit like grass leaves, so now I tend to plant the variation "Blue Melody," which has variegated leaves that less resemble grass. Some related species like wood camas are also good to eat, but probably reproduce much more slowly under most circumstances.

Clove currants, "Crandall," *Ribes odoratum*

Currants and gooseberries both make excellent additions to the beautiful, bountiful garden. But there is one variety that deserves special mention for its ornamental and edible qualities, North American native Crandall clove currant. This has sweet, Clive-scented yellow flowers in spring, and grape-like, tart, sweet fruit.

Comfrey selections, *Symphytum species*.

Comfrey is one of the great gardener's friends. A fantastic medicinal plant and universal companion plant, it has been written about extensively on the internet and in other books. But I'd like to introduce a few beautiful variations, like this variegated comfrey, which are gorgeous in the landscape.

The most vital piece of information to know is that species comfrey, *Symphytum officinale,* is an extremely aggressive plant with large, eager seeds. If you have a market for comfrey products, such as an herbal lotion business, then like me, you will not regret planting the species plant. Almost any other gardener may regret it.

However, the selections are much less aggressive, and in most cases, do not produce viable seeds. Most of these have "Bocking" and a number in their names (from the research program that created them) but many are sold as just "Bocking comfrey." You can only guess which selection they actually are. Dwarf comfrey, *Symphytum ibericum* is a related species that is native to Europe and less aggressive.

Zones: 3-10

Soil and aspect: Yes. Will be happiest in full sun and garden soil. May flounder on very sandy or shady sites.

Garden uses: One of the most powerful ecological allies we can recruit. It is loaded with minerals, is an amazing insectary and pollinator plant, a mulch-maker and vigorous fortress plant. Chop and drop, slash-mulch, and used as a fertilizer tea. Fill a bucket with leaves and water, then wait for it to smell like god-awful death. It is rich in nutrients and microbial diversity. Dilute 1/10 and spray on plants.

Daylilies, *Hemerocallis fulva, etc.*

This is another indispensable plant in the edible ecological garden, one that can add a wide variety of beautiful colors, and can make a certain statement against plant prejudice. Some scoff at this most useful plant genus as representative of a plant that only offers beauty, instead of non-aesthetic benefits. It's often used as a symbol of "garbage plants" by native plant extremists in the U.S.. I've had utterly indignant garden guests scoff "DITCH LILIES!" at me. The shocked looks on their faces when they eat their gourmet flower out of a ditch is priceless.

And yet, daylilies thrive without inputs, can defeat weeds and grasses on their own, are more drought-resistant than all but a few common native plants, AND virtually every part of these plants are not only edible, but truly gourmet.

Soil and aspect: Full sun if you want flowers, but they can thrive in a fair amount of shade without flowering. Best in rich, well-drained soils, but highly adaptive and seem to stand wet feet well, hence the name "ditch lily."

Kitchen uses: This will be a long one, since nearly every part of this plant is gourmet. Let's start with the flowers, eaten fresh right out of the garden, they are sweet, crispy, and nicely perfumed. I have made an entire salad of these flowers, since the texture is like a fine lettuce and they are available in quantity. The flowers can be stuffed, lightly steamed, added to stir fry at the end, or used as a garnish on almost any dish.

The pods before flowering are another remarkable vegetable that competes well against peas in many dishes, having a similar texture, but a sweeter flavor and sometimes a hint of aroma. Most people I serve them to prefer them to peas.

The flowering stem with immature pods make a nice "asparagus," so long as they pass the "universal snap test" (they snap clean when you bend them, rather than creasing. If they crease, they are too tough.) These make an otherworldly looking addition to stir fry dishes, or sea food. These can be tempuraed for a very interesting effect.

Young shoots and leaves can be grilled like a grilled lettuce for a warm salad (I prefer the texture this way to eating them raw.) The flavor is outstanding.

Perhaps the best and most interesting part of the plant are the corms, the tuberous roots that are starchy and sweet like potatoes. I have often utterly wowed people by serving chips made of these, with sea salt and pepper. They can be roasted in butter like potatoes and compare very well.

Garden uses: these are the stars of the extensive gardens, hedgerows, thickets, edible meadows, and young forest gardens. Swaths of classic orange flowers are a marker of summer beauty. Other colors can be used and may not be as vigorous, showing well in garden beds. I'll have a word of warning later, not all are equally edible.

Propagation: Easily propagated from root division in the dormant season.

Varieties and variations: First, this entry deserves some cautionary tales and warnings.

Most importantly, daylilies are not lilies. They are a different genus, and many lilies will be poisonous.

Next, there are now an astonishing number of daylilies in different colors, from different hemero callis species, and while most report these are not poisonous, I've had a few specimens give me a bit of a temporary burn or tickle in the throat (beets also often do this.) So, I taste test, and keep around stands that do not have that effect for eating. The most reliable good eating are the common "ditch lily" *Hemerocallis fulva, and H. lilioasphodelus*, the now ubiquitous small yellow lily. Virtually any of the ornamental versions in fascinating colors and forms are worth giving a taste, though the quality and this throat tickle may appear in some more than others.

Fuki, *Petasites japonica*

This exotic-looking, perennial, woodland plant thrives in shade and can produce a LOT of edible biomass in deep dark woods. Perhaps too much biomass, if one isn't careful. DO NOT PLANT THIS WHERE IT IS NOT NATIVE WITHOUT CAREFUL MANAGEMENT.

And I will admit, this ample biomass is not my favorite to eat. Yet, I am including this plant because there are so few great vegetables for shady ornamental gardens.

Soil and aspect: Deep shade to light shade. Good in fruit tree guilds with mature trees.

Kitchen uses: Flower buds. I like to slice these and use them at the end of a dish to lightly cook and eat. Stalks, sliced against the grain to keep them from being stringy. Like celery, but with a bland flavor. The best about this vegetable is that it soaks up whatever flavor is being used. Good in a dish with battered and sauced chicken, for example.

Garden uses: An aggressive spreader in shady guilds, can be used as part of a "terminator guild" to eradicate plants like periwinkle and English ivy.

Varieties and variations: There are giant versions and a nice variegated version on the market.

Hostas

Another all-star of shade, this is one that has challenged my own distaste for the common HGTV landscape gardens with the same six plants. Unless they are eaten, I consider them an affront to good taste. When eaten, they taste pretty darn good, so they get a pass.

Zones: 3-9, possibly 10

Soil and aspect: Full shade, rich, woodland soils with coarse woody debris mulch.

Kitchen uses: Flowers, spring shoots. These spring shoots are incredibly versatile and I have added them to stir fry, tempura, pasta dishes, vegetable medleys and so on. The only limit is your creativity.

Garden uses: I will say I only ever recommend these in fenced-in gardens protected from rabbits, groundhogs and deer, who seem to prefer hostas to all other plants. They do not offer much for a guild, other than to fill in beautifully in woodland guilds.

Varieties and variations: I have experimented around eating peoples' hostas wherever I go for over 15 years and I will report a very wide variety of flavors and textures. Some are bitter and strong, and also tough. Others are light, crispy, and sweet, making a very good vegetable.

Japonica Maize

This lovely variegated maize deserves a special mention for its beauty in the garden and versatility for the homestead kitchen.

I don't think there is any plant I have grown in 20 years that I spend more time looking at and feeling fascination for. As I write this, out the window there is some planted along the busy sidewalk, and almost daily I watch people startled by it as they pass. Often people, even corn farmers, have asked me what this amazing looking… thing is.

Soil and aspect: Full sun, rich fertile garden soil.

Kitchen Uses: This is a very versatile maize for home gardeners. The medium-sized, dark-red ears of corn make a particularly rich, aromatic flour for polenta and corn bread. Nixtamalized with ash then ground into masa, there is no corn I've eaten with a richer flavor. And yet, in the milk stage it makes an excellent sweet corn, though the soil must be well-watered prior to harvest and the window for harvest is small before they become starchy. Bring the water to a boil first, then go pick the corn and drop it right into the water, or right onto the grill in the hulls.

What it lacks in sweetness it makes up for in genuine, old-fashioned corn flavor. If you store it in the fridge overnight, it will be chewy and starchy and only fit for the compost pile.

Garden uses: Planted in drifts, planted in blocks for ornamental effect, or in an ornamental Three Sisters guild as we'll discuss later.

Lemongrass, *Cymbopogon species*

When I'm doing design/install work, clients often ask to have these plumes of grasses that have become popular. And while some of those grasses make first-class mulch-makers, I like to include edibles, and lemongrass deserves at least a brief mention. We can get the look and still get a high-quality edible out of it. It will perennialize in **zone 8** and up, but I have always potted up a small hunk to overwinter, which works well in cold climates.

Lupines, *Lupinus species*

Lupines feature heavily in this book, so they get their own profile, even though they are a common ornamental.

There are lupine species that are native over much of the world, and all of them fix nitrogen, helping to fertilize the soil. But did you know there are also edible varieties?

Or rather, all varieties are edible with leeching and cooking, but a few species are edible without preparation, most notably the Andean-cultivated varieties of *L. Mutabilis*, which have larger peas and a better texture.

In the right conditions, these may naturalize in guilds to create nice, edible ecosystems.

Mallows, *Malva species*

There are a variety of mallows that make beautiful ornamental edibles, including musk mallow, marshmallow, *Malva silvestre*, the hollyhock mallow, and rose of Sharon. It's a shame to lump them all in together, but I could write a small book about the culinary and garden uses of these plants. Besides, all are worth getting to know on your own.

All are naturals in the cottage garden or jardin de curé. Musk mallow, *Malva moschata*, is an excellent salad green and edible flower, while marshmallow is the vegan's best friend, providing a thickener and vegan meringue that can be turned into, yes, marsh mallows. The greens of most make a decent cooked spinach, and marshmallow leaves can be used to thicken soup like okra.

Milkweed, *Asclepias syriaca*

Everyone knows milkweed is important for monarch butterflies, but for me, it's my gateway foraging vegetable. This is how I get friends addicted to my hobby. It's so foreign to anything sold at the grocery store that it takes some real buy-in, and then it is so accessible and delicious that everyone is sold on it. One taste of my milkweed jalapeño poppers and you'll be a forager for life.

Yes, this plant is actually several good vegetables all rolled into one. The shoots are edible as an "asparagus" substitute, though they have a strong milkweed flavor that is very distinctive and takes getting used to. The advantage is that they are usually available much later than asparagus, thereby extending the season for shoots.

The flower buds are edible and also the flowers. I like to fritter these in a beer batter.

And then finally, there are the pods and the pre-silk inside the pods. They have to be harvested before the silk forms, usually when they are under two inches in length. These can be eaten whole or separately.

To make jalapeño poppers, dredge them in flower, salt, and jalapeño (or other pepper) powder. Then fry in shallow vegetable oil. Test one, sometimes they require a full tempura treatment depending on how much moisture they are holding. Done right, the outside should get crispy like a popper, and the inside should "melt" into a cheese like substance.

That milkweed cheese is also useful on vegan pizzas and is pretty convincing.

So plant some for the monarchs and make sure to eat some for yourself.

Zones: 4-9,

Soil and aspect: Full sun to part shade. Highly variable soils, can thrive in sandy soil to swamp.

Kitchen uses: spring shoots, flower buds, flowers, and young seed pods.

Garden uses: Deer-and-wildlife-resistant, attracts monarchs, extremely versatile in guilds and doesn't tend to outcompete other plants. Tolerates grasses well.

Miner's Lettuce, *Claytonia perfoliata*

This is another annual, but one that can so easily become self-sowing in the right contexts, that is is practically an perennial. Wherever it grows, it is a key player in a "guild matrix, " which we'll cover later. It makes a beautiful over-wintering ground cover, and I prefer it in salads to almost anything else.

Easily grown from seed. Let it grow to set its own seed and hopefully you will never be without it again. The picture by Permaculturist Christina Colis Perma Dumbrava shows it as part of a thriving self-sown polyculture with lettuces, garlic, and other delights.

Mints

Mints are a much-maligned and under-utilized garden powerhouse. The biggest "problem" people have with mints is that they're so afraid to plant them that they only grow a teensy pot, and so they never use them!

But mints are incredibly varied, and some are much less aggressive, AND can be used in quantity as a green. Here we have a highly flavorful, nutritious green that grows very aggressively, which is why there are many world cuisines where it is used as the base for salads, teas, and soups. There are mints which are so mild they are actually cultivated as a salad-leaf crop. And there are mints that are so strong they're used as the prime ingredient in various soaps and fragrances (even the word "cologne" is associated with a mint that was once used as the star in this perfume.) My favorites are the *mentha citratas*, which have fascinating blends of aromatic and flavor compounds that run the bill from orange, lavender, oregano, musk, peppermint all mixed together in different ratios. There are even perennial mints that taste very similar to basil, easily grown with no pests, while gardeners sometimes struggle to grow enough basil to can some pesto for winter.

Moving from the *Mentha genus*, we have a whole family of mints that make amazing greens, and potent medicinals, and all of which are easily grown. There's even a mint which makes a GREAT water-chestnut-like tuber, crosnes (shown above pickled in seasonal colors, see runners-up section!)

All of these are absolute powerhouses of ecological function, too, often being star pollinator and insectary plants, easy ground covers, fortress plants, and universal companion plants. If the goal is a low-work, high-output garden, having a solid basis in mints is a good strategy.

The only problem is that these cannot be grown true from seed, and so are difficult-to-impossible to source within the U.S., where they are nevertheless nearly always grown from seed anyway.

Monardas

Monardas (**zones 3-10,**) especially *fistulosa and didyma,* feature prominently in this book.

Both may be aggressive, useful plants that are amazing insect attractors. *M. fistulosa* tends to do better on sandy soils and full sun, while *didyma* thrives in heavy soils and more shade. They are shallow rooting and do not tend to compete with deeper-rooted perennials, while holding their own against grasses, making them useful in many guilds.

They are also very useful in the kitchen. As a dried cooking herb, they can be used much like oregano, they are good in tea blends and the flowers are a colorful addition to salads.

Ostrich fern, *Matteuccia struthiopteris*

Another selection for the shade garden, this North American native fern is a gourmet vegetable and attractive as an ornamental. Spring "fiddleheads" are one of the earliest vegetables and I look forward to them each year. Ferns provide a beautiful interest in the woodland garden and give off a true "forest" effect.

Zones: 3-9, possibly 10

Soil and aspect: Full shade, rich woodland soils with coarse woody debris mulch.

Kitchen uses: Spring fiddleheads, lightly cooked, don't turn them to mush!

Garden uses: Deer-and-wildlife-tolerant addition to woodland guilds into full sun.

Peas, *Lathyrus species*

Peas are another common vegetable with some very beautiful varieties available, including both vegetation colors and flower colors.

Tuberous sweet pea, *Lathyrus tuberosa*

This is a particularly ornamental perennial edible with a choice edible tuber for a root. It makes a great perennial and edible alternative to aggressive annual sweet peas and is less likely to become an obnoxious spreader in most landscapes. The biggest problem with the species is that it is so delicious you'll want more, whereas it tends to be a low-productivity vegetable. Perhaps it is a candidate for selection for high-productivity specimens.

Soil and aspect: Full sun, well-drained soils. Can do well in poor soil to rich garden soil, provided it is well draining. At home in wild thickets and "weedy" old fields. One stand I visit appears with grasses, wild carrots, asparagus, and lots of milkweed. It appears as a nitrogen-fixing weed in agricultural fields where it can bind up equipment, so it loves somewhat depleted well draining soils. Very much at home as a rambling flower in hedgerows.

Kitchen uses: Primarily the tubers, used like potatoes. Starchy, and sweet, an excellent alternative to potatoes if you can grow them in large enough quantities.

Garden uses: Nitrogen-fixing climber. I imagine uses as a slash-mulch crop as a nitrogen-rich mulch-maker plant.

Turkish Rocket, *Bunias orientalis*

People who spend much time with me in the kitchen might get the idea that Turkish rocket is my favorite vegetable. And while that's not exactly true (not *exactly*) I do really, really love this plant. And though some think it looks like a "weed" due to their prejudice against yellow sprays of flowers, I hardly know of any plant that can bring so much joy and color to a garden as this one. And the wildlife absolutely love it, too.

Zones: 4-8, possibly higher.

Soil and aspect: Full sun, in a wide variety of soils.

Kitchen uses: one can eat the leaves in a famine, but the real show is the "broccoli," or "rocketini." These have a broccoli rabe texture and a zingy arugula-like, nutty flavor I just become addicted to, sautéed simply in butter or roasted with roots and shoots. They are at their best when grown to nice large stalks, then sliced very thin and cooked over high heat for a brief time with garlic and butter. The root is also edible as a substitute for horseradish.

Garden uses: Another plant that can tolerate grasses very well in a meadow, or with company can act as a grass-repelling fortress plant. A top tier pollinator plant and mulch-maker, as well, means it is an excellent multi-purpose element in guilds.

Sea kale, *Crambe maritima*

One of my absolute favorite plants in any garden I have. I would not garden without this gorgeous multi-use plant.

I once had a professional journalistic photographer come to one of my ornamental edible gardens to take pictures for a magazine article on the project. I expected him to romp around taking pictures of the 60+ species of edible plants that happened to be flowering at the time, but instead, he spent almost the whole session taking glam shots of sea kale, captivated by the light on the blue-silver, translucent foliage and homey spray of flowers covered in beneficial insects. I mean, I get it! It's a gorgeous plant.

Zones: 3-9,

Soil and aspect: Full sun, can live on sandy, well-drained soil, and tolerate salt, but prefers rich rich, garden soil if given some protection from grasses.

Kitchen uses: This will be another long one. Depending on the treatment, a wide variety of forms and vegetables can be drawn out of this remarkable relative of the brassicas.

In spring, the emerging plant can be "blanched" by putting a bucket filled with sand over it. Pressing up through the sand it will produce a crispy vegetable like a cabbage. It was this vegetable that is the sought-after gourmet treat in its native range.

The flowering stalk can be harvested and eaten like broccoli. In my taste tests, many people prefer both the flavor and texture to actual broccoli, as it lacks the distinctive "cabbagy" flavor. Blanched, this will exhibit more like cauliflower.

The young greens can be eaten like kale, though a bit meatier. A friend makes "kale chips" out of these that are crunchier and more fulfilling that those made with actual kale.

The flowers are edible and certainly choice, and the young seed pods are like cabbage-flavored peas! Even the root is edible and makes a very interesting addition to roasted root medleys.

Garden uses: This one requires the protection and care of an intensive garden area with rich soil and protection from weeds and grasses, but is more than worth it. It probably acts as a beneficial insect attractor.

Varieties and variations: Lillie White is one variation on the market, noted for being large. *Crambe cardofolia* is a related species with some similar characteristics, being much larger in the garden, but probably less palatable for most uses.

Shungiku

It's simply impossible to do a chapter on edible ornamentals and not mention the annual edible chrysanthemum, shungiku. High-quality, edible greens and flower buds with a great texture and a slightly pine-like taste I find addictive. Some have said this plant may become biennial.

Soil and aspect: Sun to part shade, garden culture and soil.

Kitchen Uses: Flowers, buds, leaves. Usually steamed or in dishes with sauces. The strong flavor is unpleasant to most people raw or un-dressed, but holds up well to sauces.

Skirret

What on Earth did Europeans eat before the colonial period when corn, beans, squash, peppers, and tomatoes came from the Americas? Skirret. They ate skirret.

This perennial carrot-like vegetable has won over everyone who has tried it in my taste tests, roasted in a little butter for a vegetable not too far off the french fry, but with a carrot like taste. Leaves can also be used to garnish a Bloody Mary, with a fascinating carrot-celery-parsley flavor.

According to medieval herbalist Gerard, Gerard, that skirret is "meanly nourishing, but they are something windie, by reason thereof they also provoke the lust."

TMI, Gerard, TMI!

And also, Gerard seems to have found it scandalous that some German women would often "prepare the roots hereof for their husbands and know full well wherefore and why!" Oh, these poor German men to be so taken advantage of by their lusty German wives!

I am worried less about the windie lust than I am about poor Gerard who seemed to conflatuate the two conditions. I do hope the poor pious man's wife never got ahold of any Jerusalem artichokes, as that would surely have been a raucous occasion. Personally, I've never found them "windie," but I eat a lot of raw vegetables compared to the average medieval European worried over vegetable-induced moral peril.

Zones: 5-9

Soil and aspect: Full sun, a plant of near-wetlands in rich mucky soils, standing clay and high organic matter well. In the garden they like a deep mulch and rich soil, and plenty of water.

Kitchen uses: Primarily the tubers, used like potatoes or carrots. Starchy and sweet, an excellent alternative to potatoes, or parsnips. Can be mashed, fried, roasted or eaten raw. Very good! The side shoots that form as new propagules are an unusual and delightful vegetable of their own, with a texture like a very crispy pac choi and a flavor of celery and carrots.

Garden uses: Can become self-sowing, seems to repel grasses well. One of the best insect attractors in my garden of 300 species. Would be at home in a rain garden, so long as the water was filtered: I wouldn't want to harvest roots out of polluted soil.

Sorrels, French, blood-veined, silver, etc.

As a kid I used to love to pick this "lemon grass" in the back yard at my grandpa's house, and I could spend all day sitting around munching on the stuff while painting pictures of hawkweed. As an teenager, I'd see a much larger version of the same exact leaves at the farmers' market, sold as "lemon spinach."

These days, the sorrels are among the most-used perennial plants in my gardens, filling a few powerful ecological roles, being versatile in the kitchen and beautiful as well. A field of the wild sheep sorrel in flower will give a magical red glow to an entire landscape.

Soil and aspect: Full sun, on a wide variety of soils. Garden sorrel and French sorrel thrive in garden soil and know how to repel grasses and weeds. Wild sheep sorrel likes depleted soils low in nitrogen, which means it can thrive in places other plants can't.

Kitchen Uses: Leaves; tender and spinach-textured, but with a rich, lemon flavor. I have made whole salads from flowering sheep sorrel found growing on old agricultural fields. What a treat! A little oil and honey and some goat cheese to cut the acid. Sorrel soup is famous, especially cold, and I love to add its lemon flavor to gazpacho. The main drawback is that when heated the leaves brown and lose their lemon zing, but maintain their spinach-like texture.

Garden uses: Larger plants can act as fortress plants, often repelling grasses from beds. These may also act as excellent mulch-maker plants. Blood-veined sorrel in particular can produce nearly as much biomass as comfrey. Together a collection of different sorrels can form a fortress planting that gives an effect like a hosta garden, but thriving in full sun and with better eating qualities. Despite being so good to eat, these have been among the most wildlife-tolerant plants in my gardens.

Yarrow

Not the most palatable of plants in this book, but one of the most useful for medicine, beauty, and wildlife habitat. Perennial yarrow comes in a wide variety of colors and there is probably no easier plant that can make a big statement and long-lasting, insect-attracting blooms to the garden.

Zones: 3-9

Soil and aspect: Full sun, to a fair amount of shade, highly variable of soils, coexisting with grasses in rich soils in meadows and persisting on sandy sites.

Kitchen uses: Leaves are fairly bitter, but I will get a hankering for them after a long winter. They can be used to add a bitter component to ciders, ales, and teas.

Garden uses: native to nearly all of the Northern hemisphere, yarrow is a welcome native plant in a broad range, though many native plant enthusiasts scoff at it and weed it out of their native plant gardens. Very useful in the meadow garden as an insect attractor and mulch-maker plant.

Sun chokes, *Helianthus tuberosus*

A definitive multi-use plant: an edible landscape plant with the beauty of wild sunflowers, that's a hard worker in the garden, a high quality medicinal, a great vegetable, and also an extremely productive "carb" crop.

Yes, yes, the sunchoke, or Jerusalem artichoke, is all that and a bag of chips. Literally. They make really, really great chips.

But that's just the start of their culinary potential. While I'm personally not too fond of the cooked, unprocessed vegetable, the thinly sliced, raw tuber (in small amounts) is a great, crunchy stand-in for water chestnuts in salads. And when processed, it becomes nearly as versatile as the potato. I use it plentifully throughout the winter as one of our main "carbs," as the base for soups, in breads and pancakes, as chips, pasta, gnocchi, and latkes.

The only caution is that the tuber contains inulin, which can cause flatulence, especially in those unused to digesting it. However, according to a few sources, and our own experience, the inulin can be broken down with long cooking, yielding an ingredient better-suited for a party. (Unless you're planning a "farty party.")

But for homesteaders, permaculturists, and home gardeners who want to maximize the value of their space, the Jerusalem artichoke is even better, because it's the perfect example of a hard-working multi-function crop.

How do I love thee? Let me count the ways:

1. It's a perennial "carb crop" extraordinaire. Since grain agriculture (the world's main source of calories) is a major driver of climate change, soil loss, erosion, fossil fuel use and a host of other evils, perennial calorie crops are often considered the "holy grail" of sustainable agriculture. Yet, in the cold temperate climates the sunchoke calls home, there are few easy-to-grow perennial crops that are calorie-dense and can be the backbone of diet. Cue the potato, though it is typically grown as an annual on rotation to avoid pests and diseases. But the Jerusalem artichoke has roughly the same calories as potatoes (26 g of carbohydrates per cup for both) yet has few pests or diseases. As such, they can (and are generally recommended to) be grown in the same place every year, as a perennial.

2. It's often cited as one of (if not THE) highest-producing vegetable crop/acre we can grow in many cool and cold temperate climate areas, likely out-yielding potatoes in most regions. While this is frequently repeated by Permaculturists, agronomists, extension offices and other industry sources, some readers may be interested in digging a little deeper to see what data that backs this up. While output numbers are somewhat dependent on season and region, we do have some data to provide decent comparisons. For example, a study in the 1980s in Oregon found an average of nearly 64,000 lbs/acre of Jerusalem artichokes, without irrigation. The authors report that outcome was in keeping with research conducted by Washington University. Meanwhile, though Oregon is one of the highest-yielding potato regions in the world, with modern high yields (actual yields, including some irrigation) at 53,000 lbs/acre. Purdue University's numbers are more conservative, citing University of Minnesota trials with an average production of 30,000 lbs/acre for sunchokes, while reporting statewide yields of 20,000 lbs/acre for potatoes. Although there is some disagreement across sources (with some placing average potato yields higher, and others placing maximum corn yields higher) the general consensus is there's rough calorie equivalency between potatoes and corn, meaning that Jerusalem artichokes can produce more calories/acre than maize, the world's most important calorie crop. While sugar beets and sugar cane outproduce corn and potatoes by calorie when converted into sugar, Jerusalem artichoke may compete or even out-perform these in terms of sugars and total fermentables.

3. That's just the tuber! Yield increases if you count the stalks, which are being studied as an animal feed, an ethanol source, and a compost crop. In GrowBioIntensive gardening, it is both a "carbon crop" used to grow fertility and create a sustainable farm system, and a "special root crop" used to maximize the nutritional value of the garden. (Jeavons, "How to Grow More Vegetables.")

4. We consider it an excellent mulch-maker plant. In Permaculture and forest gardening, mulch-maker plants are used to grow mulch right in the garden. Because Jerusalem artichoke is hardy, perennial, and produces large amounts of mass, it is an excellent source of mulch in the garden.

5. It's also a great "fortress plant." These are plants that exclude other plants or potentially even pests. In the wild, I consistently see Jerusalem artichoke shading the ground and producing stands without other weeds. Often at the forest edge, it succeeds into grasses expands the forest. It does the same thing in our garden.

6. It's a beneficial insect attractor, wildlife plant, and companion plant. According to the Ladybird Johnson foundation, it has value to birds, pollinators and other wildlife. It attracts a wide variety of native pollinators, moths, and bees, including specialist bees that rely upon sunflowers.

7. As a common plant of thickets and forest edge, it can fill in gaps in hedgerows and dead hedges while woody perennials get established, then it tends to fade as these larger plants grow in. I've used this feature at Lillie House and a few of our related client and student sites.

8. It's salt-tolerant.

9. It's medicinal. It is a well-known source of inulin, used to help diabetics and others with

problems related to insulin. As a prebiotic it may improve the composition of gut microflora. But while Jerusalem artichokes are a great source of inulin, most inulin available in the U.S. is imported. Studies have documented Animal studies have found positive results on rats pigs, and even fish. One study found that a Jerusalem artichoke diet improved the digestion of pigs and made their farts smell "sweeter."

Ah, the sweet smell of pig farts.

Considerations

To Cut or Not to Cut?

Cut out cutting out the flowers of crop plants. The internet is filled with old farmer's tales about cutting the flowers off Jerusalem artichokes to improve yields. As with most other crops, where it's either completely false or economically disadvantageous, removing the flowers dramatically reduced yields in both the M.U. study and the Oregon State study. In fact, yields were more than halved by late-season removal of flowers.

Weediness?

While Jerusalem artichokes have few pest or disease issues, are salt-tolerant, drought-tolerant and may require less fertilization than other crops, the common warning is that they may become "weedy" or "aggressive" in some situations. Some even call them "invasive" even in ranges where they are native plants!

One may wonder how, with such an "invasive" plant growing natively across much of North America, anything else ever managed to grow here! Or why it's so relatively uncommon when compared to so many plants that don't bear the burden of the scarlet I.

Or perhaps it isn't so surprising when considering that many sources, including the Illinois-Wildflower.info noting that it more commonly persists on disturbed soils, with other plants succeeding in time without tilling.

We can learn the same thing from the dozens of reports online of gardeners failing pathetically in their attempts to pull, dig up, till or otherwise "weed" Jerusalem artichokes. This does NOT work. It only makes them happier. However, I have been monitoring several wild patches for years in my area, and can report that without the encouragement of "weeding" them, they show no signs of spreading. Some patches appear to go into decline after a few years without disturbance. The same has been true at our site, where a few of our prize patches have disappeared completely.

Other patches have been killed by animal pruning. Recall the huge negative impact of late-season removal of just the top third of the plant? Imagine what happens when you remove the whole stalk earlier in the season. In fact, it is a little *too* easy to kill plants completely by poor timing of cutting. Mowing is a sure way to control their spread.

My advice is to plant them into naturalistic systems like forest gardens, where they will be controlled by the process of ecological succession, or plant them where they can be controlled by mowing. A linear food forest or edible hedgerow may be the perfect place for them. And for heaven's sake, if you're trying to get rid of them do not try to remove them by "weeding!"

Storage

The best place to store Jerusalem artichokes is in the ground. They tend to go rubbery in refrigeration or even cellaring. But patches can be dug throughout the winter where soils do not freeze, and maintain a good texture and flavor all winter.

Harvest

We leave most of our sunchokes until January, as I believe the sugars are highest by then. Tubers harvested later are considerably sweeter in my experience, than those harvested in the fall.

Growing

Part shade to full sun. Very adaptable. Mulch on sandy sites.

Zones: 2-12!

Very adaptable to different soils, but thrive in rich well-drained soil.

Plant disease-free healthy tubers during the dormant season, spring or fall. The Oregon study found maximized yields at 17 inch spacings, but closer spacings did not significantly reduce yields, making them ideal for perennializing or inclusion in naturalized ecological growing systems like food forests or hedgerows. Yield was best in full sun, but chokes can grow and produce in fair shade, yielding moderately in fairly dense shade where few other vegetables would have produced.

In my gardens, they have persisted and produced well within the dripline (under the leaves) of black walnuts, suggesting them as potential juglone-tolerant crops.

Sunchokes can be found producing even on depleted soils though they produce best with fertilization. One inch of compost or four inches of organic mulch is generally recommend for organic growers, along with minerals where soils may be depleted.

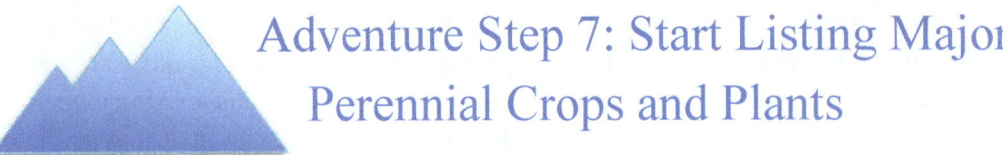
Adventure Step 7: Start Listing Major Perennial Crops and Plants

Next on our adventure of transforming the landscape is to start getting an idea of which plants are a priority, starting with woody perennial crops, trees, and shrubs. In the next few chapters we'll explore plant "guilds" so there's no need to list every plant, unless you want to. It's better to think in guilds.

If you're starting a new garden, getting a collection of high-value perennials should be a top priority. As we explore guilds, add those to your list.

Later, we'll suggest two ways to compile plants. A sort of geek path, and a more intuitive path.

The geek path involves studying the materials on guilds in this book, and compiling plant lists that fill many of the guild characteristics and "roles" we'll discuss.

But if that doesn't seem like the right approach for you, there's a second more intuitive path. Which is to just load up on biodiversity. If there is enough biodiversity, nature will probably have the tools to make a guild.

A general best practice that has emerged out of the Permaculture community is to aim for at least 30 species in a small garden, of say 300-500 square feet.

For better results, use a mix of annuals, perennials, and plants that thrive in different seasons.

And for even better results, if you're only going to think about one issue, study the section on guild matrixes. A planting with a solid guild matrix has a high likelihood of becoming both stable and very useful.

But for now, it's good to just start a plant list for the garden if you haven't already. Then we'll have something to develop as we work through the rest of the book.

Some Starter Edible Perennials For a New Garden

For more information, search www.pfaf.org

Perennial Alliums (the onion and garlic family):
Egyptian Walking Onions
Allium unifolium
Chives
Garlic Chives
Welsh Onion
Welsh Red Onion
Perennial Garlic
Ramps
Perennial Leek

Perennial Vegetables:
Asparagus
Perennial Bulbing Fennel
Lovage
Turkish Rocket (perennial broccoli!)
Jerusalem Artichoke
Ground Nut *(Apios Americana)*
Air Potato (Chinese yam)
Fairy Spuds *(Claytonia Virginica)*
Camas bulbs
Fiddle-Head Ferns
Crosnes (Mint Root)
Sea kale

Greens:
Star Herb Minutina
Perennial Kales
Sorrel
Blood-veined Sorrel
Peach-leaved Bellflower
Perennial Endive
Salad Burnet
Marshmallow (a perennial cooked "spinach")
Anise Hyssop

Herbs:
Oregano
A large collection of Thymes and Creeping Thymes
Yarrow
Valerian
A large collection of Mints
Comfrey

What to order: seeds, starts or plants?

Since we will be ordering a lot of plants, things can get expensive quickly. Seeds can save money but only if they germinate and grow well. Buying plants may save time, but in other cases seeds actually grow faster! So, what is best for each crop?

Establish easily and grow better from seed in site. (These are probably far more cost effective from seed.)

Annuals:

Arugula	Cucumbers	Rutabega
Radishes	Wall rocket	Almost all root veggies
Lettuces	Tomatoes	Mache
Cabbages	Bac choi	Nasturtiums
Calendula	Beets	Spinach
Beans	Chard	Sunflower
Corn	Kale	
Squash	Peas	

Perennials:
Turkish rocket
Anise hyssop
Salad burnet
Sorrel and blood veined sorrel

Buy "bare root"

All named fruit and nut trees
Most seedling fruit trees

Grow from tuber, root cutting, or thongs

Perennials:
Sunchokes
Crosnes
Potatoes
Sweet potatoes
Skirret
Garlic
Walking onions and other Alliums
Most Perennials
Sea kale

Home starts from seed (easy enough for home growers)

Early tomatoes* (research says usually more productive grown from seed, but earlier from plants.)
Peppers
Basil
Annual herbs
Eggplant
Broccoli
Cauliflower
Brussels sprouts
Onions
Leeks
Sunflowers?

Probably just buy potted plants

Anything cool you want to try.
Peppers
Early tomatoes?
Perennial herbs
Chives
Monarda, mints, and other spreading perennials

Any named perennials as these must be grown from cutting

Chapter 27: Patterns for Sacredness

When I say "sacredness" I do not mean it in any religious sense. One doesn't have to have religion to have entered a space and felt the presence of its sacred place, and neither do we need any belief to invoke such powerful magic in our own landscapes.

There are probably many reasons why we feel places as sacred, and I believe some are hardwired right into our DNA.

Many birdwatchers have marveled at the complex patterns of birds flying in the sky, and how they seem to move in a predetermined unison which creates swirls and gyrations wrapping like ribbons through multiple dimensions. And yet, all that apparent complexity is the result of a few brief rules of flight programmed into their DNA: keep wing space at a certain distance, keep to a certain speed, if wing space deviates, change direction, etc.

And mice are programmed genetically to keep to the walls for safety, programmed with a survival mechanism that keeps them safe from brooms and helps them quickly disappear into dark crevices. Those dark crevices must just feel good and safe. Cats on refrigerators, and rabbits in cubby holes must feel a similar feeling.

What do humans feel when they come into a new valley, with cool, clean water bubbling up from a well, or perhaps healing hot springs, mountains standing guard to our backs and a long open view to easily see any threats that approach? Uphill, lush forests trickle down nutrients into a valley filled with the thick grasses that speak of rich soil, and an overflowing with flowers that hint at an abundance of edible plants? Something silent but clear reverberates through us, "here, this is it, you have found what you have been seeking." Have you heard this voice yourself?

Invoking the magic of intention

Another sense of sacredness speaks to us of how to tend place. It's our instinct for how to build a secure home. It is the feeling we get when we see rich, "chocolate cake" soil, the clear waters that we get from a well-balanced ecosystem, the healthy growth of plants in fertile soil with deep mulch around.

It is said that Americans in the SouthWest farmed in areas where the soil was practically sand, and rain was scarce such that Europeans wouldn't even attempt it. How was this possible? By invoking the magic of everyday sacredness. Each seed of corn is sacred. Each plant is sacred. The earth, the place itself, the preparation of the soil are all sacred. And with that sense of sacred attention, abundance can grow in the desert.

If you invoke the magic of everyday sacredness and truly believe that your dining ware is truly sacred, you will never have dirty dishes pile up again. When your living space is sacred, it will always be kept, and the effort will not feel like a burden but a ritual.

Sacred garden tools, heirlooms with the signs of human craftsmanship and wear, will last well for generations.

The Two Directions Art Can Move the Mind

To invoke sacredness of place in the landscape, we need to understand the two directions art can move the mind in. This sounds complicated but it is one of the most practical things in this book.

These two states are quite different, yet both can be sacred, and both can be found in the same space or same work of art. These two directions are inward, into our fantasy world, or outward, into the present moment of the real world and our perceptions. Thought, memories, fantasies of the future are all inward. Direct experience of what's going on around us is getting "out of our heads."

"Induction" or hypnotism

With most art, say when we go to a movie, what we are going in for is a bit of hypnotism, hypnotic induction. We are being inducted... out of our bodies, out of the present moment, and into the fantasy world on the screen. We enter the theater, with the sticky floor, find our seats, hear the chatter and laughter of voices around us, and then the lights dim. If the movie is half good, the theater, the annoying guy talking to his friend three rows up, the crackling of the speakers and smell of popcorn all but disappear as we are absorbed into another world. We will feel the feelings of the characters, perhaps even smell what they smell, weep for their losses, and rejoice at their victories.

And many paintings move us in the same direction. We look at the painting of the cabin in the woods and have a little fantasy, enter into that painting, imagine quitting our jobs and going off to live in the mountains at high altitudes.

Music, poetry, and yes, a garden can accomplish this same act of hypnotism. When we enter a garden we might go on a bit of fantasy… perhaps to that cabin in the woods, or perhaps we imagine what it was like to be entertained in the gardens of Versailles. This type of fantasy is common in historic gardens and landscapes, but is also sometimes invoked in ornamental gardens.

Many people have told me that as they entered the garden gates at Lillie House, they felt they entered into a vision of what a sustainable, ecological civilization could look like. Others said they entered immediately into a fantasy of what their own sustainable futures could look like. That garden gate was an induction into a powerful state of potential!

But… in the garden
As we sit, the cool breeze blows,
The spell is broken.

Wakefulness

In traditional haiku poetry, there is a "cut," the kireji, a change between topics, which is said to wake the mind out of fantasy, out of hypnosis, and into the present moment. This little break out of hypnotism into the present moment is one of the "tricks" that good haiku can accomplish.

This is the other direction art can move the mind in, out of the head and into the body, into the cool breeze, into the warm sun, into the smell of cherry blossoms, into the present moment here and now.

This direction of movement was common in Asian art, in the tea ceremony, and in Japanese gardens, and caught on in Western art after being introduced from the East.

One might look at a painting by Norman Rockwell and think "what a beautiful fantasy world." But one might look at a piece of Buddhist art, then walk away and think "the rough character of the sidewalk is beautiful. Just like this."

We can invoke that direction in the garden as well. Through direct senses. Through experiences that wake us up to the basic goodness of life. That we have a body and we can be here and experience the freshness of the roses, the taste of fresh strawberries. Simple textures and forms tend to wake us up, rather than to invoke a fantasy. (By the way, suburban conformity and conspicuous consumption are always a fantasy, always hypnotism, but never a very helpful one. The conformist gardens of HGTV with cheap factory materials hypnotise us into accepting a dysfunctional world with fantasies that we, in our McMansions, are fabulously well-to-do, and that whatever ills the world faces, our lifestyles and concrete pavers are worth it.)

In a great space, we may have a whole experience of moving inward and outward as we pass through various garden gates, invoking a fantasy, then waking up to the present moment again. It is possible for us to invoke a fantasy of a better world, wake us up from the fantasies of consumer culture, wake us up to our own power, create a fantasy of a better life, wake us up to the beauty and potential that we have, right here and right now, in a garden.

The Qualities of a Sacred Place

If we have a biologically programmed sense for a supportive environment, what are some of the qualities that we sense?

The Sacred Grove

Trees are a sign of decent soil, that there will be game and other species, that there is adequate water, and protection.

Trees make humans feel good. The presence of trees in a neighborhood has been shown to increase human happiness. Trees enhance our behavior. Perhaps because they make us feel secure and supported, they shift us out of a scarcity mindset of struggle, reducing crime, and increasing the value of property. And trees heal us. Time in a forest has been found to enhance the immune system, reduce stress, and heal illness.

Many of us are especially drawn to old, wise, tree-beings. There is perhaps an evolutionary reason for this sense of kinship to these eldest of our living relatives.

Trees take care of each other. They have been found to share resources with individuals who are lacking. This seems to go against the idea Darwinian competition and survival of the fittest. But trees rely on a healthy system, since they can't just leave for a new one. They evolved an understanding that if one member of their community goes without, they may become a risk for pests or diseases to take hold and become a danger.

Old trees have been demonstrated to possess a library of experience for dealing with pests and disease issues that may occur. There are few problems ancient and venerable trees haven't encountered, and used their chemistry skills to solve. They can concoct chemicals to poison or weaken dangerous organisms, or failing that, to attract allies that might attack the problematic species. Trees are master chemists.

Then, these ancient trees have been found to pass on these chemical tricks to younger trees. So a space with ancient trees is sure to be one of greater stability and resilience, Perhaps why we humans are instinctively drawn to them. A treed ecosystem will be a healthy and resilient one, and we will have powerful friends watching over us.

Therefore, we should create treed landscapes wherever possible, and honor ancient venerable trees, never removing these powerful protectors from our ecosystems.

Patterns related to tree-crop systems will be helpful in making the most of our tree friends. Nurse logs and mulches can be tree crops in addition to fruits and nuts.

If you have ever come upon a grove that is thick with ancient trees rising far above their usual height and blocking the view of the sky with their cover of intertwining branches, the loftiness of the forest, the seclusion of the spot, and your wonder at the unbroken shade in the midst of open space will create in you a sense of the divine. Or, if a cave made by the deep erosion of rocks supports a mountain with its arch, a place not made by hands but hollowed out by natural causes into spaciousness, then your mind will be aroused by a feeling of religious awe. We venerate the sources of mighty rivers, we build an altar where a great stream suddenly bursts forth from a hidden source, we worship hot springs, and we deem lakes sacred because of their darkness or immeasurable depth.

—Seneca the Younger, Letters 41.3

Water, the Sacred Chalice

My Permaculturist friend, PJ Chmiel, has said that the small pond he added to his forest garden became one of his favorite spots, and that he enjoyed just watching the wildlife and insects that showed up to partake of the water. Never underestimate the impact of water.

Water is life. It belongs in every landscape.

It's no wonder that water is one of the most common and earliest additions to the garden, probably because water is the single most important thing humans must have in a secure place. And wells and springs have been sacred around the world for all of human history. Of course, that sense of sacredness would call us to tend and protect the well, taking care not to pollute the source of life.

Perhaps if we can reinvoke this sense of sacredness of water in our own landscapes, it will echo out through our culture. We can once again become tenders of sacred wells.

Bring water into the garden, even if only bird baths, and small ponds. Make it a conspicuous feature and put it where it can be seen and appreciated. Even the sight of water alone can be refreshing and cooling on a hot day. Not just for us, but for the many beneficial birds and insects that will gather around this source of life.

In a later chapter, we will explore patterns for designing water features and ponds, including herb spirals, ponds, dieu ponds, sacred wells, and natural swimming pools.

The Mountain Protector

Mountains, too have been considered a source of security, giving us the higher ground in times of conflict. **We humans feel most comfortable when we have such a source of strength behind us, opening to a V-shaped view in front of us.** Even research on office dynamics has found this is true, and chair positions with security behind can put people at ease or without that, put them off guard.

So if we wish to feel at ease and secure in our gardens, then **we should position seating with a secure symbolic mountain or wall behind and flanking us, opening to a beautiful view with a focal point.** This protected feeling will make us and our guests feel more relaxed and at ease.

Enclosure

Similarly, enclosure gives us a sense of ease and security. For early humans looking to protect crops, the garden enclosure was a necessary defense and pest-prevention—certainly our most important pest-prevention!

In fact the word "garden" comes from this "garde," or enclosure protection. The latin name Hortus Conclusus referred to an enclosure with plants. The European mosaic landscape with hedgerows is another example that makes people feel secure and is naturally beautiful to most observers.

This probably also mimics the feel of swidden agriculture, the way our ancestors gardened in forest clearings. Enclosures create a garden of rooms, which also provides us with a feeling of exploration and adventure.

Therefore, the best gardens have zoned enclosures to protect crops and provide for a feeling of security. Forest gardens, tapestry hedges, and hedgerows are particularly useful patterns for creating this sense of enclosure.

Flowers

Flowers signal food. A landscape of grasses may not be of much use to human animals; and neither will one of rocks or scrabbly stubble. But plants that flower in a gregarious way must store energy for that display. This is usually in the form of large, and often edible seeds, or in root storage organs like tubers and bulbs. Humans coming upon a show of flowers must have known that there would be plenty to eat. Often such plants developed relationships with humans, calling to us with fruits and vegetables to eat, tempting us to save and sow seed.

Humans are not the only beings cared for by flowering plants. Most of the hardest working beneficial organisms in the garden, the birds, pollinators and predatory insects are drawn to feast on their nectar, so a landscape with flowers is also one in which our crops will be more secure.

Plant flowers even in the most utilitarian productive landscapes and the humans, wildlife and crops will all be better off for it.

Plant a season of flowers for the best result. Plant them in guilds for ease of care and integrate showy ornamental edibles throughout.

Biophillia

Biophillia is a fancy term for the love of nature expressed in design. To share the love of nature through art, we may try to include "biophillic" elements, such as nature images, plant images, tree or leaf designs, and so on.

These can be expressed in garden art and statues, or in things like **symbolic path layout**. Many Permaculture gardens are made in the layout of a tree or leaf shape, which can be both aesthetic and functional.

Therefore, consider including elements that depict nature as beautiful in the garden, in art, in ponds, in paths, etc.

Sacred Geometry

The alchemical symbols for transcendence and adventure. You'll find these glyphs hidden throughout this book. An approach based on transformation and adventure will help you make realistic, lasting, positive change in your life, landscape, and world.

Among biophillic elements are **signs of sacred geometry, certain proportions and geometric features that seem woven into the fabric of nature itself.**

While some see these as the fingerprint of a creator, those of us who don't worship a creator instead marvel at these as a part of the base code of the universe, something that you and I have in common with our dearest loves, our greatest enemies, with the largest tree, the smallest bacteria, the alien-seeming life in the depths of the sea.

Some of the most recognizable proportions are the spiral, such as that in the tiny nautilus and in the spinning of galactic forms, and the hexagons of the honey bee hive, and the cell walls of plant tissues.

Sacred Symbols

Sacred symbols have long been used in garden layout to add spiritual depth and resonance to places. This goes back to Persian pleasure gardens, perhaps the oldest Permaculture gardens, which were laid out with a long symbolic central aisle. These usually led to water or a simple alter. Christians adopted the form, but added an aisle to make a cross. Most medieval gardens were laid out in a cross so that heaven was always close at hand.

One of my recent gardens, the garden at Lillie House, was laid out in an ankh, a symbol of Earth-worship and eternity, which for me translated to the idea of a permanent culture. It still immediately invoked a formal garden, but standing at the garden gate and looking up to the house, many guests exclaimed "an ankh!"

Beauty in Abundance

Sacred Symbols for Path Design

Several spiritual symbols in forms that would make nice paths and garden beds: the turtle of the world, another symbol of endurance and permanence; anarchy; secular humanism; the ankh; the Universalist flame; the goddess; the Celtic cross; the Star of David; and the crescent of Islam.

The Deep Resonance of Stacked Symbols

We can create a great depth of meaning by using multiple related layers of symbols and experiences.

This is what gives much of great literature, art, and music its emotional depth and significance to us. The great master artists of the past used this technique, creating paintings with layers of allegory and secret messages, political and emotional commentary, messages about the joys and tragedies of life.

So deep were these messages that we can discuss them still hundreds of years later. Great novels do the same thing, quoting and echoing other great works so that the emotional depth of those works gets added in for the reader to this new experience.

The symphonies of Mahler, and most Romantic composers were like this, too, quoting from pieces of popular songs, folks songs, other symphonies, to connect with the listener on these levels. Of course, modern popular music and hip hop in particular has become fond of this technique, playing little games with the listener to see what layers of reference we can catch.

And of course, all of these used spiritual and religious symbols to connect with the audience on the deepest level of all.

Of course gardens can do the same. In fact, these layers of meaning are an essential part of how we experience any place. We may know the politics of the town, the universities there, the battle that was once fought here, a famous well or monument nearby. These get added into our experience of the spirit of the place.

By doing the same thing in our gardens, we can connect to our own landscapes on this deep, powerful level. We can build an experience for ourselves that connects us to our spiritual goals, our values, our loved ones, people we've lost...

I can't tell you what will have that resonance for you. But I can suggest you seek it.

And as with those great novels, works of art, and music, don't make it too obvious. Most people find overtly religious yard art to be tacky, for example. But if these messages are somewhat hidden, then we get to have a little adventure of seeking our and uncovering meaning as we appreciate the landscape.

Right View: Visual Mandalas

Painting, Rebecca Stockert

According to Chogyam Trungpa in "Darma Art," and Guessepi Tucci in "The Theory and Practice of the Mandala," the symbolism of the mandala is the flow of energy from a central point. As a meditation tool the use of line to draw the eye into a center point, and perspective to create an illusion of depth, drawing the eye on a journey through levels or dimensions....

If the effect is good, we may get a whiff of zen from it.

The view of the garden from the street at Lillie House (along with several other views) were based on the mandala effect.

Knowing we were on a busy street, the goal was to make sure the garden had a strong visual impact, and that all who passed by got a shot of permaculture and of zen. I made several sketches of the garden from the front, and the view from the moon gate.

The trees were used to create false perspective, making the garden appear deeper than it actually is. Color and form echoes and near symmetry draw the eye inward.

From the porch, the view opens toward the garden (**Mountain Protector**) because of the false perspective of the trees, a pattern that makes people feel safe, and looks down toward a moon gate as a second mandala.

Stacking these sort of aesthetic and spiritual patterns—color, form, the ankh, the jardin de Curé, the physic garden, the mandala—into a garden creates a space, and that is rich in meaning and resonance.

While a mandala garden is a symbol of a mandala, we can actually create garden views that function as mandalas. and give ourselves a little whiff of zen if we hit the mark.

Beauty in Abundance

Chapter 28: Patterns for Garden Layout

Working too hard in the garden? Learn about zones.

One set of terms we talk about a LOT in this book is "intensive" and "extensive." We talk about it a lot because for the whole history of humans producing food this has been the number one way of dealing with pests, with irrigation, with weeds, with diseases, and overall, for reducing work. Still today, it is probably the number one thing we can do for all those problems, yet we moderns usually design homesteads and gardens without giving this most powerful tool a single thought!

If your whole family runs away when there's weeding work to do, this could be the number one way to restore family harmony!

This is a way of talking about how much energy, resources, and work we put into a garden or farm system. Intensive systems get a lot of energy and usually resources into them, but hopefully get a lot more out. Extensive systems are low-energy, and hopefully low maintenance, but probably will have less produce harvestable for humans. This last bit is important, because almost always these systems will have more total calories. But most of these calories may be going to non-human species, and we're just getting a fair share. This is GOOD because these types of gardens help protect biodiversity and sequester carbon while still feeding humans.

Intensive systems include biointensive gardens, tilled gardens, most vegetable gardens and most orchards.

Extensive systems are mostly nature managed, and include food forests, hedgerows, and edible meadow systems. Basically any edible ecosystem. The best gardens, farms, and homesteads I've seen all use a mix of intensive and extensive systems to maximize productivity and minimize pests, weeds, diseases, watering, and labor.

If you are feeling overwhelmed by garden work, you need more crops in extensive systems, and less in intensive. You'll certainly decrease your work, and likely increase your productivity, too.

What traditional gardeners knew to do was to put the crops that required the most energy in one place, and then protected them. For example, they took the plants that deer would really love to eat, pulled them out of the field where the deer were, and put them up close to the house in a small area of enclosure. It's easier to protect a small intensive garden than a whole field! Things that were sensitive to weed pressure they put in a spot for easy weeding. Then they grew things that were wildlife tolerant out in the field. They grew things that could grow in weedy thickets in weedy thickets and things that could tolerate forests in food forests that required very little work.

And this is especially important if you want to grow with less plastic, petrol, and poison in the garden. This is also the most important tool for reducing work in ornamental gardens. It just requires some knowledge of these systems and what crop goes where. Getting these things nailed down is the best thing you can do to reduce your garden work time, period. This is the key to how I grow more food than a family can eat on a small space with just a few hours of non-harvest work per week.

The Fine Stopper Starter Pack

One of the big reasons for this book is the large number of clients I've had who deal with complaints from neighbors and city fines as they try to create ecological landscapes, food gardens, or regenerative farms.

I'll tell you from my experience, getting a fine usually causes indignant homeowners and gardeners to double down, insisting that nature itself SHOULD be seen as beautiful, or that it's their property and they should get to say how its used. I have never once heard of this winning the case or stopping the fines from coming.

Instead, I recommend looking at it as a design constraint that will make our designs better and more effective at communicating a love of nature to people.

Keep in mind, most people have been brought up in this long war against nature. To them, the order of nature looks "chaotic!" And there may even be some good reasons for it. If humans are nature, and the home garden pattern is how humans naturally human, then we as humans may intrinsically recognize some signs of human use as a sign of safety and even ecological health. These may communicate that there is a healthy transition point from human management to natural management. It may communicate abundance, and safety. So, why shouldn't we be kind and attempt to show some of that to our neighbors in our human habitats? We can do so without ruining "nature."

And yet, for busy farmers or those busy with the work of transforming a landscape from lawn to something healthier, there may already be too much work and too little time to waste making things look "pretty." Therefore, we need patterns designed to actually help save time while creating beauty. In the early days especially, we may need to prioritize these.

Fine-Stopper Patterns:

Planting bulbs in meadows, guilds, and other wild areas. A fall planting of spring bulbs before letting the grass grow wild can be one of the best ways to jump start a meadow into looking like an aesthetic feature instead of blight. First thing in the spring, the neighbors will see flowers.

Signage. Signage like the Landscape Transformation Recognition sign will help show the neighbors what you hope to accomplish ethically.

Formal paths, wild plantings. This is a classic gardening pattern. The formal layout will make even pure wild meadow look like intentional garden. Geometric patterns and clean lines can help.

Plant flowers out front. This is a suggestion from Permaculture Designer, Jenny Pell. Putting flowers in the most easily visible spaces can help signal that there is indeed a garden happening and not just blight. In a Permaculture sector analysis, we should identify human traffic areas where we want to take special care as "aesthetic sectors."

Aesthetic zones are a related concept. If we want to make a big visual impact yet do so on limited work hours, taking a zoned approach is absolutely vital.

A color scheme or unifying feature. Possibly the most immediate and recognizable pattern will be a color scheme of sorts throughout the garden. A bold color block repeated around the garden, or a charismatic plant will allow a viewer to look over a large area and take it all in as "a garden."

Edible, ornamental hedgerows. Enclosure can be powerful when working in human areas and neighborhoods. We can hide the messiness of wild nature behind a beautiful flowering hedgerow.

Hard and soft edges. This one pattern again can make a huge visual impact while actually saving us time, instead of adding more burden. We use this pattern to decide where we want to keep edges tidy and neat, and where we want to let nature take over more.

Use a theme. If all else fails and the neighbors still aren't happy, start using one of the themes in this book. Most humans look at the "big picture," and if what they see conforms to some idea, then they will see THAT, not the slightly messy details. If they see a "romantic cottage garden" you can get away with a lot of natural wildness.

Formal Layout, Wild Beds

For many landscapes (and families) the major aesthetic problem is balancing tidiness against ecological health and ease of maintenance.

Despite the tendencies of some of us wilderness-loving hippies, there is a good reason for tidiness. To some extent, it is the "signs of care," the visible impact of human energies on the landscape that make us feel safe and at ease, help us know that dangers have been taken care of. These signs of tidiness also tell your neighbors you have dealt with the danger of fines from municipal planning staff.

Yet, tidiness comes with a price to both human labor and ecosystem health. What looks "ordered" to humans is chaos to nature, and vice versa. So one of the best patterns for solving this problem is to use formal layouts in places, but to allow wildness and natural order within the beds. These can be straight lines, or curved so long as some are defined by "**hard edge**s" so they look like clear signs of human care. This gives us "order" to human eyes, yet also gives us the benefits of ecosystem health and ease of maintenance of natural order.

Hard and Soft Edges

As we look to strike the right balance between human tidiness and natural order, one problem we encounter is knowing where to put our energy. A nice clean, well-tended edge is a clear visual sign of care and neatness. Yet, maintaining them can be a massive amount of work!

Hard edges are those that are maintained by humans. These can be dug with an edging spade, edged with bricks, or cut clean with a weed whacker. Soft edges are those maintained by plant communities. Usually these will be a wild mix of plants, and perhaps an edge is simply mown or scythed next to it.

The idea of zones gives us one of the best tools for solving this problem, recognizing that areas that people will see most often will have the biggest impact if they are tidy, while there is little need to tidy up a line no visitors will ever lay eyes on.

If the guests pull in the driveway and see nice clean lines, and then walk up to the door along nice clean lines, they will look around and think "what a nicely maintained landscape!" It won't matter if all the rest of the edges are left to the plants to tend!

So to solve the problem, choose areas where guests will see the most for nice, clean edges, and let the rest go wild.

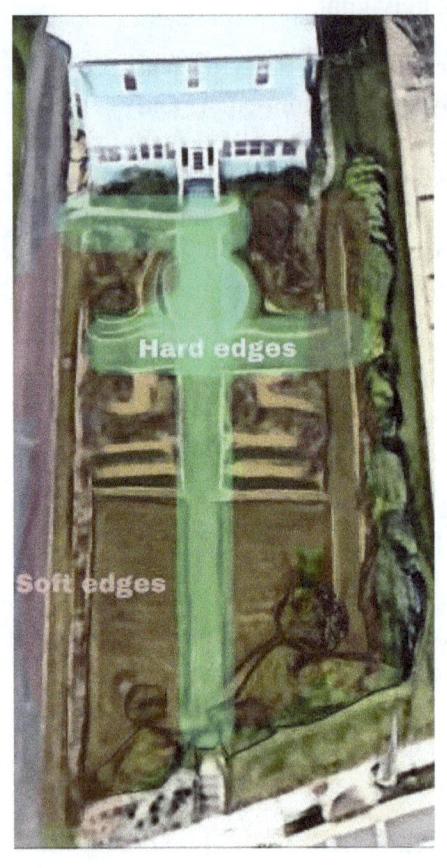

The Herb Hedge or Edging

A classic and instantly recognizable feature of formal, tidy gardens is the box edging in geometric shapes around garden beds.

There is probably no more instant sign that you are in a "garden." Yet for the ecological gardener, these box hedges have a lot of problems of maintenance, and little utility to pay for that upkeep.

In many climates, box performs very poorly, is prone to disease, is slow to fill in, and then requires a lot of upkeep labor to keep it pruned! It is prone to being overrun with weeds and grasses, and once weeds and grasses have infiltrated, it is incredibly difficult to hand weed them out.
What do we get for that work? Nothing.

However, we can get the same look with much less work by using herbs like oregano. Oregano is beautiful, with correct timing can look like formal box, and is great bee forage. It is a great "fortress plant" that keeps grasses and weeds out of garden. It fills in quickly and can be easily maintained. And each time you trim it up, you get to fill the freezer with ice cube trays of oregano pesto.

Other herbs can be used to a similar effect in the right conditions, and can include thyme, mints, and germander.

Chapter 29: Basic Patterns for Garden Color

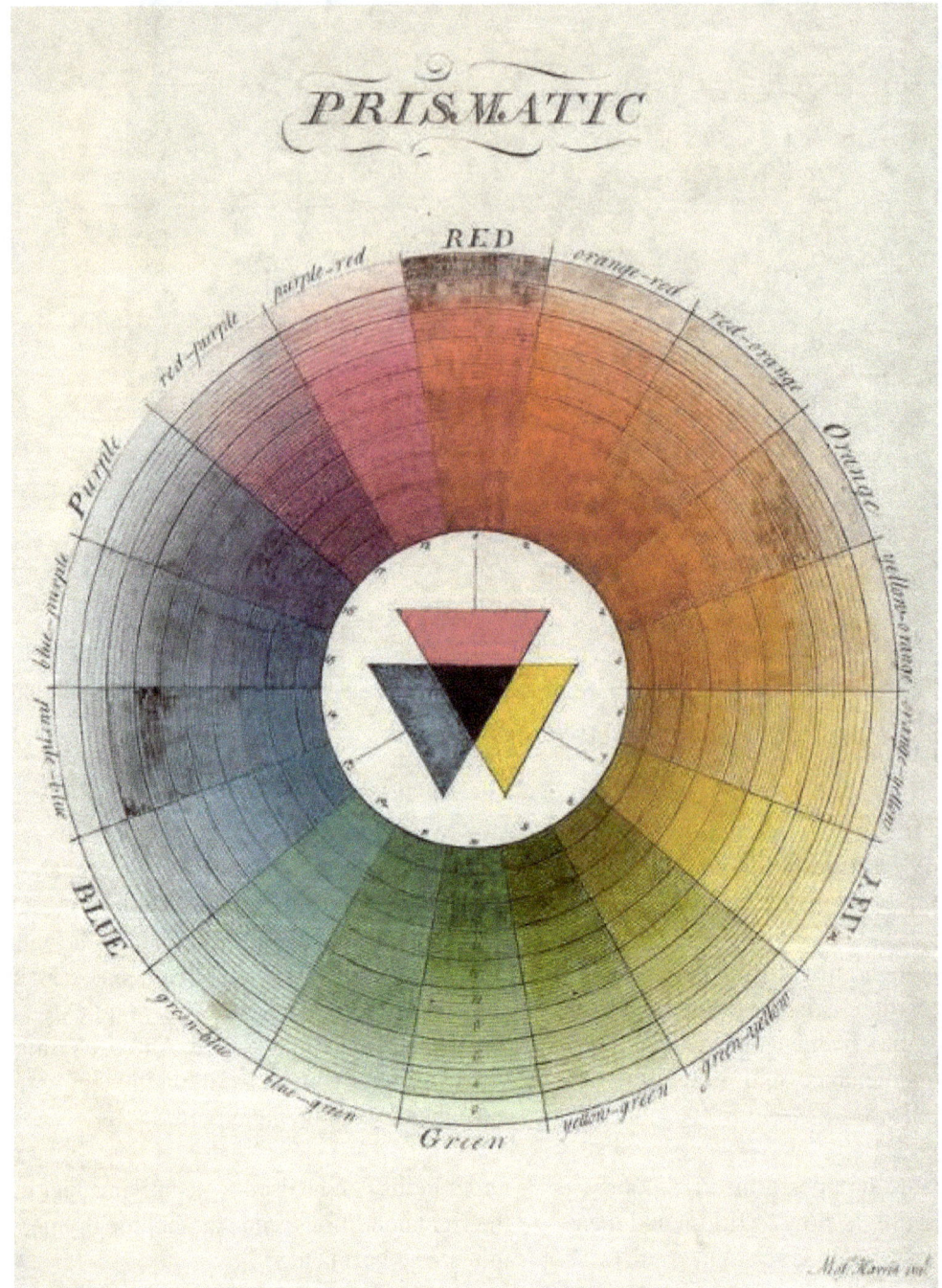

Moses Harris, The Natural System of Colours (1776)

For a lot of people doing natural gardening, farming or Permaculture, or even professional landscaping, it comes as an absolute revelation that they could make intentional use of color in the garden and landscape. And it's an even more profound revelation to discover that this can be a powerful tool for affecting health, performance, community, healing, crime, spending and a whole host of other behaviors. Perhaps surprisingly, color has even been well demonstrated to even effect the behavior and well being of livestock and wildlife.

And because color is so immediately recognizable, it's one of the top elements on our "**fine-stopper starter pack.** A garden with a good color plan will never look like a passive-aggressive attempt to meet the bare minimum requirements to avoid fines. It will look like it was conceived with beauty in mind.

Oh what joy awaits! There is a whole world of expression and effects that open up when we start learning about and applying color in the landscape.

Imagine the art of painting without the intentional use of color, or even theater or movies. A color palette makes the whole difference in how a movie can feel. Just imagine a Batman movie in the upbeat colors of a Disney cartoon, or vice versa. Color has a huge impact on our emotions.

In the Buddhist tradition of "Maitri," color is sometimes used to promote different emotional and mental states, and modern science has backed up this old practice.

For the most general example, cool colors, blues, greens and purples tend to invite a calm, rational frame of mind and overall healing. Warm colors, blends of yellows and reds tend to cause an energetic mood. This can be both positive and negative. It can give us energy when we are low or make us feel happy or friendly. Red can even cause people to react with greater speed and force. Or, it can make us feel agitated or angry. Colors can cause us to spend more, be more patient, heal faster, feel more romantic, or perform better on tests. We'll go into a little more depth on color and emotion after we go over some important basics of color theory.

Some Color Theory Basics

As seen in the color wheel above, there are three primary colors: red, yellow and blue. All the other colors are mixes of these.

In painting, we can also create different "tints" of a color by adding white to them, and different "shades" of a color by adding black. In general use, most people will simply call these different shades of colors. We can also talk about the "saturation" of color as we move away from a light tint to a pure, bold color.

Different shades and tints of blue, wikipedia.

So when we look at color wheels, they will usually show different tints and shades moving in and out from the center.

Picking color combinations

There are actually many theories of how to use the color wheel to create color combinations that "make sense" visually to us. Luckily, the range of flower colors and the ever-present backdrop of green mean that there are just a few most common color combinations used in most gardens. Keeping things simple, those are the combinations we'll cover in this book.

> It's important to keep in mind that no garden needs to stick strictly to a color scheme to give a good visual effect. Think of the color scheme being the dominant colors, and including some pops of other colors will have a nice effect in most cases.
>
> Another important note is that we're usually not looking to create an entire garden on one color scheme. The goal is to create views and vignettes of color. As we look around the garden we might see an area of blues and whites, and in another, an area of reds. Perhaps the front garden is bold, warm colors to make the house seem fun and accessible, while the back garden is blues and purples to help us relax after a long day. Maybe we have an outdoor gym and we want to use those warm colors to help our workouts. We can even create a garden that draws the eye around the color wheel, with beds progressing gradually from blues through to purples and then to reds.
>
> And again, remember these are tools, not rules. There's no wrong way to use color in the garden! But these tools might help create more harmonious and professional feeling landscapes.

Monochrome and analogous color schemes. Monochrome is what we get when we pick a single color, like blue, and build an entire garden off it. A monochrome garden may still use different tints and shades of blue. I have usually tried to keep my opinions to myself in this book and provide tools, but I'll admit that I think it's hard to pull off a monochrome garden well. This is partly because it limits plant choices too much and partly because… slight variations, such as are standard in nature, can tend to ruin the effect.

One classic exception is the all white garden, which is of course really a green and white garden. This is particularly useful to Permaculture and sustainable landscapes, because we have so many great edible plants that flower primarily white. In a garden dominated by white flowers, a few colors will shine like beacons! And the white garden will absolutely glow in the dusk, as the green foliage fades to black and the white comes off bold in contrast.

Analogous color schemes expand the colors to nearby colors on the color wheel, giving a very harmonious feel. There are two main analogous color schemes used in gardens, warm colors and cool colors, though I've seen amazing gardens that simply used the palates of greens found in foliage plants. I suppose that is the effect we usually get when entering into a forest.

Beauty in Abundance

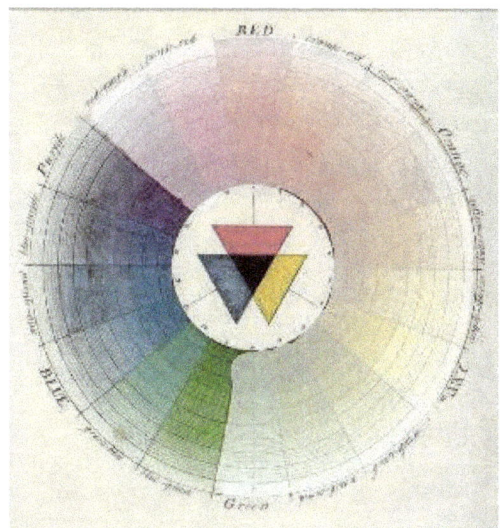

Cool colored gardens are a classic color pattern. They tend to feel restful and subdued.

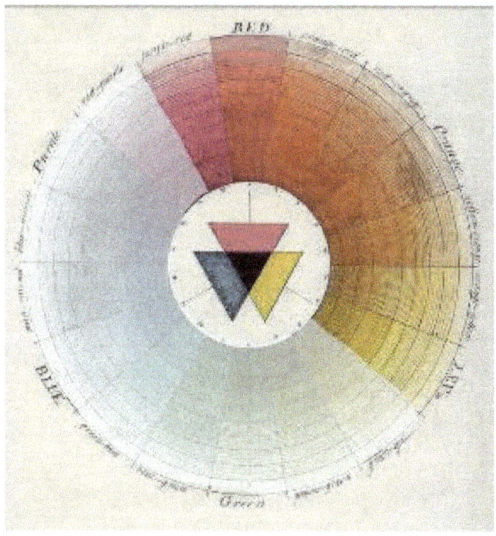

Warm Colored gardens are another classic, and tend to feel lively, friendly and energetic.

Complementary Color Patterns

If an analogous color scheme does not feel bold enough, or seems to be lacking something, adding a direct contrast can kick the whole thing up a notch. So we call these colors on the opposite ends of the color wheel complementary colors.

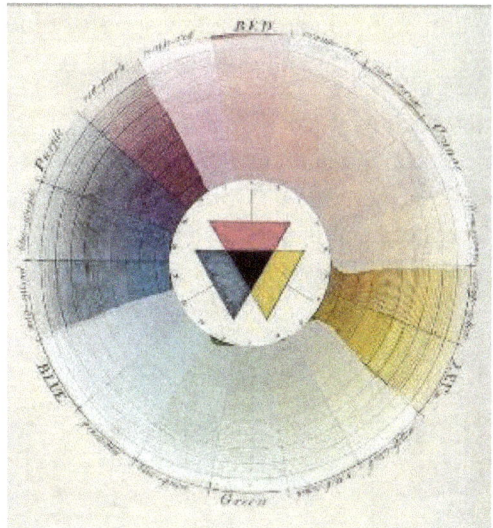

For example, it can work very well to complement a set of analogous colors, such as blues, greens, and purples with a bold yellow. This is one of the classic color combinations in gardens.

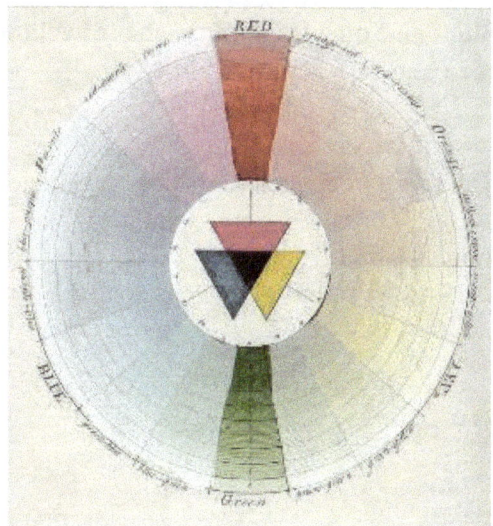

Another classic complementary color pattern is red and green. This is the foundation of the red rose garden.

We can add whites to any of these to tint the garden a little. The Jardin de curé color palette is analogous blues, purples, and whites. A few sprays of yellow, such as brassica or mustard flowers, will compliment the garden well.

345

In a warm colored garden, a few spots of blue or blue violet will stand out beautifully.

A tinted pastel garden. These colors are classic to cottage gardens and give a harmonious, naturalistic, romantic feel to a garden. These can include soft yellows, pinks, pinks and lavenders.

Some Classic Garden Color Patterns

Monochrome or analogous,
—**The all white garden.** A bold statement at dusk and dawn.
—**Warm colors.** Energetic, friendly, but possibly anxiety-inducing.
—**Cool colors-** Relaxing, healing, calming.

Complementary color patterns:
—**Blues and purples with touches of yellow.** Relaxing, yet fun and energetic.
-**Green and red.** Romantic, lusty.

Jardin de curé:
—**Blues and purples with whites.** Healing, restful, intellectual, looks great at dawn and dusk.

Cottage garden:
—**Tinted soft colors in pastels.** A restful, relaxed, romantic feel.

Color and Emotion

Now let's look in a little more depth at how color might affect our emotions in the garden and how we can put this to use.

Remember, the strongest effects go with the general idea of cool colors rather than warm colors. The more bold the colors are, especially the warm colors, the greater the effect.

Warm color combinations:

Warm colors will generally feel energizing, friendly, and exciting. Bold, warm colors can help us feel energized and can even help us exert more force. Bright reds and yellows (McDonalds, Coke) appeal to emotional, impulse buying, but may put off more careful, thoughtful shoppers.

Cool colors:

Blues, greens, and purples together will give a calming, and rational effect. It can move us away from the emotional mind towards the intellect. These can improve our test scores, help us deal with crises and may aid depression.

Red; passion, libido, pink; romance. A garden of pink and red roses has long been associated with sex and romance. One of my first communal living experiences was in a house with a red room we all jokingly called the boom boom room. Now research has found that these colors do indeed enhance libido for men and for women.

Blue has the biggest positive impact on our overall health and feelings of well-being. It can help us regulate emotions, triggers intellectual performance and even helps us reset our circadian rhythms. There is evidence to show that it helps to treat depression and cope with grief. No wonder that cross culturally, the color blue has a strong inclination to be the favorite color. While reds and yellows may cause impulse buying, blue appeals to the intellect, and more considerate shoppers.

The colors green, white, and pink have a tendency to support positive emotions, as do soft colors and pastels. A garden in these colors will feel welcoming and comforting.

Related Patterns:

This book includes several examples of well-functioning guilds by dominant color patterns.
Insectary plants by color and season.

Notes: Color effects perceptions of stress vs relaxation. https://cornerstone.lib.mnsu.edu/cgi/viewcontent.cgi?article=1172&

Blues: Use key: *edible, ^medicinal, +insectary,

Blue is the most common favorite color in the world.
It is calming, peaceful and draws us towards the rational mind. It has been demonstrated to promote healing, help with depression and processing grief.

Spring:

Grape hyacinth* (edible flowers) Columbine* (edible flowers), Camas* (edible root) crocuses+, blue flowered primroses* (edible)

Summer:

Ageratum+
Perennial flax*+^
Alkanet+^
Sage and other salvias
Blue sage*+^
Sea holy+
Blue vervain+^
Spiderwort*+
Blue Lobelia
Borage*+^
Veronica (edible, medicinal)
Bugloss+
Violets+*^ (edilbe)
Campanulas *(bellflowers, all species.)
Comfrey+^

delphiniums+^

Flax*^+

Flox^

Globe thistle*+

Lavender, Hidcot blue
Lupines*+ (N-fixer, some edible)

Monarda fistufosa*+^, blue moon, etc.

Monkshood+^
New England Aster+

Shrubs: Rose of Sharon*, lead plant, blue lilac

Purples:

Purples have a similar emotional effect to blues, usually appearing as a cool color to us in the garden. They may have a cultural association with royalty.

Spring:
Alliums*, Crocuses^, Camas,* Columbines*, lilacs*, Violets*, Hyacinth, Irises+,

Summer:
Alliums *, All heal*+^, Anise hyssop*+^, Blue false indigo, Comfrey*^, Dame's rocket*, Lavenders*+^, Liatris+^, Lunaria+*, Lupines*, Monarda fistulosa*+^, Nepatas, New England Aster, Purple prairie clover+*, Spiderwort*+^, Sweet william, Verbascums, Veronica (speedwells), Violets*^+, Vervain

Purple vegetables: (Purple varieties available)
Beans, eggplants, tomatoes, peppers, mustards, peas, carrots, broccoli, cauliflower, brussels sprouts, cabbage, kale

Reds:

Red is one of the strongest colors in term of its human impact. Red can increase emotions, making us passionate, angry, fearful, and alert. Red may give us energy and even increase strength. Sports teams wearing red are statistically more likely to win.

Many fruits will give a red color to the garden, including cherries, cornelian cherries, goumi, apples, pears, and many others. Redcurrants may be the most startling.

For trees, sumacs are a nice red and echo the flower shape of amaranth.

Spring: Tulips, red Primroses.

Summer: Amaranths*+^, Anise hyssop "raspberry," Cardinal flower+^, Cockscomb (an edible amaranth) Coneflowers,+ Daylilies,* Gerber daisies, Hollyhocks*, Hummingbird mints*+^ (agastache, edible) Knotweeds of various species*+^, Monarda didyma*+^, Empress of India Nasturtium*, Poppies*+^, Roses*+, Rose of Sharon*, Sorrel, blood-veined*+^, Sumacs*^+, Sunflowers*+^, Yarrow.*+^ , zinnias

Shade: Cardinal flower, Monarda, Blood-veined sorrel.

Red Vegetables:
Rhubarb, oxblood beet, chard, peppers, tomatoes, amaranth, red cardoon, blood veined sorrel, dragon's eye knotweed, chicory, red lettuces, Redcurrants, hibiscus, Scarlet runner beans

Pinks

Pink lacks the strong emotional effect of red. It is a calming color and a pacifying one. It may be romantic or even sentimental.

A prison study found that housing prisoners in pink rooms reduced violence and aggression.

Another found that pink rooms tended people towards happy thoughts and emotions.

I like pink and black together, and pink flowers by black garden shed with dark green foliage does it for me.

Spring: Tulips, peonies, hyacinth*, allium unifolium and other alliums*,

Pink Flowering Trees, SPRING: cherries, plums, apples and crab apples, redbud, rose of Sharon

Summer: Anise hyssop "avocado pink," azalea, beards tongue, begonias, butterfly bush, calla lily, carnation, chives*+^, dahlias, day lilies*, hollyhocks, lavender "rosea" *+^, lilies, lotus *, meadowsweet*^+, mints*+^, peas, including tuberous sweet pea, pink monarda, primroses*+^, musk mallow*, rosemary*+^, snapdragons, thyme*+^, wild marjoram, yarrow, zinnias

Oranges:

The boldest of the warm colors, it increases energy and happiness.

It will be seen as friendly and accessible, though perhaps with more aggression than yellow.

It may help with healing. In one study orange placebo pills had a greater effect than blue or white. Another found an orange environment helped recuperate after injury.

In the warm colored garden it will make a bold statement.

Summer: Amaranths*, Butterfly milkweed, California poppy, Calendula*+, Coneflower, Coreopsis, "crazy cayenne," Chrysanthemums, orange, Daylily*, Hummingbird mint, "tango," Indian blanket, Marigolds^, Nasturtium, Red Sunflower, Sunflowers*+^, Yarrow, "firefly peach," Zinnias

Orange fruits and vegetables:
Orange eggplant, orange peppers, pumpkins, gords. Oranges.

Yellows:

A classic warm color that brings happiness, friendliness, and energy. In one study it was found to increase appetite, and it has become a common color in dining rooms and kitchens.

Spring: Allium "Molly," Daffodils, Tulips, Primroses

Summer:
Brassicas, Brown eyed susans, most of them, Calendula, Compass plant, Coneflower, Cup plant, Evening primrose, Goldenrods, Marigolds, Mustards, Sunchokes, Sunflowers, Variegated comfrey, Tomatoes, Yarrow,

Yellow Vegetables: lemon cucumbers, Turkish rocket, Arugula, Lettuce, Patty pan and other summer squashes,

Whites

White colors play well with others, blending and softening other color schemes. White flowers surrounding a particularly vibrant color will help that color pop.

Many useful and beneficial plants flower in white.

White has a nice quality of shining out in the evening, at dawn, dusk and at night. Blues and whites are the classic color scheme of the jardin de curé.

The one drawback of white is that alone, it can tend to look… not like a garden, except at dawn and dusk.

Emotionally, white may be neutral, though in one study, a white environment increased creativity.

Spring: Snowdrops, crocuses, tulips, daffodils, alliums, lupines

Tree blossoms: apples, pears, cherries, plums, kousa dogwood, Camelia (tea,)

Summer: Alliums, Archangel deadnettle, Baby's breath, Borage (white,) Black cohosh, Carrots, Chamomile, Comfrey, Clover, Daisies, Feverfew, Day lilies, Garlic Chives, Jasmine, Hollyhocks, leeks, Marshmallow Meadowsweet, Mints, Mountainmint, Oregano, Poppies, Roses, Skirret, Sweet woodruff, Tobacco, Valerian, Yarrow, Watercress, White lavender,

Chapter 30: Patterns for Attracting Natural Beauty to the Landscape

"In a biodiverse garden, you expect (some pest damage.) We're feeding the insects and the rabbits too. That's okay. Enough for all. The multiple species of native bees and other insect pollinators all over the garden, in addition to honeybees, tell me that all is going well."

—Corin Pursell, Permaculture Activist

Yes! Believe it or not, if we start with realistic expectations and with wildlife habitat as one of our priorities, seeing some critter damage can actually be a great joy—in the food garden!

So many of the people I know who have transformed their landscapes find this to be the most gratifying and rewarding part of the whole adventure. Seeing a huge diversity of insects come back to a previously dead landscape is hugely fulfilling. The spectacular colors and forms of these creatures is endlessly fascinating. And while some may be worried about being "bugged," the ecological truth is that a balanced, biodiverse ecosystem will have fewer obnoxious pests of humans.

And for this reason, books on wildlife gardening now abound.

But what of the productive landscape? Can we actually balance the needs of human food production and wildlife habitat?

Absolutely, yes we can. And in fact, it is not only possible, there is an abundance of research that shows that this can be profoundly beneficial to the production goals, reducing pest and disease pressure to the point where, in some studies, chemical interventions are no longer needed at all.

Since this is a book on productive landscapes and beauty, and since so many great resources exist for creating man-made habitat helpers like birdhouses and bee boxes, we're going to focus on holistic efforts to naturally increase habitat and doing so in ways that are beautiful. And we'll focus on the base of the food chain, the plants, soils and habitat that support healthy diverse insect populations, which go on to feed everything up-system.

While having the right birdbath or bird house can make a big difference in certain cases, research has shown that natural habitat is often better for wildlife and avoids certain downfalls of manmade habitat. And in most cases, all that is needed is to stop poor human management choices that are destroying habitat, such as removing hedgerows and mowing too much.

If we stop being stupid, they will come.

The garden at Lillie House is a great example. With only holistic efforts to increase total habitat, the landscape was absolutely transformed into a wildlife and beneficial insect sanctuary. Many guests, even professional biologists who visited state that it has the highest insect biomass and biodiversity of any single place they have ever visited. Guests are often absolutely thrilled to see this amazing show of insect life.

One major insight is that we must put up with a few years of pests in some cases to get to where there are many beneficial pest predators.

It takes a few years of pests to get to a balanced system where biodiversity is the pest management, but I have seen it happen at multiple gardens of mine now.

So let's look at a specific example of how this natural pest control works when we stop doing stupid stuff.

Beneficial Insects: How I get FREE Nematodally Awesome Pest Control

If you've been gardening a while, you've probably heard about "beneficial nematode" sprays as an organic, poison-free alternative for controlling insect pests. Nematodes are tiny round worms that have all sorts of ecological niches. For example, some help stabilize nitrogen, some eat our plants, and some lovely species eat the things that eat our plants.

For a few decades, folks have been selling beneficial nematodes to apply to garden pests, and research has found they are tremendously effective against a huge range of pests. Depending on the nematode species, this includes most borers, most caterpillar pests, most beetle pests, including Asian beetles, Japanese beetles, flea beetles, potato beetles, lawn grubs, Northern Masked Chafer, most fly larvae, most weevils, fungus gnats, most root maggots, etc. Basically, they especially like insects that feed on plant material, so nearly anything that eats our plants MAY get eaten by nematodes, depending on the genera and species. (Note: while they generally prefer insects that eat plants if you spray on a gazillion nematodes, these will also impact beneficial insects.) They are tremendously effective and kill their targets within a few days.

Now, here's the cool FREE part. The two main commercial beneficial nematode genera Steinernematidae and Heterorhabditidae are endemic to pretty much everywhere, and commercial strains of both have come from Asia, Australia, Europe, North America, South America, and Africa.

If you happen to live in one of those places, you probably already have species of beneficial nematodes in your soil, without having to buy them.

BUT, conventional gardening practices destroy their populations, so they don't exist in high enough numbers to do their job.

The good news is that there are now LOTS of studies showing that with healthy habitat and good garden practice, beneficial nematodes can exist in populations high enough to provide a high level of pest control, enough to help keep pests from getting out of balance.

That brings us to my picture of tilled soil (in the chapter on soils) with lots of compost added (on the left) and untilled, mulched, and continuously planted soil (on the right.) Soil nematodes primarily live in the thin film of water that coats soil particles. The soil on the right has great "structure," lots of little nooks and crannies and surface area that holds a lot of that thin coat of water for a long time.

The picture on the right, in addition to holding water and nutrients better, is nematode habitat. The picture on the left has very little nematode habitat.

Think about that. It means if we plant a crop susceptible to plant-eating nematodes into soil with minimal habitat, those populations will skyrocket, attacking all our plants and filling all the habitat that beneficials COULD live in.

This is why studies have found that good management both controls harmful nematodes and increases beneficial nematodes. If one goes up, the other goes down.

So, how do we create beneficial nematode habitat?

First, here are some things that have been found to DESTROY beneficial nematode populations:

Spraying poisons.

Nitrogen fertilizer.

Tilling.

Monocultures.

No grasses.

Bare soil.

Solarizing the soil with silage tarp.

Using plastic mulch or landscaping fabric.

If you do those things but buy nematodes, you will forever have to buy nematodes, because they will all die without habitat.

Here are some things that have been found in studies to bring nematode populations up to effective levels (I start to sound like a broken record):

No-till, no spray, no artificial fertilizer.

Deep mulch.

Compost.

Constant soil coverage.

High diversity. Sophisticated intercropping and crop rotations greatly increases beneficial nematodes and decrease harmful ones.

Having grasses in the garden. At least one study found the highest populations of nematodes under grasses, even under lawn. They don't travel far so lawn paths or meadow areas near the vegetables may provide good nematode habitat. Ironically, one study on Asian beetles found the populations were lowest in orchards with integrated lawn, where nematode populations were highest.

Since nematodes live in water film, regulated water, such as through a swale system should theoretically help to stabilize beneficial nematode populations.

Have gardens next to or surrounded by natural habitat areas, these can be meadows, hedgerows, or food forests.

If we do those things, there are multiple studies that show that nematode populations will be high enough to provide good control without having to buy imported nematodes.

And in some cases, research has shown our native nematode species may be more effective at control than imported exotic ones! Purchased strains will usually be more specific feeders. The true key, then, is to create good habitat and have many beneficial nematode species, so that the soil has the tools it needs handy to battle any pest that threatens to get out of control.

And you don't have to pay Jeff Bezos to deliver that sort of "natural pest control" to your house by drone.

Now, here's what's really nematodally awesome. This doesn't just work for nematodes. Nematodes are just one trendy natural pest control method. These same techniques will work on a whole variety of critters that eat the critters that eat our garden plants. Again, it is a perspective of creating total good system health that can have a huge impact on preserving biodiversity, and having an easier garden.

Beauty in Abundance

Can a yard be beautiful and ecologically healthy?

This is probably one of the most common questions I hear, and one of the reasons for this book. Often people think there has to be a choice, it's one or the other, beauty, or healthy.

It doesn't have to be so.

In fact, in many cases, a landscape that is more beautiful can be of much higher wildlife value than a yard that has simply been let to go to an untidy mess. I've visited many "wildlife gardens" or "certified habitats" that were simply overgrown lawns with a few low-value native pollinators next to the house. Such a landscape certainly has more value to wildlife than a mown lawn, but it's not really very useful. In fact, one study in the UK found that tended gardens had far more value to wildlife than unkept messy yards. A landscape with a variety of flowers and forms of habitat will be far more useful than a simple lawn let go.

And it will certainly be much more beautiful. A transformative landscape will have a variety of flowers all season long, in addition to trees and hedgerows, water and different types of habitat. Not only will it be more beautiful, more ecological, it will have more useful space and utility for humans. An unmown lawn will have little use to humans, other than as a thing to look at.

And when it comes down to it, a transformative landscape need not take too much more time than an unkept lawn to tend, if we use the landscape transformation strategies in this book.

....

Patterns for Natural Wildlife Habitat:
So, with our perspective of creating natural habitat for insects, what are some high-value patterns we can use to create a beneficial insect sanctuary?

1. Mosaic Landscapes. The diverse landscapes discussed earlier, with lots of edge, and lots of different ecosystem types has been demonstrated to be one of the most important factors in creating a sanctuary for other beings.
2. Water in the landscape: Probably no single feature will increase biodiversity so much as having water access for wildlife on the property. Preferably in multiple places at multiple scales. This can include small ponds, large ponds, greywater wetlands, and birdbaths. See the next chapter for more ideas.
3. Some periodic disturbance. It seems counter-intuitive, but "just let it go back to nature" is not the best way to steward wildlife, especially in a modern world where we can't let forest fires to create disturbance. A landscape with some areas of grassland, bare soil and even desert will have more niches for wildlife. So, the kind of disturbance we create when vegetable gardening is good for wildlife.
4. Hedgerows. Research has shown that perhaps no single landscape feature will do more in most climates than creating hedgerows, which are naturally highly diverse ecosystems.
5. Heterogenous textures instead of homogenous ones. Homogenous textures refers to all the spacings and plants being the same. For example in the conventional orchard or vegetable garden. Heterogenous textures refers to the randomized mix of plants and spacings we get in a forest. No rows. This has been found to increase niche and habitat and to dramatically increase biodiversity and crop health.
6. Natural mulches. Natural mulches imitate ecosystems and can be one of the most important things we can do to increase beneficial insects and biodiversity.
7. Rock piles can provide excellent pest control by giving snakes and lizards a warm place to sun and call home.
8. Nurse logs are another excellent feature for increasing wildlife habitat. For extra habitat value, drill holes in the nurse logs of different sizes. Different pollinators will use these to make homes.
9. Standing dead wood. We're often hasty to cut down dead trees. But if it's possible to leave them standing, they can do a great deal to increase habitat for insects, birds, and mammals.
10. Avoid garden clean-up.
11. A Season of Flowers

Useful Insectary Plants for Pollinators and Beneficial Insects

	Plant and color	Latin name	Native region	Early Spring	Spring	Summer	Late summer	Fall
1	Crocuses	Crocus species	Eurasia					
2	winter aconite	Eranthis	Eurasia					
3	grape hyacinth	Muscari neglectum	Eurasia					
4	snowdrops	Galanthus species	Eurasia					
5	dead nettle	Lamium species	Eurasia					
6	strawberry	Fragaria species	North America, Eurasia					
7	golden alexanders	Zizia aurea	North America,					
8	angelica	angelica atropurpurea	North America,					
9	lanceleaf coreopsis	Coreopsis lanceolata	North America,					
10	swamp milkweed	Asclepias incarnata	North America,					
11	common milkweed	Asclepias syriaca	North America,					
12	valierian	Valeriana officinalis	Europe					
13	borage	Borago officinalis	Europe					
14	wild maroram	Origanum vulgare	Eurasia					
15	field forget-me-not	Myosotis arvensis	Europe					
16	mountain mints	pycantheum species	North America,					
17	Anise hyssop	Anise hyssop	North Amierca					
18	thyme	Thymus species	Eurasia					
19	yellow coneflower	Ratibada pinnata	North America,					
20	purple coneflower	Echinacea purpurea	North America,					
21	boneset	Eupatoreum perfoliatum	North America					
22	wild burgamot	Monarda fisfulosa	North America,					
23	meadowsweet	Filipendula ulmaria	Eurasia					
24	bee balm	Monarda didyma	North America,					
25	mints	Mentha species	Europe, North America					
26	skirret	Sium sisarum	Europe					
27	fennel	Foeniculum vulgare	Europe					
28	catmints	Nepata species	Europe					
29	rattlesnake master	Eryngium yuccifolium	North America					
30	globe thistle	echinops species	Eurasia					
31	sea holly	Eryngium giganteum	Eurasia					
32	sunchokes	Helianthus tuberosus	North America					
33	New England aster	Aster novea-angliea	North America					
34	anise goldenrod	solidago odora	North America					

A selection of the highest quality plants for pollinators and beneficial insects, by color and flower time. Plan for a whole season of uninterrupted flowers. Use as many native species as you can fit.

★Flowering times are approximate and depend on climate conditions.

Citations: xerces.org, Michigan State University, Penn State, Royal Horticultural Society

Notes:

1. Bhat, A.H., Chaubey, A.K. & Askary, T.H. Global distribution of entomopathogenic nematodes, *Steinernema* and *Heterorhabditis*. Egypt J Biol Pest Control **30,** 31 (2020). https://doi.org/10.1186/s41938-020-0212-y
2. R. E. Berry, J. Liu, G. Reed, Comparison of Endemic and Exotic Entomopathogenic Nematode Species for Control of Colorado Potato Beetle (Coleoptera: Chrysomelidae), Journal of Economic Entomology, Volume 90, Issue 6, 1 December 1997, Pages 1528–1533,
3. Strategies to Enhance Beneficials, SARE Outreach, 2005
4. I.J.Porter P.R.Merriman , Effects of solarization of soil on nematode and fungal pathogens at two sites in victoria, Plant Research Institute
5. Nethi Somasekhar, Parwinder S. Grewal, Elizabeth A. B. De Nardo, Benjamin R. Stinner,, Non-target effects of entomopathogenic nematodes on the soil nematode community, Journal of Applied Ecology Volume 39, Issue 5
6. Shin Woong Kim, Walter R. Waldman, Tae-Young Kim, and Matthias C. Rillig Environmental Science & Technology 2020 54 (21),

Chapter 31: Aesthetic Patterns for Ponds and Water

As we follow a Permaculture design process, the next step we consider is water as a mainframe element. If perhaps we want to work to transform the landscape in a more intuitive and incremental way, this would be a good time to start a few adventures to bring water into the landscape.

This topic could be a very "deep well" indeed —too deep for the pages of an already long book. So we'll just introduce a few very basic patterns for pond implementation, then look at aesthetic patterns we could use.

Including water in the landscape itself is a nearly miraculous pattern for beauty. No other single element can do more to transform a landscape into a place that is healing for both humans and wildlife. Water alone can make a garden "pop" with life.

The most basic aesthetic pattern we have, or rule, I suppose is: don't screw it up. Meaning, don't screw up the natural beauty of water with artificiality.

Patterns for simple ponds

Make ponds that collect water, instead of ponds that require water.

The first and perhaps most important pattern of a pond that aims to be SUSTAINABLE is that it is designed to catch its own water, rather than to be filled with fossil water or city water. So, one of our first goals is to figure out where this water will come from. Usually that involves carefully siting the water feature or pond.

Therefore, site a water feature such that it collects its own water, rather than requiring water to be filled up. The best ponds in small landscapes will have access to water from rooftops or structures, or paved surfaces, which can be a plentiful source of water. In the landscape, look for low areas and keypoints where water naturally stands. These will be the easiest places to make ponds that stay full through the dry months.

The Kiddie Pool Pond Adventure
When it comes to simple implementation, it can't get better than this. Kiddie pools often sell for as little as $10 and are made of at least acceptable materials safe to humans and wildlife. I have seen unpretentious pools of this kind converted to absolutely beautiful edible/wildlife ponds, even without digging them in. When dug into the ground, no one would even know it is a kiddie pond!

If you're confident about the location, you can dig it in. If not, it's great to just get a small pond started.

Lay a branch into the pond to allow stuck wildlife a foothold to get out.

Ask folks from your local Transformative Adventures or Permaculture groups if they have aquatic plants to share. Many local small nurseries also stock aquatic plants and appropriate small fish.

Building a Small Pond Without Plastic
Since there are many resources for digging ponds and using plastic pond liners, I will include a type of simple pond I have created now on a few sites using bentonite clay.

This type of build is probably best for a small pond under 15' * 15'.

Steps

Step 1. Site the pond. The ideal pond will be in a low spot or keypoint where it will naturally fill with water. An experienced Permaculture designer should be able to help with this. Swales from the downspouts can help fill the pond.

Step 2. Dig the pond. Set aside any clay material you find to return to the pond. Dig about 6 inches deeper than you want the pond to be. The pond will work best if it is fairly shallow at the edges and gradually deepens. Such ponds will appear to stay full better. If you dig too deep, you'll go over the "slump angle" and the pond will collapse in. A small pond could be a couple feet deep at most. A 10' diameter pond could safely be 5' deep on most soils. Return any clay material back to the pond. It will help act as the liner.

Step 3: Test the Soil. Find a 5 gallon bucket. Put a few holes in the bottom then fill it with 8 inches or so of soil. Add water and see if it drains. Now, add the clay in a layer about $\frac{1}{5}$ inch at a time and test again. Keep adding the clay until the water doesn't drain. This is how much clay you will need to add to the top few inches of your pond soil.

Step 4: Add the clay, work it into the top layer of soil. Then compact the soil. It is good to wait for a rain where the ground itself has some water in it but the clay can stay dry. When it is right, it should compact easily into a firm, holding clay.

Step 5: fill the pond with water to test it. If it drains let it drain and add more bentonite clay until it holds. In many cases a few pounds of bentonite clay cat litter should fill a 10' pond.

Step 6: line with rocks or blocks to provide access to the edge without a danger of damaging the lining. Consider linking the pond with pebbles or rocks. If the edge is too deep, make sure to put a stick into it to allow wildlife to climb out.

Now, if the pond was sited well to fill on its own, you should have something much like a natural pond. If it still drains in dry weather, cover it with grass clippings, and let these rot into the pond. The bacterial that feed on the grass will help create a natural liner.

Your natural wetland will start attracting appropriate wildlife, but you can help it out by importing aquatic plants.

Getting More Sophisticated, Natural Swimming Pools

Perhaps the ultimate home water feature is the natural swimming pool, which relies on plants to clean and filter the water, rather than chemicals. And many of these plants can be edible. Such pools are so effective that they can produce water that is safe to drink from! This is especially a useful technique for existing ponds. I have known some gardeners to buy homes with in-ground pools and fill them to have a garden. The pool might have been a more productive growing space than the new garden, and it could have still acted as a pool!

Natural swimming pool, Clear Water Revival, Creative Commons.

Greywater Wetlands

No book on sustainable landscaping would be complete without at least mentioning greywater wetlands. These are small wetlands designed to process and clear greywater from the house, such as from the washer and dryer or sinks. Again, this is a big topic and covering it could be the topic of a whole book. In fact, it's the topic of several very good books, including:

"The DIY Grey Water Wetland," by Caroline Kloppert and Stephan Kloppert
"Gaia's Garden," Toby Hemenway

Edible Pond and Wetland Plants

Skirret (Perennial clumping parsnips)
Cattails, celery, willow, wasabi, wapato
alliums, blood-veined sorrel, chocolate root (avens)
meadowsweet, marsh marigold, swamp Saxifrage?

Marginal:
Aronia, Pears, paw paws, highbush cranberry

Water Guild Plants

Water cleaners: cattails (every part is edible) purple loosestrife, sedges (pennywort,) wild rice

Water oxygenators: hornwort, elodea, dwarf sagittaria

Deep water plants: hardy water lilies, lotus

Patterns for Pond Beauty

"There was a grove, dark with holm-oaks, below the Aventine, at sight of which you would say: 'There's a god within.' The centre was grassy, and covered with green moss, and a perennial stream of water trickled from the rock."

– Ovid, Fasti

There's probably nothing that does more to invoke the "spirit of place," the "Genius Loci" than the presence of water in a landscape.

Like other "wildlife," we humans respond on a deep intuitive level to water, sensing how it nourishes all things in the ecosystem, cleanses, and restores us with the song of deep tranquility, a song we can so easily lose ourselves and our cares in. It's a consistent feature of the places we flock to for vacation, retreat, sanctuary.

In the ecological garden, water gives a flowering of wildlife, insect and plant diversity–the "fount of life."

In the edible landscape, water allows us to grow some truly exciting and nutritious food plants, and aquatic ecosystems are the most productive habitats on earth, in almost every measure. Water supercharges the productive garden.

But I like to start every creative process by connecting to the deep wisdom of culture, the subconscious and intuition, rather than letting the conscious logical brain call all the shots. The conscious brain is really good at things like taxes and train schedules, but when we let it lead the creative work, we end up making stuff that's about as sexy as, well, taxes and train schedules.

Speaking of taxes and train schedules, the accountant brain tends to make a pond like a math problem:
Flowers, check;
Block border, check;
fountain, check;
waterfall, double check;
Just keep adding more elements to it, then wrap that pond up with a schedule D and send it to the IRS!

Because, if you're doing "pond math," you just add up a bunch of elements until they equal "awesome pond." So if you add even more elements, you'd obviously get an even awesomer pond, right?

What does it say of our view of nature when we build a mini waterfall with Hawaian volcanic rocks in our backyard in Iowa?

But we're certainly looking for something different, patterns that will lend a spiritual depth to our home. We need a backyard to act as a sanctuary, an oasis where we feel secure that all of our needs are met and we're helping to heal our dysfunctional relationship to nature.

But surely, there are patterns that point to a simple, genuine approach appropriate to a humble human habitat, one that will look "in place" in my backyard food forest garden, without trying to fool anybody.

Gudmunder the Good's Pool, Iceland, photo: Bradley Rentz

Rie Cramer, "The Goose Girl at the Well," Grimm's Fairy Tales, 1927

From the earliest times, humans have lined natural pools and wellsprings with stones and rubble as a sign of sacredness and importance, the need to simply decorate the important landscape features. And also to protect them from animal digging and erosion. These simply lined "sacred pools" are a deep archetype in our art and depictions of sacred places. They are a "natural" feature of human habitats.

Throughout the old world, these ancient wells and pools are the stuff of sacred myth. Preceding Christianity, pagan Europe held these pools as the dwellings of the fae, and of the "spirits of place."

As Christendom came to reign, these sacred wells became associated with saints, especially ones who had a habit of getting decapitated! Often these saints, baptized in holy wells, would lose their heads to heathen scoundrels, then go parading through the country with their noggin's tucked under their arms like footballs—testament to the new god's power.

Another common pattern was the cairn, built to shelter a spring and protect it from run-off and rain contamination.

These often flowed to a pool, where pilgrims could bathe in sacred waters.

Such sacred wells are often depicted in art. It's easy to sense that there's something holy in such patterns.

And still today, many are drawn to the healing powers of such ancient places like "Chalice Well," in England. Over the ages, gardens and orchards often rose up around them, further enhancing their nourishing symbolism. At places like Chalice Well, these gardens eventually became quite elaborate, but still seems so genuine, and far from what you'd see in any casino. And at most, the original well sites often remain in their untouched paleolithic condition.

(Pond in the food forest at Lillie House)

What strikes me in all these images is the naturalness of this pattern in human habitats, the need to simply leave our mark on something important to us with the simple, local materials available.

In most of these famous European sites the local material was limestone and slate, materials that would look out of place in my Michigan back yard, as much as I love the look. I find that in Michigan gardens, these always lack that genuine "sense of place." We know it on a subconscious level, even when we don't know that we know.

So, how might this pattern look in a new stone-age in your region?

(The Holy Well, Frederick Goodall)

In Russia's glacial till, the oldest holy wells and pools repeat the same patterns, but with wood, concrete, and roundstone adorning the sites. Many of these have had ornate orthodox chapels built around them, but inside, the wells often still bear their ancient ornaments.

Ultimately, I can't say how some mythical, terroir human society might have built your pond for you. What I can do is suggest this adventure: Make a natural pond or pool, one that will fill without being filled with plumbing. Make it feel like it is natural in your part of the Earth. You will be happier with it that way.

Chapter 32: Patterns for Beautiful Paths

The number one time-waster on most vegetable farms? Path maintenance.

I was attending a farming conference a few years ago where researchers were presenting on their study on farm profitability and labor, and they said their data indicated that the vegetable farmers in their study spent more time maintaining paths than any other farm task. Everyone nodded their heads knowingly. Sound familiar? Are you nodding now?

And yet, most recent books on market farming and vegetable gardening do not have much to say on the topic. Most don't even have a single chapter on the topic. It's usually just assumed that even in a no-till system, paths will just be maintained with hard work, usually recreated from scratch each year. This is done with tilling, plastics, or poisons. I know of no other way.

Which means that good path design is key for no-till gardeners who want to avoid the health risks and ecological damage of plastics, poisons and petroleum.

And so paths are a problem on most farms and vegetable gardens. Usually if there are tilled strips between vegetables, these open spaces are the first place that weeds take hold, and so they require regular hoeing or tilling. Many conventional veg gardeners tell me their biggest garden task is hoeing, usually paths between rows. If the paths are mulched, they'll usually need annual re-mulching. I know one chip-mulched market garden where the biggest annual task of the year is re-mulching. It's less than 1/10th of an acre, yet it takes a team of 15 volunteers half a day to do the mulching. If that labor was paid, that one day would completely eliminate any profits for the whole season.

Good paths can reduce our costs, reduce our maintenance time, increase garden productivity and function, and in the context of this book, make a huge impact on how we experience the beauty of the landscape. Paths are literally HOW we usually experience the beauty of the landscape, as we walk around the garden.

But in the context of this book, path-making is also deep magic. When we plan paths, we are platting our relationship with nature, we are orchestrating the energy flows through the landscape, and we are literally planning the future of our lives and landscapes.

These patterns start with mimicking the efficiency of nature, which uses "orders," vessels of different sizes, to move resources around. We'll look at what the best materials are for different kinds of garden paths. We'll look at some efficiency measures for garden path layout. And different kinds of plant communities that can help us reduce our path maintenance time and create beauty and function.

Path Hierarchy and Orders

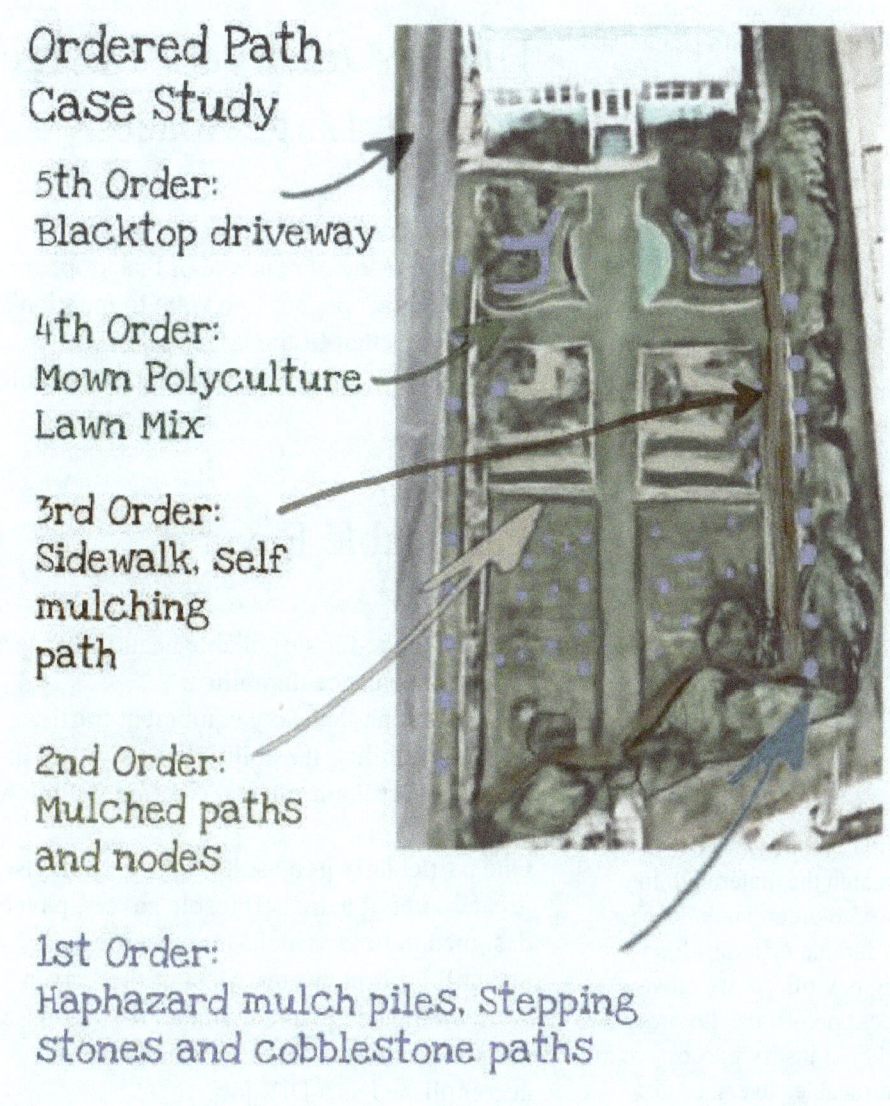

Ordered Path Case Study

5th Order: Blacktop driveway

4th Order: Mown Polyculture Lawn Mix

3rd Order: Sidewalk, self mulching path

2nd Order: Mulched paths and nodes

1st Order: Haphazard mulch piles, stepping stones and cobblestone paths

Nature uses "paths" of different sizes to transport resources through systems, which we call "orders." We can see this in the branching pattern of a tree, in the human circulatory system, or in a river system. We can emulate this pattern to create a system of paths that is easier to maintain and more functional. It will also feel more natural.

Creating a system of ordered paths can be our biggest tool in balancing path maintenance and cost, which is our big problem.

For example, when it comes to maintenance, the ideal solution to paths would be concrete. It is very long lasting and can be very attractive, though difficult to work on our enjoy a picnic on. But a garden using concrete paths with permanent beds would require very little maintenance work and could last (with appropriate care) for decades.

But there are downsides. The most obvious is cost. It would be prohibitively expensive for most landscapes. And it would contribute to a hot environment, probably not be great for the soil, and concrete is a major driver of climate change. So, the idea of ordered paths helps us keep concrete down to the places where it is really worth it, such as places with frequent access or heavy equipment.

Generally speaking, the materials and techniques we can use to create and maintain paths will either have a high upfront cost which reduces maintenance, or they will have low cost, but require more maintenance. Our task is to choose the best system for each use on our site.

As we balance cost with durability, we can use five orders to help us match the material to the job. For example, with "5th order" major access points the cost of very durable materials like concrete will have a big pay-off. As we move down to secondary access points, the 4th order, the infrequent use might not justify the cost, so a solution like gravel, permeable pavers, or just mown grass mix might be better. 3rd order paths are main access to gardens or the house on foot or for small equipment like wheelbarrows. Finally, 1st and second order paths are the small working paths inside or between gardens and different landscape features.

Higher-Order Paths: Large Paths in Large Spaces

The higher order paths, orders 4 and 5, will usually be made of concrete or blacktop, are beyond our scope, but I do want to mention a few patterns that fit the large scale, particularly mown grass paths and permeable pavers.

Permeable Pavers

The challenge for sustainable beautiful access roads is to balance durability,
 Ability to handle heavy equipment traffic, and permeability, the ability to soak water in instead of letting it run off, creating pollution.

One particularly good solution that could use greater attention are permeable pavers, pavers designed to hold vehicles, protect the soil, and still infiltrate water. Most of these even work well when planted with a mix of grasses and flowers. And these can usually be accomplished as a DIY job.

Therefore, consider permeable pavers for large access areas and parking in residential and commercial landscapes.

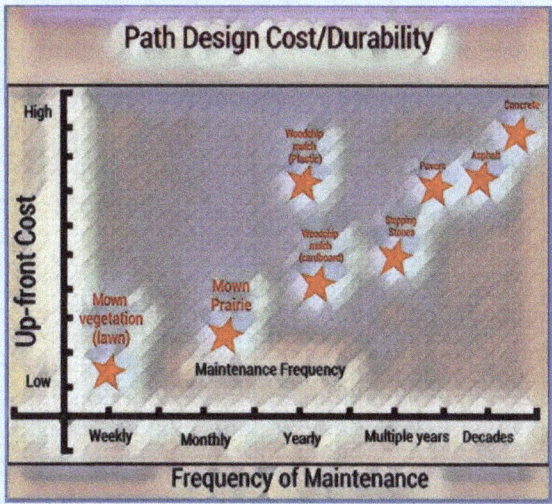

5th Order: Main site access points: frequent access for cars, trucks, equipment.
- Concrete: Highest cost, highest durability. Probably worth the cost, due to the impact of machinery and frequency of use.
- Asphalt: Slightly reduced cost, but requires more frequent repair and replacement. May not be practical where heavy equipment is used.
- Gravel: Lowest cost, but may require yearly repair and filling of heavy equipment is used frequently. May reduce the life of equipment.

4th Order: Major site access: seasonal access, trucks, tractors.
- Gravel: More practical with lighter equipment or seasonal access.
- Grass: Not feasible with high frequency or heavy equipment.

3rd Order: Major system access, garden or field access, barn access. 3-6' wide
- Mown vegetation (lawn.) Will stand up to foot traffic and light tractor access. Least cost, best value.
- Self-Mulching: In woodland or forest gardens with dense plantings, paths may become self-mulching in time.
- Concrete sidewalk: Only justifiable where access is a concern or aesthetics a priority.

2nd Order: Garden/Farm access paths and nodes. 1-2' wide.
- Self-mulching: With tight plantings, especially in forest gardens, these may become "self mulching," or maintained by traffic.
- Woodchip Mulch: With cardboard barrier. Without cardboard, may be inoculated with mushrooms.
- Dust mulch: Hoed after use. Not feasible in wet climates.
- Cobblestones and stepping stones: Not feasible in high traffic areas.

1st Order: Small connecting paths.

3rd Order Paths

The next order of paths are for main access between landscape features or gardens. This could be the path to a garden shed, the path in the garden, or the main path through a garden. In most cases, the most appropriate and beautiful type of path is simply mown lawn mix, though we can improve on the mix by adding some flowers and biodiversity.

The Mown Grass Path

Another excellent solution for paths that will not get as much traffic, or where the traffic will not be necessary if the weather is bad, are mown grass paths. These can be surrounded by beautiful prairie and look quite ornamental and beautiful. In my opinion, this look should be more normalized even in urban and suburban landscapes.

Therefore consider mown grass paths for higher ordered paths where mud and frequency of use will not be a problem.

The Mown Weeding Strip

"How the heck do you keep sunchokes from taking over?" exasperated gardeners sometimes ask when I tell them how much I plant and use this great vegetable.

Similar questions arise over blackberries, elderberries, mints, etc.

A related pattern that is useful for containing runnering aggressive plants is the mown weeding strip. Most of these plants cannot compete with grasses when there is any kind of mowing going on.

Therefore plan a 3' plus strip of grass path around guilds with plants that need to be contained. This includes fertility strips, hedgerows, berry hedges, and guilds where mints are a player.

A Beautiful, Walkable Lawn Polyculture

This is my favorite, tried-and-true, walkable lawn mix for home and farm landscapes.

It is a major achievement of the movement for better landscapes that everyone who's remotely environmentally minded now absolutely hates lawn. But, perhaps we were a little too effective in this message.

Not only do grasses play very important roles in ecosystem health, they are the absolute best plant there is for walkable ground covers. They hold up to foot traffic better than any other plant, they feel amazing on bare feet, persist almost indefinitely, make lovely picnic and nap areas, and require relatively less maintenance than many of the "alternatives."

This is where I often get exasperated questions from people who have come to loathe grasses.

"Is there an alternative to grasses that we could use instead?"

"What if we want to set aside that grasses are so ideal, that they are natural in every reasonably sunny ecosystem type, have huge benefits to wildlife, and that they have huge benefits to the function of a garden or farm... if we simply wish to avoid grasses as a political statement, is there an alternative?" In most cases, no, not in my 20 years of experimenting with the problem or in the research literature examining it.

What about clover, chamomile, thyme, etc? These can have their uses in certain contexts, and we'll talk about these, but in most cases, these are a solution designed to fail. Ecologically, none of these evolved to succeed and outcompete grasses in most cases. Clover, for example, cannot outcompete grasses, but actually helps grasses, by fixing Nitrogen, a nutrient that grass thrives with. Every clover path or lawn is a grass lawn waiting to happen, and it will become a high maintenance task in a few years to keep the grass out. Grass will easily invade other ground covers, too.

Since grass is an excellent walkabout ground cover, provides both garden and ecosystem services, and since it will eventually show up in most cases, plan most 3rd order paths to be a grass mix. But we can improve on lawn with a polyculture:

A Good Grass Polyculture:

Lawn grass mix. Start with a regionally appropriate lawn grass mix. In many cases, this will be a mix of Kentucky blue, fescues, and ryes. Your local garden center should have proven recommendations.
White clover or dwarf white clover. Sow at recommended densities.
Add bulbs: Fall plant crocuses, squill, and snowdrops in lower traffic areas.
Additional broadleaf plants to encourage or plant: Dandelions (there are white flowering varieties) chicory, plantain, english daisy, oxeye daisy.

On Sandy Soil or Dry Climates:
Add creeping thyme, pineapple weed, oregano.

A Walkable Groundcover Polyculture for Shade

Shade is where we can finally move away from grasses. In fact, as shade increases, grass becomes the guaranteed failure. I once had neighbors who loved lawn and replanted seed every single year, which by July was completely dead under the foot traffic and competition from trees. Meanwhile, my "lawn" of violets, bulbs, duchesnea, bulbs and annuals continued to look lush and green through the summer, after a spring of glorious flowers.

Shade Mix:
Violet species. Consider trying a mix of these.
Bulbs where traffic is low: squill, wolf's bane, crocuses, snowdrops.
Duchesnea. Consider fragaria virginiana in low traffic areas.
 Lower-Order Paths

As we get down to the paths between and within garden beds, a new set of tools becomes important. This is where mulch becomes a much more sustainable option. Stepping stones can create permanent paths within garden plantings. And the "keyhole" garden can become one of our most important tools for reducing maintenance and increasing functional growing space.

Keyhole and Node Design

First, the Keyhole. In this book, I'm using the original meaning, which is simply a garden bed with a single access point, which is shaped like a keyhole. We can also call these round work areas, work "nodes."

These beds maximize growing space, minimize path space and weeding, and still have more usable work space than a row garden. These are space efficient in any garden, but when it comes to no-till, they become very important. This is because they allow us to easily mulch the path with minimal material and keep weeding to a minimum. One of the biggest reasons people swear of no-till is "I can't afford that much mulch!" As we'll see, a garden of keyhole beds will require as little as 1/10th the mulch of a row garden.

One of the most common complaints I hear about Permaculture techniques especially with polyculture and French Intensive garden design is: "sure that stuff works well for small gardens, but I've got a 5 acre market garden! On 5 acres, I have to use rows!"

A very interesting point was made well by John Jeavons in his book How to Grow more Vegetables. First, NOBODY can market garden 5 acres alone. I've visited many "5 acre gardens" that had less than an acre in crops, and usually even that could fit down to half an acre of efficient row space. And these gardens still require a lot of labor and fertility inputs because there might only be a half acre of crops, but there's still 5 acres of weeds.

But, as Jeavons said, and Grow BioIntensive proved, as land expands, per acre production goes down, so the highest productivity spaces will be small. Which is why 10,000 S/F of biointensive beds will grow the same yield as 3 or more acres of efficiently spaced row crops, on far less time and with no imported fertility.

Often a single, good, well-tended garden bed will produce more for a family than a big sprawling row crop garden that gets out of hand every season.

Don't believe it? Let's look at a 20' row garden as an example:

We'll use head lettuce because the 1' recommended spacing keeps the math simple. So a 20' area gives us room for 6 rows, or 120 plants. We'll use a standard recommended minimum row spacing of 30 inches, though I visit many farms with 3'-4' rows.

Keeping the 1' spacing and using a hexagonal planting we can fit 6 beds in our garden, with 312 plants, nearly 3 times as many plants. With wider spacing, this could be 4 times as many plants in the same space as many conventional gardens.

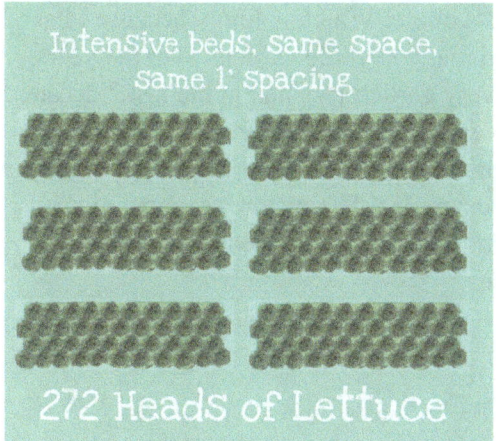

Using keyholes we can further reduce maintenance time and perhaps even cut the garden size in half in some cases.

One keyhole alone can fit 95 plants with room for companions—nearly as many in our whole row garden in less than a quarter of the size.

And a garden of keyhole beds will easily fit 380 plants compared to the 120 in our row garden, with more compact, easier to work beds with less space for weeds.

Assuming there was a standard border around the garden of 4 feet or so, usually necessary for a tillage garden, we can reclaim even more space with no till keyholes and work nodes.

On a 40'*40' garden, we gain even more efficiency, hitting 1,200 plants, while our row crop garden is at 220 plants.

So a modest 6,000 s/f of well designed permanent beds with path and node design can probably be more productive than multiple acres of successfully managed row crops, while being easier to maintain.

There may be a case for row cropping, such as if you've got venomous snakes around, but usually, I think it's a sign the design or goals could be re-evaluated.

A final complaint I've heard about this style of garden design is "sure it sounds good, but if it works so well, why hasn't anyone ever made an actual garden this way?"

That's a quote. Which is funny because I have visited many keyhole gardens over the years, and every garden I've had in the last 2 decades has been moving in this direction, and my gardens in the last 15 years have all looked like these examples. The pictures of most of the gardens in this book are beds or keyhole gardens.

How wide should paths be?

The question of path width comes back to the basic problem of paths, which is that they take time and materials to maintain, so we have to balance the usefulness with the labor and materials. For a quick example, a good general rule for grass paths is one mower-pass wide, however wide your mower is. If we add 2 passes, then we've literally doubled our maintenance time, but perhaps we add the function of being able to more easily walk with friends. If we add 4 passes, we might not get any more added use, but again doubled our time over 2 passes.

If the path is a woodchip mulched path, then doubling the size doubles the amount of woodchips and cardboard we have to use, doubling the number of wheel-barrow loads we have to carry as well. We've also doubled the area we'll need to weed. If it's a paved path, we've doubled our costs and carbon emissions, and so on.

So, our guideline here is that paths should be as wide as necessary to fill their function, but no wider than that. Ordered paths and keyhole and node design can help us again.

1st order paths: These small garden paths inside garden areas and beds can be very small, ideally just stepping stones. In the old gardening traditions, such as the best English cottage gardens, these paths were often ephemeral. They were designed to appear in the off season when plants died back for access when transplanting and work was required. In the season, the plants grow in, creating a full "border" or garden look.

This "closed canopy" garden, as Permaculturist Michael "Skeeter" Pilarski calls it minimizes weeding work and water loss, and creates a lush environment under the plants. It was a brilliant system best achieved with stepping stones or mulch piles and narrow paths around 1' wide, with larger nodes for work in key places.

This same system can be used in zone 3 and 4 plantings like 3 sisters that don't require frequent visits. The paths can disappear under a closed canopy by midsummer. Just make sure that you won't need constant access during the season!

2nd order paths: These regular daily garden access paths need to be wide enough to easily access by foot on a daily basis, around 2' wide or 1' with work nodes and care not to plant plants so they encroach on paths.

3rd order paths need to be wide enough to act as major garden access with a wheel barrow or for walking with friends and guests. As plants grow in you'll lose a bit of path in summer. My experience is that 4' is the minimum for walking with guests, and 6' is probably optimum unless you're doing regular garden tours with groups. 3' with nodes.

4th and 5th order depend entirely upon your equipment.

Haphazard Mulch Mandalas

Haphazard mulching is a term popularized by Michael Phillips in his book The Holistic Orchard. Each year a new mulch pile is placed around a fruit tree.

This pattern ensures that the tree always has materials of different ages and stages of breakdown, to supply different types of nutrients and habitat to different microbial communities, maximizing the microbial biodiversity. It also naturally creates a keyhole system of steppable work areas underneath productive trees. These mulch piles can also be used in rows in an orchard planting to create natural work keyholes and nodes. A series of these mulch pit nodes in a garden creates a natural, rustic mandala garden. Surrounded by nurse logs, they can be functional and beautiful.

Sometimes, when we think we need to make paths, all we really need are some piles of mulch.

The Walkable Stepping Stones Mix

The best solution to 1st order paths that I have found is to use a mix of repurposed or recycled stepping stones and walkable ground covers. The polyculture I use most often is a mix of creeping thyme, creeping chamomile, and white clover. In appropriate climates corsican mint works very well.

This can be very effective as a main path in small gardens.

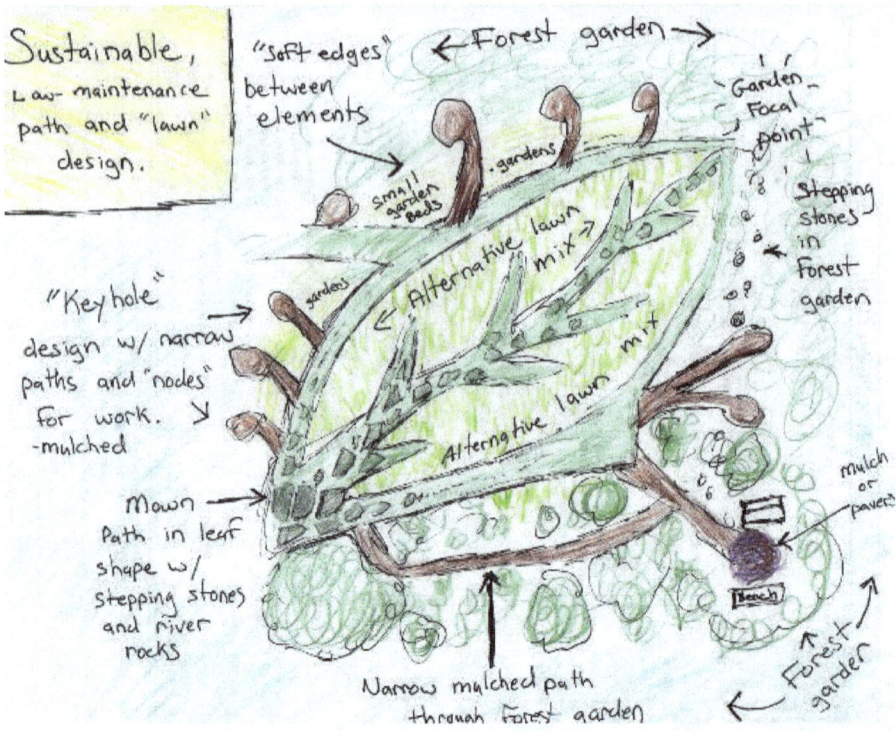

A garden design using the elements in this chapter: ordered paths, stepping stones, keyholes, nodes, walkable groundcovers...

Design by Michael Hoag, implemented by Mike Hoag and Scott Iwlew

Paths on Contour - Level Paths

One of the most basic facts of gardening is that we usually want to keep water in the garden, and that water naturally rolls downhill… so, if we put our paths on contour, it can slow down the water loss and help water soak in.

And yet, a drive around farm country in most regions will show that probably better than 50% of farms have their paths and furrows off-contour, helping water move away from their plants, and ensuring that it will cause erosion and waterlogging, too.

So, whenever possible our paths in productive landscapes should be level and on contour. This not only is good for our water, plants, and soil, it generally makes moving around the landscape easier, too.

EXCEPT of course, when it doesn't make it easier to move around the landscape. In some cases, it creates major harvesting difficulties to keep paths on contour. Water can always be captured down hill with basins, swales, or forests, but good convenient design can't easily be fixed. A big swales acting like a "moat" between the house and the main garden will be a daily inconvenience for the whole life of the swale.

Which is why for this book, we're taking a balanced approach and focusing on transformation. We can make sure that every garden area we plan will catch and store its own rainwater with appropriate tools for the job, instead of thinking that every path must be on contour. But, where possible and convenient, a level path keeps water problems from forming and having to be addressed later.

Therefore, always look for opportunities to put major paths on-contour. Always design gardens around main paths on-contour when possible.

Chapter 33: Places to Sit

Every landscape for people should have relaxing, enjoyable places for people to sit. This might sound obvious, but having toured hundreds of small farms and Permaculture projects, I'd say the vast majority have no planned seating areas. Most farms I've worked on have few or no places for laborers to take a break, or for guests and volunteers to sit and enjoy the experience. Most of the home native plant gardens I have toured also have no seating in them, symbolic perhaps that there is no room for humans in them.

On the other end of the spectrum, there's an often-repeated cliche: the garden seat no one sits in. We've all seen these inappropriately placed garden benches, sitting looking out to the road, exposed to the neighbors bathroom window, or positioned in a part of the garden rarely visited and with no view, overgrown with plants. A garden could have a dozen seats in it and still have no place for people to sit.

Yet, one of the most important factors in truly enjoying the garden is a space to sit and enjoy it from. And when it comes to work, nothing helps productivity more than productive rest in a space that is regenerative for us. The ability to sit in the shade next to work areas could be seen as a massive productivity-increasing and time-saving feature. And yet, few books on farming or Permaculture include patterns or projects for sitting.

Types and styles of garden seating are so varied that this could be a very long chapter indeed. There are many interesting books on garden seating with a goal of providing ideas and inspiration, so in this day of Pinterest and social media I will pass by that low-hanging fruit and try to give some patterns and ethical objectives on garden seating.

Ethical objectives? Yes, it would be a much better world already if mass garden seating became a cause of environmental or ethical concern. And yet, if we want the whole landscape to be regenerative, or even transformative, then we will need to put some thought into whether our garden seating is selected skillfully.

Beyond the ethical objectives, we can look at some types of seating areas in the garden and some patterns for making garden seating work well.

Ethical Objectives:

For me an ethical garden is always more beautiful.
Ideal garden seating should be made of materials that are safe, do not pollute, and do not contribute to social problems like oppression.
Materials should ideally be low-carbon, and have low embodied energy, which means they should be durable and last a long time. Locally made artisan seating and tables is preferable to corporate ones made with exploited labor and fossil fuels for shareholder profit.
Or they should be antique, used, recycled or repurposed materials.

Patterns:

Privacy. The first and most important pattern for garden seating is a sense of privacy, if it is necessary. As we said in the first chapter, the very first and highest impact project for making a garden habitable and workable is often creating a space with some privacy to work in. It is difficult for most of us to work, or to relax in a space that feels like a fish bowl where we're being watched. Fast growing screen materials might include bamboos, elderberries, sunchokes, and other sunflower family plants.

Views: When it comes to seating that provides active rest and relaxation, views are probably the next most important fact.r, and the hardest to create. If there are natural beautiful views in the landscape, these should be a mainframe element in your design thinking and marked on your zone and sector maps if you are doing a Permaculture design.

The Mountain Protector. We've discussed this pattern elsewhere, but humans tend to feel more comfortable with protection behind us and an open view in front of us, where we can easily see any threats approaching.

Higher ground feels good, generally. I suspect this is why we like to build patios, porches, and balconies.

Water: seating near water elements is a must. Most people will naturally gravitate towards the healing power of water and wish to sit next to it to gaze at the reflective surface and the wildlife that comes to visit there.

Aromatics. We've covered aromatic plants previously (page Number) and here is another example where they are very important. Most of my gardens feature aromatic plants near main seating areas. These again are spaces we naturally go to.

Fire: fire pits are another natural human element

Gradients of inside-outside. Outdoor rooms in zone 1 can feel nurturing and bring day to day life outside. A sense of enclosure is important to creating this effect. We can do many of the same techniques to create a feeling of enclosure and protection or even coziness outdoors as we'd do inside. Appropriate rugs, curtains, hangings, decorations are all welcome.

Mosquitoes and ticks and other pests. We should take natural efforts to reduce pests like mosquitoes in main outdoor living areas. We can and should do so without spraying poisons where we spend time. Some proven-effective natural mosquito tricks include:

—Run a box fan around seating areas. Mosquitoes are weak flyers and can easily be pulled into a box fan to their demise. Running a box fan near seating areas for a while before sitting out can dramatically reduce their numbers.

-Grow mosquito repellent plants near seating and living areas. Nepatas in particular have been shown to repel mosquitoes well, and being able to crumple a few leaves and rub them over our clothes can help.

—Mosquito netting is a surprisingly cost effective way to reclaim the outdoors. I spent most of the most notoriously mosquito ridden year in S.W. Michigan sitting under a sun umbrella with a mosquito net draped over it,

—Garlic spray. Many people are shocked to learn that Disney Land, notorious for the complete lack of mosquitoes in the middle of a Florida swamp, does not use chemicals or poisons to repel or control mosquitoes. The Magic kingdom uses the power of garlic spray, proven highly effective. More importantly, it has been demonstrated that the method of effectiveness is smell deterrence, not death, meaning that it does not reduce biodiversity in our ecosystems the way poisons do.

Types of seating areas to consider.

Oculus: Earlier I mentioned that most of the farms and gardens I have visited did not have seating areas for rest near work areas. The oculus is a pattern from the Permaculture world intended to remedy that. The oculus is a space where we can sit to view the work area to see what needs to be done or to rest. A good oculus is in the shade and has storage for a few tools and seeds. I keep a few basic hand tools in a mailbox near my oculus so I won't have to go back to the house if I forgot something, and a few basic seeds to sow in case I pull weeds and want to sow something in the spaces. Usually, this is the "useful cover crop mix" or the "beautiful polyculture mix."

Meditation space I always like to have a few outdoor meditation areas. I find them more productive than indoor spaces. A good meditation seating area will be in a quiet, secluded space with some privacy. A small alter or statue can provide a view and focal point. Grounding objects, objects that bring you into focus or concentration are good to have in a meditation area. . Aromatic plants are especially important here. Scent is a powerful emotional trigger, which is why the use of incense is so prevalent in religious institutions. If we practice getting into a peaceful, still frame of mind, and associate that with certain aromas, then smelling those aromas can help us get back into those states more easily. Running water for white noise, or wind chimes can help mask noises and create a more conducive environment. Most people find the sound of water naturally soothing.

Meeting and talking spaces for groups of different sizes. Most of my gardens have multiple spaces where people can sit and have a private talk. Often at parties and get-togethers these are welcome and used.

Outdoor dining and picnic areas. Absolutely vital, in my opinion. There is nothing more natural and essential than eating in the great outdoors, next to our produce.

Garlic Mosquito Repellent, Disney recipe:

The Disney recipe is a liniment mixed with water and sprayed around the periphery of the property.
Ingredients;
Rubbing alcohol or 100 proof Vodka.
10 Garlic bulbs or 15 bulbil heads.

Crush garlic (do not chop) and cover with alcohol. Let sit in a cool dark place for at least 1 week. Mix 1 part to 10 parts water and spray around periphery of property. Reapply after rain.

Garlic Mosquito Repellent, Mike's recipe:
20 whole garlic plants.
5 gallon water bucket.
Additional plants can include lemon balm, lemon grass, catnip.

Preparation: Cover with water and cold steep for one week or until strongly aromatic. Spray around periphery of property. Reapply after rain.

Gardening Skills: How to Sit in a Garden

Sitting well in the garden takes some preparation and practice to get good at.

One of the first steps is to find a good place to sit. This can take some experimentation. As we move around any landscape, we'll find there are places that naturally make us feel more at ease. This is hardwired into our biology. Just as mice are hardwired to feel safe next to walls, and rabbits always need two routes of escape to settle down. We tend to like protection to our backs and a nice open view in front of us, for example. Shade is probably good. The Buddha found enlightenment under the Bodhi tree. Sit here and there and find places that feel like good sitting places.

As you sit, take a few deep breaths. Sitting in the garden well takes some focus, so take the time to get ready to really have a good sit.

Look around. The plants you see have evolved over 2500 million years. The insects you see have evolved over 480 million years. They've had a long time to get good and efficient at being plants and insects, finding their places in the world. And one of the most important abilities to thriving in this world is to be very good at observing and resting.

The bird in the tree, the frog sunning on a rock, a dragonfly, still at the edge of a bird bath, they are relaxed yet dignified, calm yet awake and observant to the world around them. We humans have great imitating capabilities, and we can often learn surprising athletic feats simply by observing and just doing it. We feel it in our bodies when we see someone do such a thing. If you have the chance to watch and imitate one of these masters of sitting, don't pass up the chance for this masterclass. This is worth getting really good at.

If you've got other things on your mind, put them aside for a few minutes so you can do a good job of just sitting there. "Don't just do something, sit there." It might be the most important work you do all day. Many times, I've gone to the garden with an idea of what had to be done, and I was so focused on it that I couldn't see what was around me. Only when I took the time to do a good sit could I see that there were much higher priorities to be attended to. Many times I have discovered an opportunity while sitting that I had missed when I was gardening.

If you're really busy, taking the time to sit there can be the best way to make more time. Our buttocks have a special organ that creates more time when pressed against a seat. When I feel busy and stressed, my mind is too messy to prioritize well, and I waste time worrying about how little time I have.

If I can find time to sit, I find I actually have much more time than I thought I did.
Feel that flutter in your chest, there's a dull ache there... that's time's hook, pulling you ever into the future. What's it saying? I've got to email so and so back. I've got that deadline to meet. Try to relax that spot with a breath and time's hook might come out. There's nothing more important than this sitting and breathing and observing you're doing right now.

Breathe in, see the garden as though it is the first time you've ever seen it in your life. As though you were born with that breath into this brand new universe. The air you're breathing is, in part, a gift of the plants around you. The polyphenols they put in the air are proven to improve your immune system and help you relax.

Breathe out, and let go... if this was the last breath you'd ever take in this life, do it so you'd die content, enjoying the heck out of this moment of just sitting there in the garden. Breathe in, as though born, and see the world for the first time again…

Look around. *What a lovely day.*

Beauty in Abundance

Adventure Step 8: Plan your major paths, structures, and hard-scaping.

Next on our adventure of transforming the landscape is to start laying out the plan for paths, structures, and hard-scaping (any paved, concrete or permanently installed surfaces.). This is such an important element that - even if you do none of the other design sketches in this book—I highly suggest you do this exercise. Once we begin any serious landscape transformation, the first thing that needs to happen is the hard-scaping, or any we put in later risks damaging trees or plants. So we'll need at least a design sketch to start with.

Start with a print off or sketch of the site. For this, it is also helpful to have a contour map showing our contour lines. For most sites, putting as many paths on contour as possible will help naturally catch and store rainwater.

- How will path hierarchies work?
- Where are the major structures in the landscape?
- Where are any minor structures like sheds, arbors, or gates?
- Where are any seating areas? Will these be paved? Mown?
- Looking at our uses overlays, which areas will need permanent major access paths for cars and equipment? Can these be on-contour?
- Which will only need seasonal access?
- What uses will require regular foot access for groups, or for small equipment?
- Which areas will need small, mulched paths? If these are in productive garden areas, keeping as many on contour as possible will be very helpful in collecting water, and reducing erosion.
- Where can we get away with simple stepping stones and mulch piles?
- Are there opportunities for patterns in the landscape?

Part 5: Aesthetic Guilds for the Beautiful, Abundant Garden

Of all the techniques in this book, there is one which is most important for transforming the landscape, and that is the guild. As we have said before, guilds are human-designed plant combinations intended to function like natural plant communities. **Guilds allow us to invest our energy into setting up self-maintaining, highly durable, productive plantings.**

While in the Permaculture community, the word has gone on to largely mean the same as "companion planting", in its original sense, a guild is a set of organisms connected by needs and energy transactions such that they fill up niches. Since there's little wasted energy, there's little room for weeds or pests to move in and become a problem. A true guild is different from a companion planting in that a guild is stable over time. Meaning, once we become a part of the guild, harvesting from it as the insects and wildlife do, it becomes largely stable with our harvest, so that little work other than eating is required.

Mind you, even the best guild can get out of balance and need some work to pull it back into stability. But it should not need constant inputs or work to maintain or keep weeds out. If it does, it is not yet a guild and something is missing.

And this is why stable guilds are the most important technique when it comes to transforming landscapes. They bank our labor, allowing us to invest long-term in transforming the landscape in a way that is low maintenance and stable. In fact, we could say that landscape transformation is a technique of guild building.

Types of Plantings

Monoculture: A planting of just one plant.

Polyculture: A planting of multiple plants. Research has found this has many benefits to plant growth and health. This book contains many samples of polycultures you can copy to get started with learning.

Integrated polyculture: A planting of multiple plants mixed in together. Usually this is what is meant when people say "polyculture."

Interplanting: A polyculture designed to take advantage of the growth habits and timings of plants so they fit together without competition. For example a succession of peas followed by a succession of cabbages. Or corn grown with shade-loving root crops beneath.

Companion planting: A polyculture where plants are intended to help eachother grow. For example, marigolds can help repel soil nematodes. While some companion plantings have been proven through research, most have found that supposed companions do no better than interplanting. The benefits come from polyculture, not from the specific companions.

Chapter 34: Guild Design Basics

> **The 3 Types of Guilds:**
>
> 1. Guilds designed to keep grasses out
> 2. Guilds designed to coexist with grasses
> 3. Guilds designed to fail
>
> When we don't know which we have, it's probably #3.

*This is hard-won knowledge. I call this **the first rule of guilds**. It is one major thing that separates guilds from companion plantings.*

The first rule of guilds: There are three types of guilds:
1. Guilds designed to keep grasses out.
2. Guilds designed to coexist with grasses.
3. Guilds designed to fail.

When we don't know which we have, it's almost certainly #3.

There is a 4th kind, which is a guild designed to be maintained by human weeding, but again, this could better be called a companion planting, since the ecological concept of a guild is that it should have some stability (with our harvesting, of course.)

That stability is also usually what people expect, too, so they are usually disappointed if/when the guild copied from a Permaculture book is overrun with grasses and their plants choked out.

So, if we want a guild to be successful, there are many strategies we can use. Here are a few:

Type 1: install fortress plants, weed barriers, edging, a guild matrix.

Type 2: grow field plants that thrive in grasses in your region, like Turkish rocket, chicory, and asparagus.

We talk about both of these in this book.

Beyond this first rule of guilds, guilds should be a relatively stable community and resistant to invasion, so we can look at nature to find what makes guilds work.

We can never know all the roles that are played by plants in a stable plant community. These relationships are far too complex. But we can use some of the most basic roles we can observe, and we can imitate these, hoping that we will design something more likely to achieve stability.

What are some of the jobs we can observe being done in ecosystems? There are plants that maintain the edges of the community, giving it a definite, stable structure. For example, forests are typically edged by species that are good at succeeding into grasslands. We'll call these "fortress plants." Other plants increase stability in other ways. For example, some are good at maintaining soil conditions, by fertilizing the soil, or covering the soil to conserve moisture and regulate biological communities. Others increase total system health, such as by increasing insect and microbial biodiversity.

So, we can look at these roles in nature and attempt to imitate them.

We can also look at the ways natural communities minimize competition between elements and maximize competition. In this way, we can get good at fitting plants together the way they fit together in natural plant communities.

Approaches to Creating Guilds:

In Gaia's Garden, Toby Hemenway proposed 3 approaches to creating guilds:

1. Mimicking nature. We can find natural plant communities and attempt to replicate them. This book shows many examples like this.
2. Copying from books. This book includes many examples of guilds you can copy and adapt to your local conditions. They might not be perfect, but they'll be a good start.
3. Being nerdy and looking at guild roles and trying to create something that will work.
4. And I will add a 4th, just assembling together enough biodiversity. If we can fit maybe 30 species into a small space, including species that thrive at different times of the year, then it will probably turn into something stable. For all of these, there's one tool that will help ease the way towards success: the Guild Matrix.

Patterns for Dealing with Grasses
Lillie House Case Study

1. the first rule of guilds is pick one: design guilds to keep grasses out, design them to persist with grasses, or they are designed to fail.

Guilds designed to keep grasses out

Terminator guild designed to eliminate grasses

Edible meadow guild of grass tolerant plants, used for slashmulch three sisters.

Mown grass paths next to grass free areas

Little bare soil means no grass germination

Fortress planting near grasses.

Shade, woody mulches suppress grass

Deep mulch gardens with fortress plantings

Hedgerow with fortress plantings

Guild Design Roles

Guilds are plantings designed to provide multiple harvests and reduce work by getting plants to do it for us. They often are used with fruit trees, but can be used to serve annual garden beds, nut trees, or other crops.

Nitrogen Fixers: These plants usually rely on a beneficial relationship with microbes that take nitrogen from the air and deposit it in the soil. Alders, broom, clovers*, leadplant, goumi*, bladder senna, lupines*, hog peanut*, sea buckthorn*, siberian pea shrub, blue false indigo, groundnut*, perennial pea*, black locust, legume crops*.

Dynamic Accumulators: These plants bio-accumulate large amounts of nutrients and minerals in their tissues, and it's thought that these become available to other plants when the plants die back. There is controversy around the scientific terminology, but as a practical model for design it is very useful: most trees, carrots*, cleavers*, clovers*, comfrey, sorrels*, chives*, alliums*, chicory*, fennel,* yarrow*, valerian*, watercress*, willow, thistles*, sunflowers*, parsley*, nettles* borage*, lambs quarters*, good king henry*...

Insectary Plants: These attract beneficial insects such as pollinators and predatory insects: buckwheat*, bee balm*, all carrot family plants*, almost all asters and sunflowers*, almost all mints*, yarrow*, fennel*, dill*, pennyroyal*, valerian, oregano*, thyme*, comfrey*, lavender*, borage* native plants*... **see the chart of insectary plants in this book**.

Mulch-makers: These can used to grow your own mulch: Cardoon, sorrels, blood-veined sorrel, pokeweed, Jerusalem artichokes, squash family plants, sweet potato, potatoes, yarrow, marshmallow, catalpa...

Ground-covers: Thymes*, strawberry*, perennial chamomile*, white clover*, claytonia*, miner's lettuce*, wild cresses*, snowdrops, violets, duchesnea...

Fortress plants: these plants tend to fight off grasses and weeds: thymes, oregano, comfrey, woody perennial herbs, sorrels, pennyroyal, valerian, walking onions*, Turkish rocket*, sweet rocket*, chives*, daffodils, alliums, sunchokes, dwarf sunchokes....

The Guild Matrix: The Most Important Aspect of Planting Design?

Right now, this guild matrix idea seems like the most important concept in guild and planting design to me. Which makes this one of the most important elements of the most important concept in this book. So it's worth taking a few minutes to elaborate on.

This spring, I am breaking ground on a few new garden projects, so I'm really thinking about this idea of practically transforming landscapes.

I am pretty confident I will be able to transform about 1/4 acre of land to highly productive, fairly mature ecological garden in the first year with no tiller, very little digging, almost no imported fertility, with only one or two flats of seedlings, and minimal time. (Update: I did successfully achieve this goal!) How? All-star plants and the guild matrix. I started using the term matrix while thinking about how most plant communities are defined by a few key interlocking players other plants grow into. It's like chocolate chip cookies. You have the chips, and the cookie dough matrix that holds the chips together. A really great cookie requires both to be good!

All-star plants is designer/teacher Jason Padvorac's clever term for the type of plants that make great guild matrix candidates.

To understand the guild matrix, we'll compare 3 edible ecosystems.

First, imagine you walk into an ecosystem that is lush with plants, and some of them are edible. The value of this system is going to depend on how many are edible and how much work it all takes to harvest and maintain, right? If it takes more work to maintain than the value you get out of it, it's not going to be valuable! In a few years, that kind of garden usually fails. For example, people who plant only fruit trees and understory support species, or maybe the whole understory has gone to comfrey, because they're thinking only of the long term with no year one yields. Three years in, the only yield is a burned out gardener!

Second, imagine you walk into a lush forest and all the plants are edible! Thing is – that alone STILL isn't enough to make the garden truly useful! Perhaps they are low value foods. Perhaps no one you know likes them. Perhaps they take a lot of processing and it isn't economically viable. Perhaps it's all one plant, like Japanese knotweed, which produces heavily one week of the year,

producing a glut too large to harvest, and little else the rest of the season. Or maybe it just takes too much work and cost to maintain.

Last, you walk into a third forest, this one has mostly edible plants. And, here's where the magic is, this one gives exactly the amount of produce you can reasonably harvest or sell, staggered all season long. And these aren't just "you can eat that" weeds, it is highly valuable produce, desirable and useful to you in the quantities needed for meals. And because the plants are vigorous, spreading and symbiotic with each other, 90% of the maintenance that is required is just harvesting to thin the plants and keep things in balance. All we have to do is eat from the garden daily all season long.

Clearly, the third one is going to be our goal, yet most people never really set out for that, let alone achieve it. Yes, perfection may be nigh impossible, but we can always move closer to it, and the closer we can get to it, the more valuable the garden will be.

What gets us dramatically closer to that goal is a "guild matrix" of good all-star plants. If you observe natural plant communities, you will see that most are characterized by a "community" or "guild" of 2-5 players that seem to work well together, fit together in filling in niches, share similar needs, cooperate well and avoid competition. There is usually a basic matrix of a few plants that make up much of the ecosystem, with specimen plants appearing here and there like chocolate chips in a cookie. This is what we are going for.

For example, in much of the Eastern Woodland, especially under maples, you can walk into deep woods and find a matrix of ramps, solomon's seal, claytonia virginica, ostrich fern, and perhaps may apple, sweet cicely all rambling around together. And sometimes, this understory can look as dense and productive as the most dense, intensive SPIN farm you've ever seen on YouTube. That is the matrix other things grow into. Here and there tucked into it we may find currants or gooseberries, or elder, or service berries, paw paws, cherries, and mast trees in the overstory.

In moist acidic woods we may find a matrix of wintergreen, ostrich fern and blueberry. Witch hazel, currants, and mast trees may be tucked into that.

But what we also want is to make sure our "matrix" is as useful as possible, producing lots of plants, growing fast enough to handle fairly heavy harvesting, and productive over a long season, and ideally multi-purpose. Walking onions are the most useful plant in my garden, next to garlic. Hardly a day goes

Approximate Mowing Times:
First Cut: As close to the first sign of spring bulbs as possible. Mow as short as possible.
Second Cut: After daffodils have started to die back, but before summer perennials are too tall. Mow as high as possible.
Final Cut: In fall, after perennials go into dormancy, you may mow again on the lowest setting.

by when I don't use both. I can harvest them both in different forms from the time the snow melts until after the snow is a foot deep, and bring them in for storage even in the winter. And both can be used for mulch, and to create pest and disease repellent sprays. Garlic spray is famously used at Disneyland to repel mosquitos! Both can help repel grasses in the right conditions, and keep weeds away.

It would be hard to imagine having too many of such useful plants. Luckily, they self produce rapidly, too! With careful harvest, it is easy to maintain populations of both by just letting them self sow and divide. Excellent guild matrix candidates.

Sorrels are delicious, productive all season long and provide a green for daily soups or salads and act as fortress plants suppressing weeds in the garden and creating a stable matrix. Alliums can easily grow vertically among sorrels, while sorrels spread and cover the ground to prevent weeds.

A wild and mild mannered mint might ramble around and fill in spaces too, and these have been known to both prevent weeds and encourage net biodiversity in the right conditions. Mints have been shown to be particularly good at eradicating field bindweed. Yet the effects of allelopathy are minimal when the garden is abundant and the soil is healthy. But if a weed is getting out of control, these vigorous allelopaths will crank up their chemical production and keep the system in balance. I know many people who simply can't grow enough basil, but are afraid of growing mild mannered basil mint in the garden!

Crosnes are a mint that is an amazing fortress plant, useful as tea, and produces a great edible tuber in huge amounts. And the 4 of these plant friends together have a strong likelihood of creating a stable, useful groundcover throughout a garden. And, they provide tons of useful, edible biomass all season long.

Now, when you walk into the garden, everywhere you look, there is food that can be harvested, every single day of the year, and we haven't even gotten to our "featured plants" yet. That's a garden that is guaranteed to be highly valuable

A few Examples of Possibly Successful Guild Matrixes to Try

Spring bulbs, basil mint, walking onions, garlic, sorrel.

Spring bulbs, alliums, monarda fistulosa, Turkish rocket, brown-eyed susans.

Spring bulbs, monarda didyma, dwarf sunchokes, groundnut.

Edible spring bulbs, crosnes, skirret, Turkish rocket.

Spring bulbs, wild strawberries, sheep sorrel, mountain mint.

Resources on Guilds:

1. Wayne Weiseman, Daniel Halsey, Bryce Ruddock, *Integrated Forest Gardening*, Chelsea Green, 2014. This book provides us with theory and models for guilds.
2. TobyHemenway, *Gaia's Garden,* Chelsea Green, 2009. Excellent plant lists and model guilds.
3. Bryce Ruddock, Milton Dixon, *Free Plant Guild's E-Booklet,* Midwest Permaculture.

Chapter 34: A collection of Aesthetic, Functional Guilds for Beauty and Abundance

The Beautiful, Bountiful, Scatter-Sown Polyculture

A variation on the Lillie House Polyculture

The scatter-sown polyculture or "mixed vegetable garden" is one of the easiest ways to start a new garden, a great way to learn about gardening, and it's a Permaculture classic. The basic idea is to take a handful of a variety of seeds and throw them onto bare soil and a whole dynamic diverse healthy garden comes up from that.

Pretty much like magic.

As the plants come in, we harvest a whole succession of vegetables, starting with baby greens and picking and choosing things to thin the garden out to appropriate spacings as everything comes in. The result is a whole season of greens from one planting, with very little room for any weeds. Because of the intermixing of plants, disease and pest pressures are also very low and the garden conserves water and fertility well, too. Such seed mixes can be used to start new gardens or as a productive cover crop wherever the soil is disturbed.

There are a few famous polycultures including the Ianto Evans polyculture or mixed vegetable garden, available in a few places online with an internet search. Gaia's Garden by Toby Hemenway also has the Jajarkot polyculture. Permaculturist Sepp Holzer has his own seed mix that he uses wherever soil is disturbed.

This Beautiful, Bountiful, Polyculture is one I have used for over a decade now and tested on dozens of sites. It should work in most climates and zones. With different soils and circumstances, the results are always different, but it reliably yields a beautiful and functional garden. What's different about this mix is that it is designed with beauty in mind for the edible ornamental landscape, and it uses "ecological succession," mixing annuals (plants that only live one year) with perennials (plants like asparagus that live for many years) so that it transitions from an annual vegetable garden into permanent guilds of perennials and self-sowing annuals.

As the plants fill in, we can thin them, eating all the thinnings as greens, so that we keep good spacing where the plants are not quite touching, or are barely touching.
For flexibility, I've included two slightly different variations.

The Super Simple Main Mix:
SLOW IN SPRING, MARCH-MAY
1 package "mesclun mix."
Purple boc choi
Red mizuna
Optional:
1 package Garden sorrel.
Carrots
Small white radishes, don't use daikons for this.
Walking onions if available.

The ULTIMATE Main Mix:
SOW IN SPRING, MARCH-MAY
Butter lettuce (sometimes considered the best self-sown lettuce.)
Arugula, or dragon's tongue arugula
Garden cress
Carrots
Small White Radishes
Walking onion top sets, maybe 5 for every 10' of garden.
Salad burnet
Garden sorrel
Blood-veined sorrel
Optional additions:
Turkish rocket
Nasturtiums
Red orache
Amaranth, Hopi red dye
Note: Don't add brassicas for spring in this kind of garden.

Not too many of these (these are allelopathic and will keep other seeds from germinating if we sow too many. This is the usual reason polycultures fail is that if these are over used it's all that germinates. Still some can be included.)
Carrots
Bulb fennel
Dill
Mustards, especially red broadleaf or mizuna
Radishes
Parsley

PLANT IN SUMMER. (Approx. last week in May.)

By the last week in May, the garden should be filled with lots of veggies. As we pull some veggies, we can plant in some of these.

Consider adding a scoop of compost with each if available:
Bush beans
Bush Squash, like bush delicata or patty pan squash.
Bush cucumbers
Cilantro

START FROM SEED in spring (or buy plants,) Plant in SUMMER
Tomatoes (especially "Galapagos," which can become self sowing.)
Peppers
Perennial herbs like chives

ADD IN LATE SUMMER (Start these indoors in June, plant in the shade of taller veggies in late July or August.)
Cabbages
Redbor kale

SOW FROM SEED IN LATE SUMMER:
Radishes
Golden beets
Boc choi
Garden cress
Plant in fall if desired for next year:
Garlic
Potatoes
Any perennials

FALL: PUT THE GARDEN TO BED.
"Chop and drop" any plant materials such as tomato and squash vines to make a nice mulch over the garden. In the spring this can be pulled back to reveal a nice weed-free bed for next year's garden.

Beauty in Abundance

The Aesthetic Edible Meadow Guild that Keeps Neighbors Happy

An oversized lawn can be converted to an ornamental edible meadow filled with food, native plants, and flowers all season, which takes less time to care for than the lawn.

Twenty years ago when I first heard the words "food forest" I immediately wondered "can we model food systems on other ecologies, too?" And of course my first thought was to edible grassland ecosystems. This would be valuable since grasslands sequester a significant amount of carbon, provide habitat for a wide range of plants and wildlife, and since most of our grain crops grow in these naturally. So I began reading, and just as there are models of traditional forest gardens around the world, there are models of edible grassland ecologies, too. This is a large and complicated topic, and while there are human-food edible grasslands, the most common way humans turn grassland into food is by feeding them to ruminants.

That is a much larger topic than we can cover in this book, and there are already many books on it better than I could write.

But there is one very important pattern we can put to use in this book, the beautiful bountiful meadow, an edible ornamental grassland.

This has special value in this book, because many homeowners who want to go sustainable start by giving up the lawnmower, and what follows is usually hate mail from neighbors, and fines from the municipality. Ironic, since such meadows have become highly fashionable for the wealthy on estates and in rich neighborhoods, yet, middle class folks are legally forbidden to emulate these more beautiful, functional, wealth-building landscapes. Of course, such luxuries are for

the wealthy, the lower classes must feed the lawnmower gods.

Others try to establish native prairie type systems, and are disappointed to learn that in most cases, this requires large amounts of tilling and poisons on a regular basis, and still these systems end up dominated by "invasive" Eurasian grasses and "weeds." I myself worked for years helping people create such systems and studying to learn how to do it in a way that would work. As of writing this book, I discussed the topic with several university and industry experts on the topic and can say (sadly) no one has figured out how to create these systems yet without poisons and tilling.

And yet, this is EXACTLY what more of us need to start doing, putting aside the lawnmower, poisons and tilling. As soon as we stop mowing an abundance of high-value plants and wildlife start showing up to enrich our lives.

So what these homeowners need is a method for converting lawn to useful, ecological, beautiful grassland systems.

So the beautiful bountiful meadow system is a way of doing exactly that. Instead of being destroyed by grasses, it is based on the Eurasian grasses that tend to dominate as we convert from lawns. And most importantly it keeps the neighbors from hating us.

Steps to create a beautiful, bountiful meadow:

Step 1: Start with bulbs. The project begins with a fall planting of bulbs and some special perennials. If you would like to start sooner, then begin by collecting plants to use in the meadow.

I'd like to stress the importance of beginning with a good bulb order. This is the key to the transition working. Bulbs are one of the cheapest ways to start a flower show, and that show starts first thing in spring. If the neighbors see the new meadow flowering abundantly first thing in spring, it will train their brains to see "flower meadow" instead of "untended blight." It also fast-forwards succession by adding biodiversity that competes with the grasses and creates niches for other plants to get established.

Many bulbs have a second advantage in that they naturalize easily in grasses, giving many years of flowers, or even increasing in time with this system.

Good bulbs to start with are daffodils, look for varieties that are good for naturalizing) camas, crocuses, allium unifolium and other alliums, and garlic as a major edible in the system. The garlic will naturalize and become a self-sowing productive element. All of these should survive as the meadow develops.

Step 2: Include fall perennials. Along with the fall bulb planting, I add strategic clumps of a few special plants. These Perennials have been selected for thriving in lawn, tolerating a little mowing, sowing lots of seeds, and most importantly, spreading by runner to outcompete the grasses. Some plants that are especially good in this role are monarda fistulosa, mints including American field mint and corn mint, native to the Northern hemisphere, crosnes or Florida betony, short-toothed mountain mint (the only one that spreads by runner) oxeye daisy, sunchokes (dwarf sunchokes like Sunray are more desirable here) and goldenrod. Asparagus deserves special mention as one of the best edibles to start with, even if it doesn't spread by runner. Asparagus crowns are cheap and easy to plant into lawn.

By planting a few starts or clumps of these plants into the grassland, you'll be ensuring season-long flowers in year one and a diversity of spreading flowers by year three.

Step 3: Summer planting. In the first growing season, plant a variety of additional perennials and annuals that can become self-sowing and naturalize in the system. See the included list. Using a **slash-mulch system** or planting into **resident vegetation,** we can plant some seeds, but my preference is to put my money into planting starts. One successful start can produce hundreds or thousands of seeds, and sow them at the time they are most likely to establish, while I would be wasting money on seeds while getting low germination rates. Better to let the plants become little seeding factories for me.

Plant in clumps. Biodiversity makes good for biodiversity, and putting a clump of a few plants near each other will make it easier for the plants and easier for you to care for them. For aesthetics, use odd numbers and somewhat uneven spacings to make the plants look more natural. Plant near the most visible areas first, then think of creating color echoes in the landscape.

Step 4: Maintenance:
One element of maintenance will be to eat your edible meadow. Gentle occasional digging of root crops and harvest of herbs will create disturbances to help the meadow become richer and more diverse.

Beyond eating, the most important maintenance issue is cut timing. Most homeowners will still want to occasionally mow their meadow to keep it from looking too wild, and there's some evidence that one or two mows per season will create the maximum biodiversity.

Cut timing is very climate-sensitive, but there are a few things we're looking to accomplish:
- Set back grasses. This can be done by mowing short at their peak growth in late spring or early summer in many climates. Mowing right as they begin to flower is a second way to steal energy from these plants when they are vulnerable. It also prevents them from sowing as many seeds. This coincides with hay season.
- Create niches for broadleaf plants. Very short mowing or grazing in the dormant season can help open up room for broadleaf plants to germinate.

- Generally minimize mowing. When cut, grasses put their energy into running laterally and spreading, creating thatch that chokes out broadleaf plants. If the grass is let to grow, it puts its energy into getting taller to compete with the other plants. This creates an open understory under the grasses that our flowers can germinate in. This is why regular mowing was considered important to prevent "weeds" in lawn.
- Avoid harming the broadleaf plants. This means we need to avoid cutting them when they are about to go to seed, and we need to let them store up energy for next season. In spring, this means waiting until spring bulbs like daffodils die back. This is my signal that it is time to mow. There is usually a magic window where daffodils are ready to be cut while summer perennials have yet to really take off and won't be badly harmed. This summer mow I do with the mower on the highest setting.

Plants That Succeed Into Grassland

Keeping in mind our first rule for guilds, one may wonder what plants can we plant that will coexist peacefully with grasses and remain relatively productive? A drive around the countryside near your home can often be instructive, especially if your municipality hasn't done too much tax funded spraying. What do you see? Asparagus? Milkweed? Dame's rocket?

The following list will be a good start, though each region is different and mileage will vary depending on soil and precipitation. Remember a general rule is the more species there are, the

Plants that often thrive with grasses	Plants I have seen thrive with slashmulch
Allium unifolium	Corn
Asparagus	Beans
Blood veined sorrel	Squash
Breadroot	Amaranth
Camas	Quinoa
Chicory	Potatoes
Comfrey	Skirret
Crosnes	Sunflowers
Daffodils	Pea
Dames' rocket	Melons
Dandelions	Cucumbers
English daisy	
Garlic	
Mints, if there are enough other species	
Monardas	
Oregano	
Scorzonera	
Shasta daisy	
Sheep sorrel	
Sunchokes	
Thyme (usually only on sandy soil)	
Turkish rocket	
Yarrow	

Edible meadow case study: Cold Climates

This long-lasting meadow filled with edible plants is in Northern Michigan, and only required rare mowing. It had persisted for at least 50 years.

The main edible guild matrix consisted of American field mint, monarda fistulosa, wild strawberries, sheep sorrel in abundance, and ground cherries. This matrix would be happiest on sandy, somewhat depleted soils. Grasses will out-compete strawberries where fertility, water, and sun are good.

Often, strawberries will persist at a field edge, rambling through a hedgerow to maintain a population, then making sly forays out into the sunny field to set fruit when the time is right.

Case study 2: warm Mediterranean
This naturally occurring edible meadow in a Mediterranean region is absolutely filled with food, flowers, and beauty. Nearly every broadleaf plant is edible: radishes, mustards, corn mint, monarda, borage, fennel, and carrots. Historically, similar systems in the U.S. would have been filled with camas, yampa and breadroot.

A Front Yard Three Sisters Guild

This Native American guild majors in corn, beans and squash, but can be grown without tilling in an edible meadow system. Yields can be as high on corn as modern intensive agriculture, while also yielding many other vegetables. This version uses highly ornamental varieties so it is beautiful as well as abundant. That will keep the neighbors or farm guests happy.

Many gardeners today have heard of the "three sisters garden," a set of traditional ways of growing corn, beans, and squash together amongst each other, typically along with other crops, as well.

Rather than one specific system, this refers to a set of regionally adapted, appropriate systems for growing these crops. Different tribes had different systems and there were probably three main broad patterns that were adapted to local conditions.

For those interested in historically accurate and culturally informed versions, I will again recommend that you seek out local native experts who have stewarded the knowledge of how these crops were grown in your region. They will have the best information on the best variety selections for performance and may also have culturally significant wisdom to share.

But since these systems were all adapted to local conditions, I would like to offer one adaptation, an aesthetic three sisters guild that helps us hide vegetables right in our front gardens without anyone complaining. It is based on the Wampanoag-style planting used in much of the Great Lakes and Eastern Woodland Region.

In this planting, we will use some highly ornamental varieties, which I am told are highly well-adapted to three sisters plantings. In my experience of growing three sisters gardens over twenty years, this planting works as well as any other I've personally used.

Corn: Japonica maize. This variegated flint corn has beautiful foliage colors and scarlet tassels. Even corn farmers have asked me what this extraordinary ornamental plant is! The corn is an excellent corn for homesteaders, useful as a flour corn but also edible in the milk stage as a sweet corn when picked at perfect ripeness.

Squash: I recommend moschata varieties, the species most resistant to pests and diseases in most locations. Tonda podana is a particularly aesthetic squash for a semi-ornamental planting. Summer squash can be interplanted for 60% of the squash, and then removed as the moschata vines need more space.

Beans: Hyacinth beans, scarlet runners, or purple beans are all excellent choices. Hyacinth beans in particular will have excellent color echoes to the Japonica. It is important to plant these highly vigorous beans after corn has already grown to nearly knee height, or they will pull down the corn.

Extras:
Sunflowers are a traditional addition in some three sisters systems, and red-flowered varieties are a particularly attractive addition to the Japonica. Traditionally these are usually planted to the north of the corn mounds, so as to not shade the other crops.

Amaranth is another forth sister in some traditions. Scarlet-colored varieties like Hopi red dye add another beautiful color element.

Wampanoag planting design:

In this tradition, corn is planted into mounds, with one plant to each direction, the North, East, West, and South, 1 foot to 16 inches apart. These will look like one, beautiful otherworldly plant.

Corn mounds are planted 5 feet apart into a grid or possibly 6' in Northern areas or areas with a lot of cloud cover.

Squash is planted at the same time as corn, into mounds inside the diamonds formed by the corn mounds. The squash is integral to keep weeds down and suppress grass growth.

When corn reaches knee height, beans are planted into a ring around the corn, to use the stalks as a trellis.

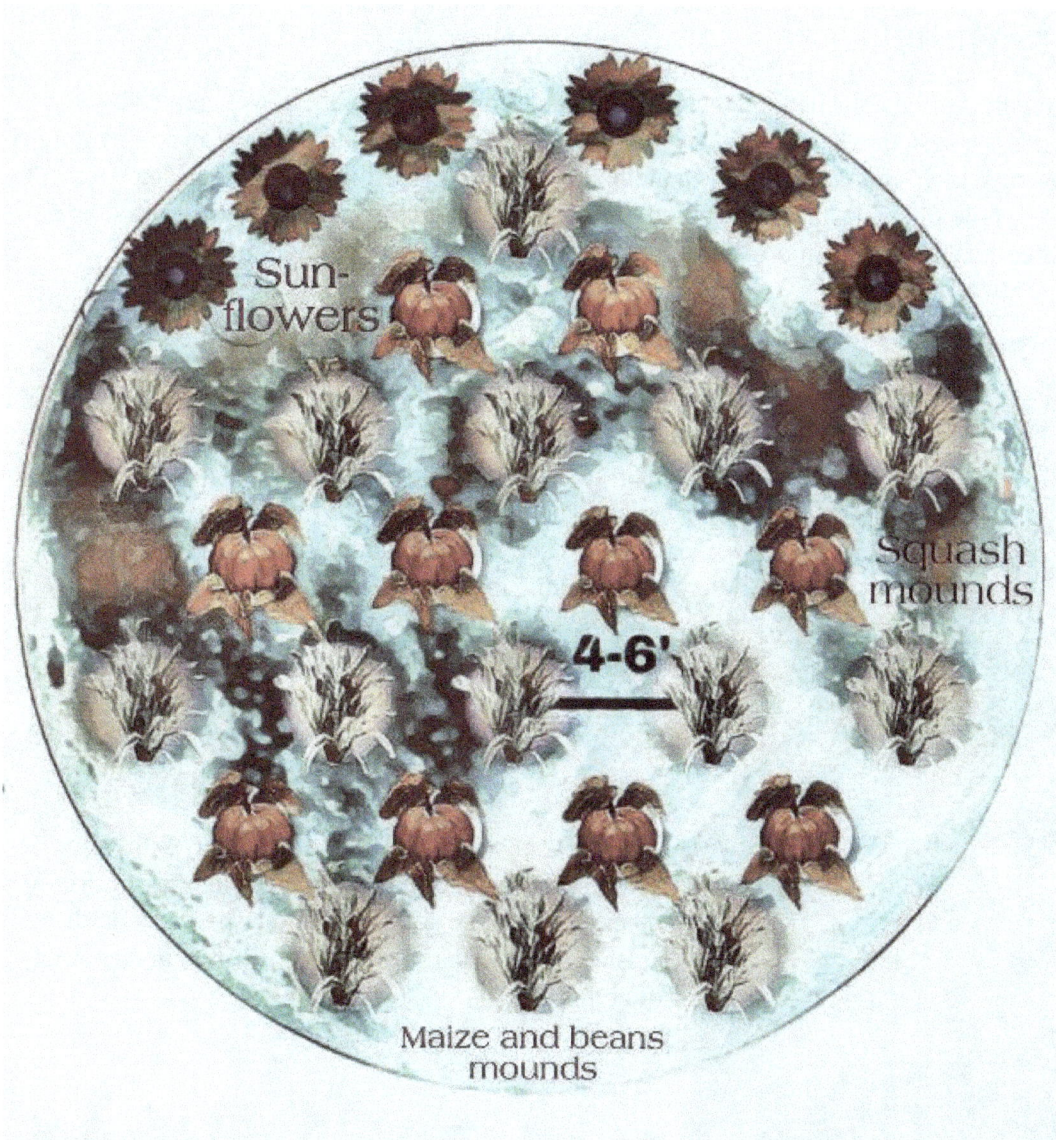

Wampanoag Three Sisters Garden

Slash-Mulch

Slash-mulch is often considered the most sustainable form of agriculture ever created by humans. It is a system where mulch is grown in place, is slashed, and then a crop is grown into that mulch. While "chop and drop" is a method for weeding, slash-mulch systems are a way of prepping whole fields for planting without tilling. I have used this system on multiple sites to grow three sisters gardens into edible meadows without tilling or removing vegetation.

In this case, I grow out the edible meadow guild in spring, until daffodils are starting to die back, then scythe the field and rake the mulch into piles. When the mulch is no longer green, I plant my crop seeds into the piles.

This image shows germination in a slash-mulch mound, with the edible meadow growing back in. Corn grows rapidly, so even just this level of disturbance may be enough to establish good, healthy crops.

This shot shows the crop in midsummer, with the squash running and further suppressing grasses. Many other crops can be seen sharing the space: monarda, daisies, garlic, salsify, amaranth, crosnes, and a variety of herbs.

Here, the corn has grown in as the meadow recovers. At this point, the corn should be established enough to produce at extremely high yields, often 4 times higher per square foot than the best conventional yields.

And in fall, the squash and corn finished for the season, Amaranth ready to be harvested. When it goes dormant I will mow low to the ground and leave all the material to mulch in place, fertilizing the soil and creating niches that encourage biodiversity.

Note: since this chapter is an element of TEK, traditional ecological knowledge, which the author has learned from various native sources, this material has been made available in free pamphlets at Transformativeadventures.org.

The Beginner's Annual Garden Make-Over Guild

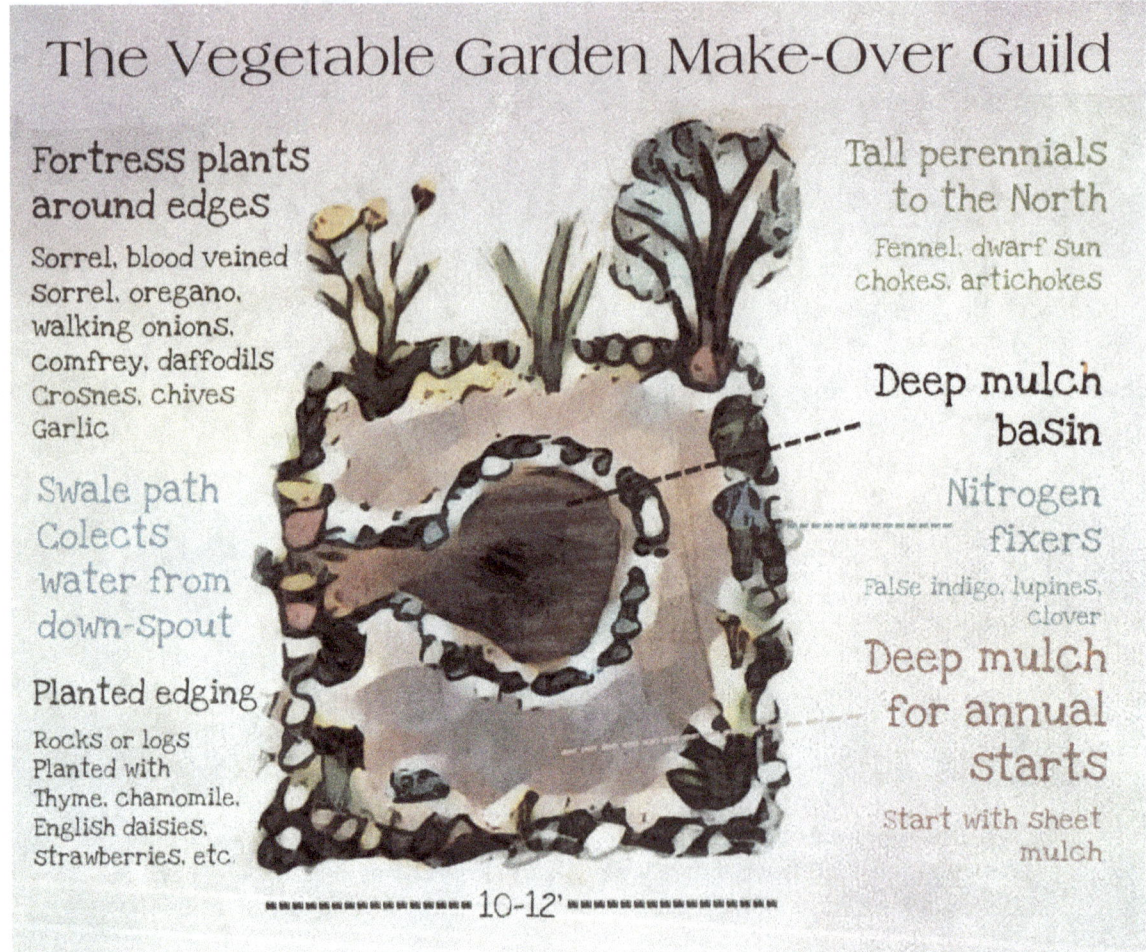

A classic Permaculture design for an easy, low-work home garden.

Most of my veggie gardens use a version of this easy, productive design.

These patterns create a garden that minimizes some of the most common draw-backs of home gardening: the work, the weeds, maintaining paths, watering, fighting pests, and getting haphazard harvest results. They're especially chosen to bring new life to that old, overgrown garden everyone seems to have in the backyard.

This single key-hole design works for a garden between 10′ * 10′ to about 15′ x 15′.

Narrow entry path: Paths are often difficult to maintain and a perfect place for weeds to invade.

Double-reach beds: at their widest (the corners) the beds should only be about 5′. This allows you to reach to everything in the bed (2 1/2′ reach from the inside and out) without ever having to step on the bed.

Planted border: everyone loves border rocks and bricks, but they also make the perfect niche for difficult weeds to invade. Plant these niches yourself with mat-rooting herbs like thyme and deny weeds a foothold.

Mulch-basin "keyhole;" the narrow path opens to a more spacious work space. This mulch basin can be dug out to collect water much like a rain garden. The heavy mulch will break down quickly to feed your plants. Perennial plants might provide most of the material you need to keep this pit mulched.

Perennials around the outside, room for annuals on the inside. This makes your garden an ecosystem that's hard to invade. Unlike most annual gardens, this one won't quickly get overrun by grass.

Nitrogen Fixers: Help your garden fertilize itself.

Flowers: Include some for biodiversity! This will be a beautiful garden, too!

Perennials: Perennials help ensure a good harvest year-in and out, even if you don't have time to "plant your garden." You might discover that you like the perennials so much you just let them take over! Just a note, I forgot to include a few of my favorites: Turkish Rocket, lovage, artichokes (need winter protection) and rhubarb.

Applying these patterns to an old under-utilized garden plot will "yield" a perennial harvest for little labor for years to come. Better yet, they will provide plant material to help expand a Permaculture design to other parts of your yard, and perhaps to your neighbors' yards, too.

The Beautiful Bountiful Mulch Strip

A strip of beautiful, vigorous plants, grown to create mulch for garden beds. Beautiful, functional, and also provides extra yields.

Since one of our goals is to create each new garden so that it grows fertility and mulch for itself, this is a very useful planting, and also one which has been designed with flowers, foliage, and beauty in mind. It will create a long season of flowers and ample biomass for mulch and fertility all season long. One strip placed between 5 foot beds should provide all the mulch necessary to keep both beds tucked in and snug all year round.

Some suggested plants for this guild include: sunchokes, groundnuts, Turkish rocket, Bocking comfrey, variegated comfrey, sorrel, monarda fistulosa and didyma, lupines, blue false indigo, daffodils, daylilies, brown-eyed susans, Maximilian sunflower, and tuberous sweet pea. Cup plant, milkweed, colorful yarrows, and mints can be used in places where there is a strip for mowing to keep them contained.

Because we'll be cutting these plants once or twice a season for mulch, we can use particularly tight spacings, cutting recommended spacings nearly in half. I suggest planting in 2 or 3 staggered rows to minimize the width. As a "good enough" rule, we can use these spacings:

Center row, pictured from top to bottom in image:
Sunchokes, Maximillian sunflower or cup plant: 3'
Goldenrod: 2'
Variegated comfrey: 2'
Bocking comfrey: 2'
Sunchokes, variegated comfrey: 2'
Baptisia plants can be tried in the center row for more nitrogen at 3' spacings, though in some cases these have not survived frequent cutting.

Left row, Support plants 1 ½' (top to bottom)
Yarrow *2
Lupines
Blood veined sorrel
Sorrel
Yarrow*2
Monarda
Lupins

A Useful Ornamental Cover Crop

With French Intensive gardens, or wherever there is bare soil, this beautiful cover crop will help regenerate the soil and look good doing it.

I don't often use a cover crop, but when I do, it's often this one. Using tight no-till succession plantings means that a cover crop is rarely needed. When a cover crop is called for, but something beautiful and functional is required, this is my go to plant combo. These are widely available seeds, but are easily collected if desired. The plants are useful and can be harvested in small quantities without bothering the cover. This cover has a little something for every soil issue, giving nature the tools it needs.

Lettuce leaf poppies, available in many colors
Crimson clover
Red broad leaf mustard
Daikon radish (has surprisingly pretty flowers.)
Zinnias can be added and fair reliably grow from seed when scatter-sown.

Fruit Tree Guild in Cool Colors

Path
stepping stones and walkable ground covers

1. Oregano
2. Yarrow
3. Sorrel
4. Monarda
5. Valerian
6. Anise hyssop
7. Strawberries
8. Day lily
9. Chives
10. Sage
11. Baptisia
12. Hollyhock Mallow
13. Sweet cicely
14. Dame's rocket
15. Sea kale
16. Lupine
17. Walking onions
18. Bell flowers

Brick lining ----------

A Tree Guild in Cool Colors

Cool color guild, oregano hedge fortress planting

This fruit tree guild flowers in cool colors, and it is designed with a guild matrix to become stable and low-work.

While many of the textbook fruit tree guilds seem to assume weeding or re-mulching, the guild matrix here should fill in the gaps and create a dynamic plant community where things may come and go, shifting and dancing through the seasons. But with regular, thoughtful harvest, it should become mostly stable and will hopefully require very little non-harvest maintenance. The pictured guild required its first weeding, some spot mulching for grass-encroachment, after four years and a particularly hectic summer where not enough harvesting was done.

Another common complaint I hear about guilds is that the fruits are the main harvest, but that the understory plants are mostly just there for support, and not very useful.

So this guild is optimized for the home landscape, doubling as a culinary herb garden, and salad garden, especially for the temperate winter. Some calorie root crops and medicinals are included for stability.

For color effect, choose blue and purple varieties where available.

The edible Pearl Lupin, "tarwi," Lupinus mutabilis, is an ideal lupine for this guild. There are some variations in flower color but most will be appropriate to this guild. Native lupines will also put on a show. Sundial lupines may have the most visual effect and hold space best in the guild, and with careful leeching the seeds are edible but not as large as tarwi.

Yarrow, "appleblossom" has soft lavender flowers, while "firefly amethyst" has bolder lavender-pink flowers that fade to a softer lavender. For this guild I recommend a mix white with one or two of these color variations.

There are a huge number of daylily, hemerocallis options that are fine edibles with scene-stealing color that will be absolutely vital in creating a low-maintenance guild. These will persist well as the trees mature into shade. Don't under-do the daylilies, as you'll want to use the buds and flowers liberally in the kitchen, and when it comes time to thin, you get the rare treat of the potato-like roots, too. Good tasting purples include "purple de oro," "purple rain," and "wild horses." The white "Casablanca" is one of my favorites for eating. The soft peach "Naomi Ruth" is a confirmed edible that would look great in this guild. *Though of course, always test these carefully to account for potential food allergies!*

An edible substitution for valerian would be skirret, which has a similar growing season and habit, though it won't get quite as tall.

This guild could easily be decked out with a variety of edible alliums as well.

The **walkable stepping stones mix** is used here, but in appropriate climates (zones 7+) corsican mint works very well and will give a nice cool color echo when flowering.

Tree Guild in Bold Warm Colors

1. Bee balm
2. Day lily
3. Yarrow
4. Sorrel
5. Rhubarb
6. Cone flower
7. Bulb fennel
8. Asparagus
9. Daylily, d'oro
10. Daylily, orange
11. Sun chokes
12. Turkish rocket
13. Blood-veined Sorrel
14. Walking onions
15. Anise scented goldenrod
16. Yellow baptisia
17. Variegated comfrey
18. Clove currant, "Crandall"

Spring plants:
(Not pictured)
Allium Molly,
Wolf's bane
Daffodils
yellow crocuses

Fruit Tree Guild in Warm Colors

This is a highly productive and functional tree guild or mini forest garden in attention-grabbing bold warm colors. This was one of my favorite guilds at Lillie House, which achieved stability and high yields rapidly. Here, I've made a few changes to make it more broadly appealing, but this basic guild was tried and true, and then replicated in part for a few other projects.

One visiting farmer said of this guild in spring "this one guild has more food in it than my entire market garden," and it hadn't had any planting or weeding work in 4 or 5 years.

The original guild was based around an Aromatnaya quince for bold yellow ornaments in late summer, and with pussy willow, golden raspberries, and color echoes from cornellian cherry and staghorn sumac across the path. While this guild would work beautifully with any fruit trees, especially red-colored apples and crabs, my recommendations for this guild are:

Tree 1: Aromatnaya quince
Tree 2: Cornellian cherry
Tree 3: Variegated elderberry

Other options could include cherry guava, cornus kousa, sumacs, and citrus.

The guild can include coneflowers and yarrow and in both red and yellow. Variegated comfrey shines out brilliantly in this guild (as pictured in its plant profile.) Any goldenrod would work, but there are a couple like the anise scented variety which make the best use for humans. Milkweed would be at home in this guild, as would other sunflowers. Rosinweed was happy in this actual guild at Lillie house, sharing space with sunchokes. Golden raspberries were an especially nice yield next to the fruit trees in the early years, but I left them off this image because I don't want to hear complaints about spreading or thorns. Allium Molly is the only yellow culinary allium I know of, and it makes a charming addition to the guild.

Annuals including mustards and brassicas look great when flowering in this guild, too. At planting, this guild was started with the aesthetic three sisters guild, with red amaranth, too. Color echoes between the amaranth and the staghorn sumac were a daily delight.

If you plant this guild you will find that pollinators and birds absolutely swarm to the warm colors. Hummingbirds will love you for this guild, as will hummingbird moths. Add milkweed and the monarch will come and add its warm colors to the display.

Weeding Like a Jedi: Terminator Guilds

If we're following a strategy of planting in guilds that will persist a long time to transform the landscape, one question people ask is: what do we do about weeds?

The key insight here is that if you do weeding labor —if you get into the garden and manually pull weeds out—then you have worked weeding permanently into the garden. You will have created the situation where the weeds came in in the first place, and they will inevitably come back. That work has become part of the system.

But, if you try to solve this problem by understanding how the weeds got in in the first place, and you can plant to fill up that niche, then you have weeded the garden like a freaking Jedi: WITH YOUR MIND.

Then you will never have to weed it again. Terminator guilds are aggressive, useful assemblies of plants that are useful for outcompeting problem plants.

When it comes to this strategy of weeding with your mind, Permaculture authors have suggested a variety of "control" mechanisms based on research and observation of nature and ecology.

1. Rather than focussing on eliminating or combatting a species of plants, focus on eliminating the harm they supposedly cause by increase biodiversity, abundance, and overall ecosystem health. Most invasives only invade damaged or disturbed ecologies. By repairing and restoring the ecosystem to health, we should be able to dramatically reduce the harm being done by "invaders."
2. Follow the 80/20 principle. In most cases, 80% of the benefit of reduction, will come from just 20% of our efforts. This is the case with most agricultural weeds, where the goal should be to keep weeds down to levels where they are not economically impactful, not utter elimination.
3. Solve problems by increasing diversity instead of reducing it (add plants instead of spraying and "moonscaping.")
4. "Fast forward to a later stage of succession." This recognizes that "invaders," like every other species, has "niche" where it evolved to thrive, and if we can work with nature's process to move past the stage of its niche, it will no longer be a problem.
5. Let grass do the weeding for you. Grass (especially when mown a couple times each year) will out-compete many plants, including many weeds and "invaders." Converting an area to an edible meadow may defeat many perennial plants like Japanese knotweed.
6. Use minimally harmful, naturally modeled control methods like flooding and fire when possible, which holistically increase ecosystem health, rather than targeting "invaders" and recreating the disturbance which introduced them in the first place.

Putting these to guidelines to use, Permaculturists often discuss planting to "outcompete" weeds and invasives. Our terminator guilds combine all these strategies together.

Keep in mind, if we're dealing with aggressive, opportunistic plants, then the plants we'll use will often also be aggressive and opportunistic. But ideally, they will be more useful, and easier to control through simple interaction. And following the guidelines above, we'll try to typically use native plants and communities when possible.

Key aggressive plants for full sun, average to fertile soils, sandy loam to clay:

Mown lawn or pasture. Lawn is a very special ecology, and only a very few plants can survive a regimen of repeated mowing, in addition to the competition of grasses. This treatment has even been found to be effective on notorious invaders like knotweed and bamboo, which I have never heard of invading or persisting on mown lawn. Research how appropriate pasture plants for your region and an appropriate organic maintenance plan. The **lawn mix polyculture** in this book could be seen as a great terminator guild for some plants.

Sunflowers, including Maximillian sunflower, cup plant, rosin weed, swamp sunflower, giant sunflower, sawtooth sunflower, and Jerusalem artichoke. Sunflowers are among the strongest allelopaths (they poison other plants,) and are North American native plants. The perennial sunflowers often form dense colonies and shade out all but early over-wintering ephemerals. For that reason, they are best paired with ephemerals to dramatically reduce opportunities for invaders. Jerusalem artichokes add an opportunity for digging (which removes weeds) as well as a high quality vegetable.

Monarda didyma, bee balm. Aggressive, especially on slightly wet or heavy soils. Beautiful flowering plant beloved of native insects, and useful for humans as a tea and herb. This often establishes dense spreading colonies which outcompete "weeds," yet has a shallow, week root system which is easily dug or hand-pulled, and is easily controlled by cutting.

Olympic mullein is an impressive perennial that makes a grand statement as a focus plant and is also a useful herb.

Goldenrod. Generally, this plant's upright nature makes it good at co-existing, rather than outcompeting. But combined with other aggressive plants, can add beauty, density and fill up remaining niches. Fragrant and anise-scented goldenrods can add another useful herb.

Joe pye weed is a favorite native of mine, and its flowers are useful as an herbal flavoring.

Brambles. Brambles are often one of nature's first choices for colonizing new forest edge. Especially useful with other plants, and with fruit tree guilds. Brambles are excellent nurse plants for young fruit trees, and they protect young tender bark and low fruit.

Virginia jumpseed, American knotweed. Has highly ornamental varieties. In part shade, in dense plantings, or open woodland, this plant can be as dominant as comfrey, and produces ample biomass as a mulch-maker or chop and drop plant. Unlike comfrey, spring greens and shoots are edible.

Milkweeds, showy milkweeds. While these usually don't outcompete other plants, they have a way of rambling around and filling up any niches left over in guilds. So they can be useful in terminator guilds.

Not native to North America:

Comfrey and dwarf comfrey. Non-native, but highly useful and aggressive. Sterile varieties will hopefully avoid becoming weedy, but the self-sowing species plant can will freely spread - a good thing if it's welcome and being used, a bad thing if not. Ornamental varieties like variegated comfrey are less aggressive, but will not come true from seed.

Turkish rocket. Another stout plant which tends to resist the invasion of even aggressive grasses on fertile soil in full sun. It also happens to be an excellent vegetable deserving of more attention from top restaurants.

Sorrel and blood-veined sorrel. These tend to evade grass encroachment and persist into very competitive environments, so long as there is enough density of broadleaf plants.

Daylilies. These can be useful in addressing plant "invaders," weeds and grasses. Well, unless daylilies are the "invaders" you're hoping to get rid of. Beautiful, 5 delicious vegetables in one plant.

A Warm-Color Terminator Guild

This warm colored guild features the same key players as the warm-color fruit tree guild, which is why it has been so useful on many sites and why it achieves stability so well.

While the color scheme may still seem unnecessary to some of you pragmatists out there, understand that the recognizable color scheme is what makes this terminator guild look like a garden, instead of like a weedy thicket. When you outcompete weeds with this guild, the neighbors will see a strong, attractive color painting in its place.

Bee balm, Jerusalem artichoke, butterfly weed, common milkweed, licorice goldenrod, naturalizing daffodils, camas, New England aster, Turkish rocket, yarrow in red and yellow, sorrel and blood-veined sorrel, variegated comfrey, brown-eyed Susan, asparagus. Together these will turn into a rambunctious, highly competitive, beautiful guild.

Plant Jerusalem artichoke, bee balm, and goldenrod at the corners of the planting with plenty of room to spread. Spread other plants out in between and around edges, then let them all "duke it out." This aggressive community may exclude most other weeds, or at least hold them in check while maintaining biodiversity. It can also be planted as an aggressive guild around most fruit trees, along with a few shrubs like goumi and serviceberry, though these will require some protection.

The sunny forest-edge community terminator guild

This is the plant community nature often "uses" to convert weedy lawn and degraded ecologies into forest. It's a vigorous, aggressive, but useful plant community that dominates the ground and generally excludes other plants. It naturally forms a dense hedgerow or thicket. Brambles, elderberry, Jerusalem artichokes, pokeweed, sumac (on dryer soils) prickly ash, grape, groundnut. On many sites it can also include Turkish rocket, and asparagus on the sunny edge, and good king Henry in shadier areas. Often these grow with wild apples, pears, hazels, etc.

Terminator guild ideas for dry, sandy sites

On sandy soils, it can be a challenge to get any high-value plants established and thriving. Typical invasive weeds might include horse nettle, spotted knapweed, thimbleberry, various Eurasian grasses, etc. With acidic soils, Indian strawberry could be considered invasive.

Because sandy soils tend to support less dense growth, there typically remains ample room and niches for well-adapted "weed" plants to persist, so the best strategy in my opinion is to seek to increase the value, usefulness and diversity of the plant community, rather than focussing on eliminating specific "weeds."

Shorter/ground cover perennials: (Plant at approx 1' apart in sets of 3 or 5. Listed in order of utility, with more useful plants listed first) Egyptian walking onion, common chives, oregano, thyme and creeping thymes, allium unifolium, allium strentorium, camas, musk mallow, mastic thyme, Mountain mint, sage, white sage, ornamental salvias like sylvestris "May Night," lily leek, Ornamental nepatas and cat mints, lemon mint, lavenders, daffodils, saffron crocus, yarrow, poppy mallow, red valerian, daylily, lance leaf coreopsis.

Additional plants might include: baby's breath, grey rabbitbrush, lambs ears, santolinas like lavender cotton, yucca, prickly pear cactus, musk mallow, feverfew, autumn sage, prairie smoke, miss Wilmott's ghost, rattle snake master.

Useful self-sowing annuals: bread-seed poppies,

"thickets" of trees at 1'-2' spacings may establish more easily than single trees, as cooperation in retaining water and resources outweighs the competition between them. In such a case, we may use "sacrificial trees' early on to provide shade and dominate the ground, then remove them for mulch as our trees get established.

"Sacrificial" trees: Oaks, black cherry, catalpa, black locust, elms.

Another guild for sandy sites: the Maine guild

Taller perennials: (interplanted with lower plants at 5-10' spacings.) Mexican sunflower, mullein, bulbing fennel, asparagus.

Useful spreading woody perennials: sumacs, wild plums, wild cherries, hazels, rugosa roses, nannyberry, currants including clove currant, goumi, serviceberries, elderberry succeeds on some dry sites but not others (presence of nearby water might be required.) Additional shrubs: ninebark, American prickly ash. Blackberries and black raspberries succeed on some sites, but not others.

Useful trees: Turkish filbert, paw paw (requires shade during first 5-10 years) redbud, toon tree, apricots (prunus species generally do better than pome fruits on sandy soil)

Research on forest establishment on dry, sandy soils indicates that dense, well-mulched plantings of mixed species

Driving through the sandy areas of the North East U.U., up into the state of Maine, you begin to encounter a specific plant community again and again dominating roadside thickets.

This guild seems to do well on sandy acidic soils and in harsh environments.

Rugosa roses, lupine, valerian, day lilies, walking onions, garlic, allium unifolium, camas, chives, tuberous sweet pea, apples, crabs apples, blueberries, and service berry can make up the backbone of this sandy terminator guild. Lupines can again include edible varieties.

With its multiple root types, this guild also makes a good slope stabilization planting.

Beautiful terminator guild for part shade

Shade usually indicates the presence of trees or forest, which is a new problem if our strategy is to "fast forward succession," since we're already in a later stage of succession!

However, as forest matures, the understory does succeed as well, and moves towards maximizing productivity of the system as shade-tolerant plants fill out niches.

In our region, common weeds that persist into such circumstances are English ivy and periwinkle, poison ivy, garlic mustard, and shrubs like honeysuckle. Again, our strategy wouldn't be eradication, but rather to increase useful biodiversity overall, to the point where the weeds are no longer "weedy." In situations where total removal is desired, sheet-mulching with cardboard and woody mulches is recommended. Do one small area, then move on to the next.

The basis of this terminator guild is monarda didyma again. A groundcover of violets, duchesnea, or strawberries can help in spring. Cup plant can thrive in a surprising amount of shade. If there's no danger of invasion, fuki or variegated fuki will do a good job filling in spaces and outcompeting undesired vegetation.

A beautiful bountiful guild matrix for full shade

This is a guild I have planted on many sites, as it imitates the understory of the Eastern Woodland region. This guild can produce a massive amount of food in a beautiful flowering garden.

Ostrich fern, variegated Solomon's seal, Virginia spring beauty, ramps, variegated fuki, monarda didyma, variegated black cohosh, yellow bellflower, and wild ginger. For a woody perennial, add Crandall clove currant. May apple is an option, but may come to dominate the other plants.

Chapter 35: Model Gardens for Inspiration

Now that we have covered a wide variety of patterns for creating a beautiful landscape, the key to transformation is to add up self-sustaining, low-maintenance patterns into an inter-connected, well-functioning whole.

The more we have chosen and created parts that are themselves self-sustaining, the greater potential for the system to come together as a whole, like a closed-loop organism. For example, an annual vegetable guild that is mostly self-mulching may be able to provide mulch for other guilds if needed. If beds have their own built-in composting systems, then it may not be necessary to have a large centralized composting system. If each guild has abundant biodiversity, then they will work together to create a disease and pest-resistant garden.

If we have adventured well, then after year one, we will have more free time to invest in creating more gardens, not less! And so we go, adding up these self-sustaining gardens and guilds, and in this way, we transform the landscape.

Why stop there? Our beautiful, abundant landscapes and plant collections can be the source for an infinite number of new gardens. And if we have done our job well, they will be.

We can go on to transform how the world meets its needs. We can create wealth, beauty, and abundance for ourselves, our families, our communities, and the world, in ways that actually raise everyone else up, rather than exploiting people or ecosystems.

The models in this chapter are all adaptations of real gardens that have worked in the real world. I don't expect them to be copied completely, but they may make a good starting point for an adventure.

An Ornamental Front Yard Food Forest

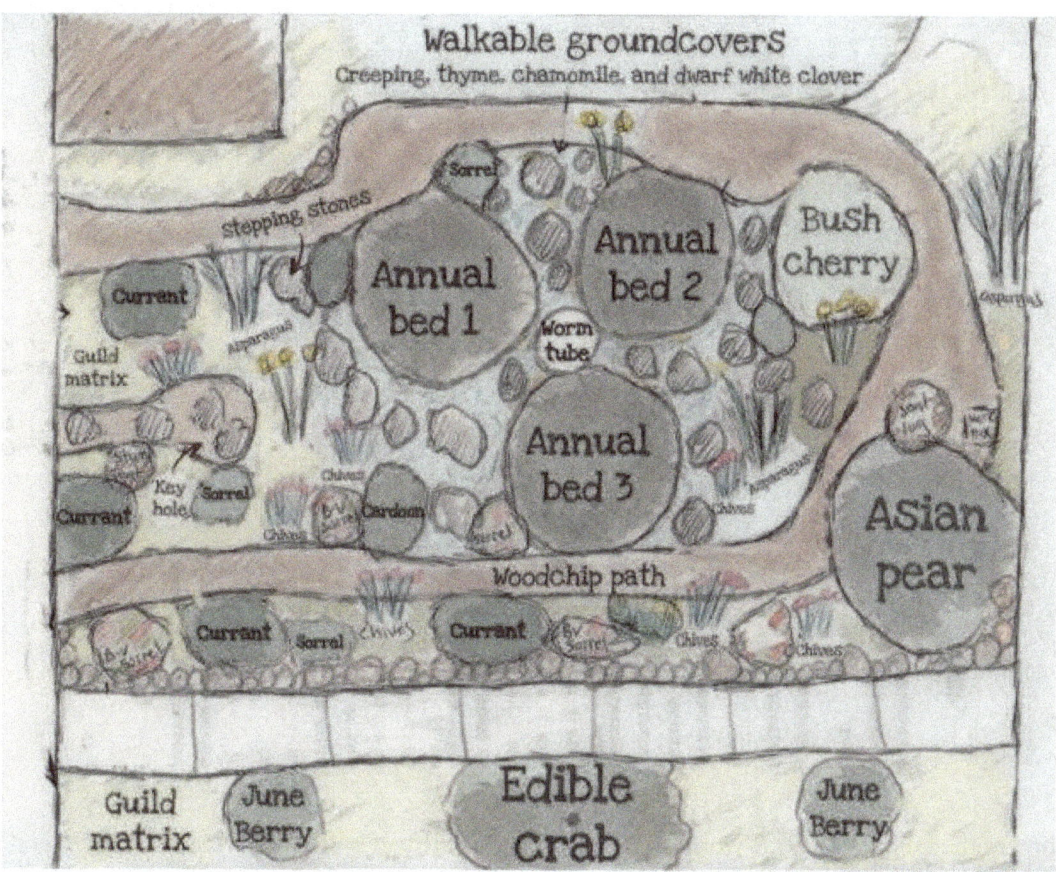

The model for this little ornamental front-yard food forest was about 30'*40' in an urban subdivision. It used the Veggie Circles Garden, surrounded by the walkable stepping stone guild, and situated in a larger guild using a guild matrix of sorrels, chives, mints, walking onions, asparagus, strawberries and currants. In spring, the garden is filled with spring bulbs, and pops of monardas and brown-eyed Susans give color through the whole season. Selected flowers from the Season of Flowers for Pollinators chart ensure high biodiversity all season long.

The veggie circles get fertility from a centrally located worm tube, as well as continuous deep mulch. The paths are micro-swales, cycling water through the garden from the roof to soak it into the garden.

A Suburban Backyard Paradise

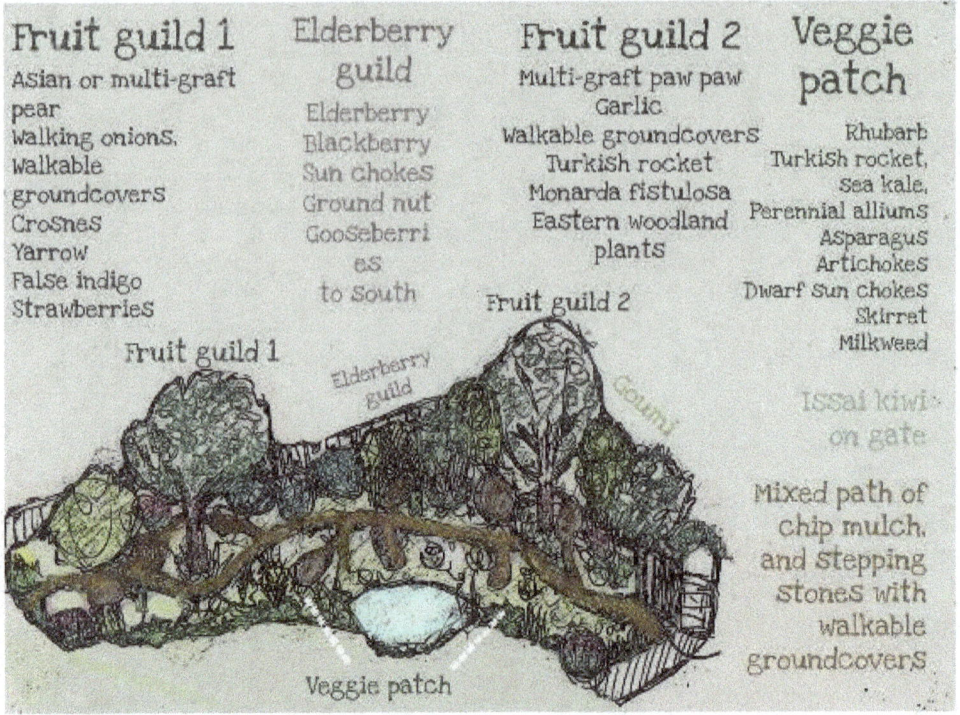

Based on another successful real-world garden, this forest garden can fit easily into a corner in any suburban back yard.

This garden uses **perennial veggie patches** with fortress plants along the front to keep a grass-free garden. Seasonal edging with an edging spade will help.

Two **fruit tree guilds** provide both fruits and vegetables, as well as mulch for the garden. The **keyhole and node** design allows us to use the minimal mulch for the maximum impact. A small pond in the front brings beauty and ecological benefits, and can be planted with **edible water plants**.

Permaculture Edible Border Garden

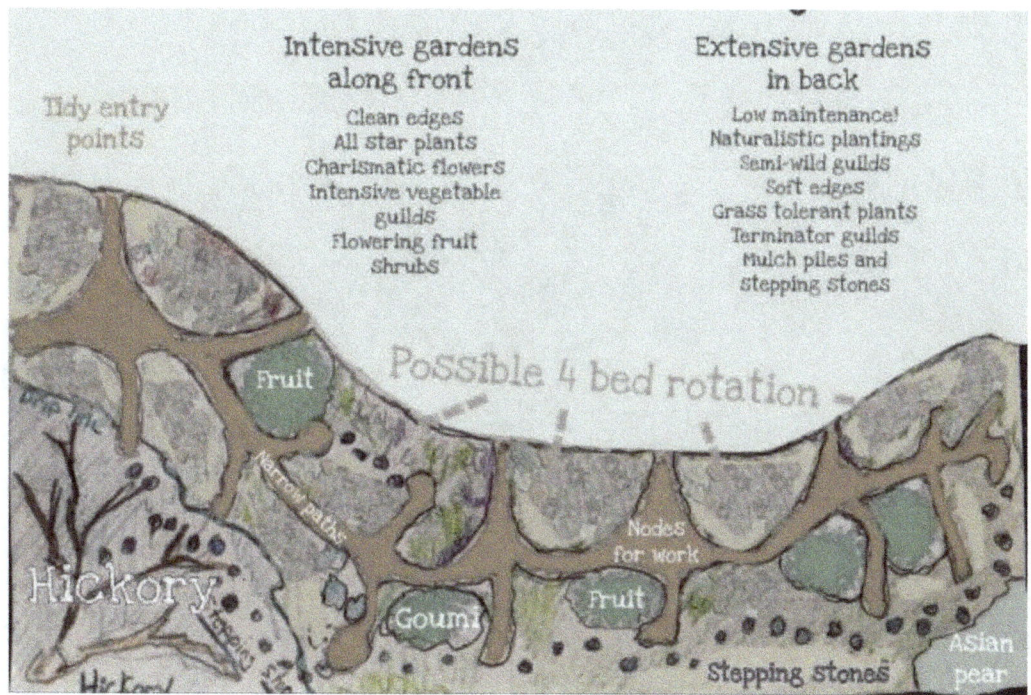

This design was for a client who was a former master gardener, who had build a few too many hard-work gardens, which had become burdensome and were scaring off the whole family. Everyone would become scarce when it was weeding time!

The goal was to convert the flower border to something beautiful, and still easy. And including fruits and vegetables would be a way to get the family more interested in the garden.

Again, it includes space for four **veggie circles**, integrated into **warm and cool color guilds**. The design is based off the **Zones for Aesthetics system**. Path and node design gives easier access to the garden than the pathless border, and stepping stones in the wildest areas provide access without maintenance.

The front of the garden uses hard edges, and guilds designed to keep grasses out, while to the back, it has soft edges, and guilds designed to persist in grasses.

A High-Productivity Vegetable Garden
With a formal look but wild ease

This design is for the gardener who wants the ease and function of Permaculture, but the look of a more formal garden. Or perhaps it is a compromise for the marriage between a "neat freak" and a "wild child." This is very similar to the intensive garden at Lillie House using rotation beds converted into keyholes.

The garden is surrounded by an oregano hedge to fight weeds and give a formal look. Cardoon or artichokes are at the corners, and in the keyholes as a repeated element and color echo. Sea kale lines the central path. In the back, the beautiful bountiful, mulch-maker guild provides beauty, perennial vegetables, and mulch for the garden.

Sun Trap Forest Garden
With a formal look and wild ease

Here we see how we can easily add elements up. The High Productivity Vegetable Garden is ringed with a sun trap of fruit tree guilds and tapestry hedges getting taller as they go to the shade side, and shorter towards the sun, to allow maximum sun into the garden, and protect it from harsh winds.

This gives us a complete, self-sustaining forest garden, especially with use of nitrogen-fixing trees in the hedge. The hedge will also provide a great deal of additional mulch, helping this garden to become nearly self-mulching.

Use of an electric chipper will help keep the garden looking neat and tidy when it's time to trim the hedges for mulch.

Beauty in Abundance

 Adventure Step 9: Putting it together and Getting Started

As the circle closes, we come back to our very first adventure: the adventure of envisioning how our landscapes can become beautiful, productive places.

At this point, we have explored many ideas and patterns, and hopefully had other adventures along the way. We have imagined uses, observed the site for zones and sectors, explored how water would be used, where fertility would come from, what plants and guilds we want, and where paths and hard-scaping will go.

How do you feel? At this point, many people find they do not even need a final design, because they see it clearly in their heads. If this is you, you're ready to begin transforming.

If that is not you and you are still uncertain, perhaps it is that you never found the seed syllable of inspiration that would guide the rest?

If it isn't clear, make three very different design sketches. Make the first your wildest ideas. Let the others flow. Don't wait for perfection, pick one and commit to it. It can develop as you go. If a consultation with a permaculture designer would help at that point, do it.

The best designs in the world aren't the ones that are perfect, they're the ones that get done.

And so it is with this beautiful world of ours. The "best of all possible worlds" is not the perfect utopia we can imagine in our dreams, it is the one we can actually build ourselves in the here and now, one step at a time.

Revisit our PLENTY checklist. Pick a project. Get it done. Rinse, repeat, transform the world.

The best adventures will be those that invest in long-term stability, but have immediate yields, decrease our labor, and increase our collections of valuable plants, food, and wealth.

And of course, most importantly, choose projects that truly fire you up. When we plan our projects around adventures that motivate us, make us jump out of the bed in the morning to get going, we super-fuel our lives. Whatever barriers arise, we have the energy to overcome them. When our vision is so clear, and beautiful that we can hardly wait to see it become a reality, each step of the quest becomes like the sun, which in burning, creates its own fuel.

If we can get that right, watch out! The whole site will transform, almost as though it happened all on its own.

If we can get this process right in just our yards, what can we make of this world?

Growing in Beauty and Abundance Over Time

And so it goes. In just a few years, if we follow a smart transformative process and invest first in projects that will save us time and money our investments will add up to profound accomplishments. A tapestry of beautiful guilds, forest gardens, vegetable gardens, tapestry hedges, edible meadows, and more, will manifest as if they had always been there, waiting to emerge. Natural guests and visitors will begin showing up and add to the beauty and bounty. With the landscape transformed, there's suddenly a wealth of plants and produce to share with others. Perhaps with a productive, low-maintenance landscape to support them, we add a few animal friends to our community, and create special ecosystems just for them.

At some point, the garden "pops," as Toby Hemenway said. It stops feeling like a garden and starts feeling like… something more. When that happens, everything gets very easy, pests and disease are mostly in balance, and the yields continue to grow.

As the productive systems add up, so does the beauty. Rich memories and experiences sink into the soil. Family pictures, laughter, and losses come and go and leave a resonance for us. Nooks and crannies grow for all the ghosts we've been trying to coax into the garden, and the land takes on a real spirit of place.

We will not be the only ones who notice this growing beauty. Any good project will grow in its impact on the community, on the positive energy that it radiates to the world around it. New projects will spring up in your community, inspired by your work, spreading relationships and connections. You'll learn of plants from your garden spreading around the town, the state, or perhaps the world. It's almost hard to create a truly transformative landscape and have it not also transform your life and even your livelihood in this way.

And so our movement of abundant beauty—and our vision for a better world—shall grow.

If we can imagine the world we want with enough clarity… if we can get that vision right, and express it with enough beauty… even just in our own back yards, there will be no stopping that world from growing in up in the rubble of this crumbling dystopia. Even today, that future is already there, just waiting for us to give it the space to grow.

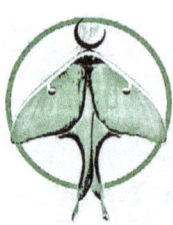

Appendix A: Discussion on Gentrification and Displacement

Since the most basic activity of this book is to take responsibility for our spot of the world and to regenerate it—transform it into a culturally and ecologically beneficial space—we cannot avoid the discussion of gentrification. Gentrification in results in displacement of communities as wealthier people move in and the cost of living soars. This process is happening in country towns displaced by corporations, in rural villages turning to suburbs, and older suburbs converting to gated McMansion zones. We especially need to be mindful in cities, since urban neighborhoods are especially vulnerable to being destroyed this way. Meanwhile, cities give us our greatest opportunities for sustainable living and right livelihoods.

Gentrification itself may refer to the improvement of access and resources in a neighborhood, and may not necessarily be a bad thing if the improvements benefit all. But the displacement that often results is indeed harmful to people and communities. Worse, it very much resembles the past generations' crimes of seizing land through colonization. Whatever justifications we may use to excuse it, this displacement may one day look indistinguishable from the methods used to justify genocide and the forced removal of indigenous peoples from their land to steal resources.

As I write this in mid 2021, gentrification remains a hotly debated topic that evades clear solutions. A few years ago, considerate people from gentrifying communities began to take personal responsibility, writing pieces about how to avoid gentrifying through more responsible behavior. To these writers, we have a personal responsibility to avoid contributing to gentrification.

Now, a new generation of writers are critiquing this attitude as a distraction from the larger systemic change that needs to happen.

For these authors, it is impossible to avoid the cycle of gentrification or participating in it. Low-income people are also entitled to places to live, and cannot help that they may be contributing to displacement simply by existing within this system that offers limited opportunities. To this perspective, all we can do is hope for systemic political change, and work towards that end. These critics understand that the gentrifiers have often themselves been economically dispossessed, gentrified out of their own neighborhoods, and are looking for a piece of earth they themselves can stick to and call home. Gentrifiers cannot help being gentrifiers, they say.

Yet on the whole, the system equates to institutionalized, justified violence based on skin color and class. Talk of personal responsibility may become just another justification for ignoring the larger political causes.

I accept this new critique, and agree we must organize for political solutions in our communities. And yet, I dislike the implication that we cannot take positive personal action to help address this problem in our neighborhoods and communities.

As always, the best path is to keep both sides in mind and seek to transcend the conflict. Both sides here indeed make excellent points.

Some things we can do:

Organize. To start with, we should not lose track that we should be investing our time and energies into broad systemic change. Just as we are organizing plants into guilds, hopefully we are organizing people into caring communities and building social capital as we build soil. At some point, it may be our time to lend our voice to those who are in struggles against gentrification and other forces of systemic oppression. If we do our "solid base" work well, (creating community connection and sovereignty) hopefully we will have effective resources to share.

And yes, we can also find opportunities to take direct action:

Set the right goals. As always, we may begin that process with the right vision and mindset. If not, we are unlikely to get the results that will be beneficial. For example, sometimes we may move into a neighborhood with the idea of intentionally "cleaning things up," displacing "undesirable" current residents. Or perhaps less horrible, importing the sort of people we would like to see, whatever that means.

Hopefully, this seems like an openly colonial mindset to you. If not, I'll ask you to be open to evaluating whether that mindset really is in harmony with your true goals as we continue the discussion. Do we want to weed people out of our communities? Or do we want to find ways to empower everyone to play positive roles?

The danger is that we think transforming the neighborhood means replacing the population. If that becomes our vision, not only will be acting violently against the existing community, and the diversity of our environments, but we may end up harming ourselves in the process. First, we may lose the very characteristics that we moved for in the first place.

And just as when weeding the garden of "undesirable individuals," we may create open niches where more aggressive species may move in, like real-estate speculators and developers keen to take advantage of the gentrification cycle. Soon, you may find yourself priced out of the neighborhood, or forced out as an "undesirable!" Or perhaps business owners may seize the opportunity and turn your neighborhood into a foul-smelling confined feedlot operation or waste dump.

Remember our Uber Principle: solve problems by building connection to nature and community. Displacing people destroys connection and wastes opportunity for diversity. In the garden, we have the goal to increase biodiversity, seek community stability, and resource our plants fully with a healthy environment, soil, and all that our plants need to thrive. Why should the ecosystem of our neighborhood be any different?

Observe, then interact. Get to know the neighborhood and the community where you move to. Don't try to remake it in your image.

Observation also means educating yourself by reading about gentrification before you move into a new neighborhood.

Act transformatively. This means grow community organically, by interacting with what's already there. We could call this the Spike Lee Ecological Theory of gentrification, since Spike Lee featured this theory that neighborhood decline can drive gentrification in his films. Natural systems grow in wealth internally over time. Neighborhoods should, too. Which means we need to grow the wealth and resources of our neighborhoods from the inside. As Spike Lee pointed out, if our neighborhoods are allowed to fall into decay, and artificially low value, then that creates an opportunity for speculators to move in and capture that instant profit! So we can help our neighbors keep the hood stable and improving, so that investors don't start swarming.

With that, we can follow terroir. Don't update your house with a bunch of trendy "gentrification starter pack" products.

Don't treat real-estate as "an investment," treat it as housing. Let's get nuanced: owning a home and transformative landscape IS the best investment we can make. It's an important part of a strategy to get free from the corporate system. There is nothing wrong with owning a home or even with providing regenerative housing opportunities within your community. In fact, I'd say these are positively transformative acts.

But if we are investing with a mindset to just flip, or otherwise drive up housing values to make money off the equity, we are in the business of displacing people. We have to ask, do we want to make a living by enabling the displacement and colonization of existing communities? It's possible to be thoughtful about investing in truly regenerative ways instead. But only if we actually do it.

notes:

20 Ways to aNot Be a Gentrifier. https://www.theguardian.com/cities/2014/feb/12/oakland-20-ways-not-be-gentrifier

We can't do anything about it but fight the system:
https://www.rewire.org/gentrifier-fight-the-system/

Tim Wise's reading list on Racism. http://www.timwise.org/reading-list/

Appendix B: Discussion of Cultural Appropriation

(This material develops in discussion with the Transformative Adventures Cooperative)

Since patterns for sustainable growing come from around the world, an understanding and consideration for cultural appropriation is important for us to not create gardens that may offend our community members.

Cultural appropriation is the theft or misuse of cultural or intellectual capital within a colonial setting.

Since cultural sharing (called "syncretism") is an absolutely beautiful and wholesome thing, the concept or appropriation can be confusing for some. Let's see if we can help clear up the confusion:

Three short sci-fi stories to understand how "cultural appropriation" is a bad thing, and different from positive cultural sharing and syncretism.

Story 1: A ragtag band of Americans discover that a secret alien society has been living amongst us, oppressing and enslaving us, and attempting to terraform our planet with Co2. The humans are able to steal some alien tech to defend themselves and begin to fight back! Who are the heroes? Who are the villains? Was stealing the alien tech bad?

Story 2: We discover a nearby alien civilization! We live as respectful equals, share our technologies, fashions, and beliefs. Human composers begin using alien themes to reflect unity, respect, and brotherhood, and alien artists come to learn about human painting techniques. Together we solve our problems and celebrate that we are not alone in the universe. Who are the heroes? Who are the villains? Did anything unethical happen in this sharing?

Story 3: A small group of alien refugees flee the death of their sun and settle in an unpopulated part of the American desert, disguising themselves to blend in with North Americans to live in peace and preserve some small part of their culture. They are discovered, and conquered in a violent raid by the U.S. military. They are tortured for their technological secrets, which are then used without compensating them. They are legally "free," but essentially sold as slaves to perform in road shows. Some of the slave owners begin to learn about the spiritual, culinary, artistic, and fashion traditions from their alien slaves, then get rich by writing books, opening restaurants and otherwise selling these alien cultural traditions to other humans while alien artists and cooks remain impoverished or in slavery. A sports franchise renames itself "The Klingons," a derogatory term for the aliens based on a superficial resemblance to an old human TV show. The term is generally used to imply that aliens are "savage and warlike." Fans dress up as a variety of scifi fictional stereotypes about violent, savage aliens. They claim this shows respect.

Would anyone see the humans as the heroes in this story? Doesn't this use of alien culture seem unethical to you? This shows how power dynamics and colonial history impact the difference between theft and free sharing between cultures.

So, when such colonial history, class, race, or power dynamics are involved, it's important to be mindful and respectful about using and giving credit for such intellectual and cultural capital.

If a technique or design comes from a colonized society, it may be unethical for people from colonizer societies to sell, market, or profit from teaching that technology. It's better to promote experts from within that culture to profit from their own cultural capital. For example, I generally would not charge for classes or materials on Three Sisters Gardens, but instead, I try to promote local experts and give any materials I make away for free.

It's also important to always credit intellectual and cultural capital. Of course, that also includes the intellectual capital of folks from colonizer societies! But too often, the capital of people from colonized societies is not considered as stolen property. Give credit!

Syncretism uses cultural property respectfully, while appropriation regards it as part primitive, often dissecting out parts for use and leaving the cultural or spiritual elements aside. Respectful use keeps cultural capital whole and in-tact, and values the spiritual and cultural elements.

Invest time in learning about the culture and connecting with people from the culture you with to respectfully share with. It's hard to hold ourselves accountable for respectful sharing if we're borrowing from a distant people we know nothing about and never interact with.

Be open to learning that your attempts to share are offensive, and don't get defensive, be open to changing.

With respectful sharing, credit, and empowerment, we can build a world where all truly are equal, and like the respectful examples above, we may grow and build a better world together as a human family.

TASK TITLE	Winter	Spring				
		Soil<40	Soil=40 F	Soil= F45	Soil=55	Soil=60F
General	Sheet mulching. Begin sowing first seeds of the new year, snow casting. Care fool tools, and for the gardeners!	Can continue to snow-cast many perennials, so long as there are a few weeks of freezing temperatures.	Planting season begins when ground is dry enough to work	Grasses running in cool season. Time to edge beds or do spot mulching.	Use dense plantings during cool wet weather to conserve water and us available water wisely.	Summer chop and drop or slash mulch for summer planting.
Lawns	Mow or cythe as short as possible, or browse with sheep.	Continue to mow as short as possible if necessary.	Infrequent short mowing is best in Spring.			
Annual Vegetables	Order seeds, store seeds	Never dig soil until it's dry! Start looking at soil temps online or with a laser thermometer.	When soil hits 40 degrees F, sow early cold hardy crops: mustards, brassicas, radishes, cresses, arugula, carrots, possibly lettuces.	Beets and chard when soil hits 45-50 degrees F.	Prime temp for lettuces, brassicas.	Sow or plant first tomatoes, corn when soil is consistently over 55F.
	Grow microgreens, sprouts, and water cress indoors.	Peas with protection, potatoes, and late onions and garlic.		Peas without protection.	Start harvesting early annuals like radishes and greens.	Plant out tomatoes and peppers after all risk of frost.
Herbaceous perennials	Dig and pot up cellaring vegetables, dandelion, chicory, asparagus, sea kale, poke...	Start transplanting and dividing perennials as soon as soil is dry and workable.		Start of prime harvest season for many perennial vegetables.		
	Dig hardy root veggies, crosnes, skirret. Harvest Jerusalem artichokes after ground is workable, for best flavor.	Spring ephemeral, chickweed, cress, dead nettle. Okay	May begin harvesting Turkish rocket. Last chance to harvest sunchokes or skirret.	Skirret, Good king Henry, sweet rocket, garlic mustard, ramps, Solomon's seal.	Blanched dandelions. Pokeweed shoots.	Asparagus starting. Rhubarb.
Woody perennials	Dig bare roots	Start transplanting as soon as soil is workable.		Ideal grafting time is a week with day temps over 60F and nights over 40F	Complete grafting before blossoming starts.	First honeyberries. May prune after blossoming for many species.
	Start winter pruning.	Collect scionwood around February.				
Habitat	Leave messy garden as long as possible.	Garden clean-up after insects have emerged.				Harvest plant seeds when ripe.
Water systems	Empty and bring in water storage tanks.	Put out water barrels and tanks to collect Spring rains.	Check microswales for sediment build-up. Use as compost!			

Summer			Fall	
Early	Mid	Late	Early	Late
Soil= 65F	Soil=75F	Soil>75F.		
Check irrigation systems. 1 inch of water per week for new plants. Do 't overwater!		If weather gets droughty, switch from dense plantings to wide spacings with deep mulch for "dry farming." Begin sheet mulching for Fall planting.	Build compost piles with Fall leaves. Mix 30-1 with grass clippings and food scraps.	Final chop and drop to put garden to bed.
	When hot dry weather hits, set lawnmower to highest setting. Reduce mowing.			Begin mowing lawn as short as possible.
Sow peppers, cucumbers, okra, beans at soil temps of 65F.	Sow eggplants, pumpkins, squash, and melons at soil temps around 75F.	Begin planting winter crops, brassicas, lettuces, root veggies. In polycultures plant these in shade of taller plants.	Last chance to plant winter crops.	Chop and drop after frost kill. Deep mulch to "put beds to sleep."	
Continue planting successions of spring vegetables until too hot.		Realistic start of tomatoes, peppers, eggplants.		
Last chance to easily transplant perennials before weather turns hot and dry.	Start dead-heading aggressive sowers.	Ramps and other spring perennials going dormant for transplanting. Daffodil fronds finally dying back can be removed.		Paw paws, pears, asian pears, kousa fruit...	
Strawberry rhubarb pie season!	Walking onion top sets forming and ready to eat! Garlic scapes!	Day lily pods and flowers.	The season of abundance! Store surplus in freezer to process when the garden slows down.		
First goumis.	Black raspberries	Cornelian cherries start, blackberries start	apples start	Start of pruning season.	Bare-root transplanting when dormancy starts.
		cherries, early plums.			
Are there flowers for pollinators? This is a gap in many gardens!		Make sure critters have access to cool, clean water!			
Careful not to water on plants in hot weather or it can cause scald or mold issues.					

Bibliography

(In the order presented in the text and chapter notes.)

1. Donella Meadows, *Thinking in Systems*, Chelsea Green, 2008
2. Lara Cushing, Rachel Morello-Frosch, Madeline Wander, and Manuel Pastor, *The Haves, the Have-Nots, and the Health of Everyone: The Relationship Between Social Inequality and Environmental Quality,* Annual Review of Public Health (2015)
3. Rachel Morello-Frosch and Edmond D. Shenassa, *The Environmental "Riskscape" and Social Inequality: Implications for Explaining Maternal and Child Health Disparities,* EHP, (2006)
4. Walter Scheidel, The Only Thing, Historically, That's Curbed Inequality: Catastrophe, *Atlantic*
5. Nicholas Georgescu-Roegen, *The Entropy Law and the Economic Process*, Harvard press, (1971)
6. Meredith Roman. "The Black Panther Party and the Struggle for Human Rights." *Spectrum: A Journal on Black Men,* 5, no. 1 (2016): 7-32. Accessed August 30, 2021. doi:10.2979/spectrum.5.1.02.
7. Cirrus Wood, East Bay food-justice movement has deep roots in Black Panther Party,
8. Garret Broad, The Black Panther Party, A Food Justice Story, Huffpo, Feb 2017.
9. Vaughn Sills, *Places for the Spirit: Traditional African American Gardens* Hardcover – Illustrated, August 31, 2010
10. Richard Westmacott, *African-American Gardens and Yards in the Rural South,*
11. Adriana Zavala, *Frida Kahlo's Garden,*
12. Maxi'diwiac (Buffalo Bird Woman), *Buffalo Bird Woman's Garden*: As Recounted by (ca.1839-1932) of the Hidatsa Indian Tribe, 1839-1932.
14. Michael J. Caduto (Author), Joseph Bruchac, *Native American Gardening: Stories, Projects, and Recipes for Families* Paperback – March 1, 1996,
15. Home Gardens in Nepal
16. Christopher Alexander, Sara Ishikawa, Murray Silverstein, *A Pattern Language: Towns, Buildings, Constructions.* Oxford, 1977.
17. Christopher Alexander, *A Timeless Way of Building.* Oxford, 1977.
18. *PermaculturePrinciples.org*. Learn about the Holmgren Principles,
19. *PermaculturePrinciples.com*. Learn the Mollison Principles.
20. Gaia's Garden, Toby Hemenway. A classic for home-scale Permaculture.
21. Peter Bane, *The Permaculture Handbook.* Chelsea Green, 2014.
22. Jessi Bloom and Dave Boehnlein, *Practical Permaculture,*
23. Zack Loeks, *The Edible Ecosystem Solution,*
24. David Holmgren, *Retrosuburbia*, 2019
25. Permaculturevisions.org.
26. Graham Bell, *The Permaculture Way,*
27. Graham Burnett, *The Vegan Book of Permaculture,*
28. *Permaculture Magazine.*
29. *Permaculture Designer Magazine.*

30. P. R. RobichaudA,C, J. W. WagenbrennerA, R. E. BrownA, P. M. WohlgemuthB and J. L. Beyers, "Evaluating the effectiveness of contour-felled log erosion barriers as a post-fire runoff and erosion mitigation treatment in the western United States," *International Journal of Wildland Fire* 2008

31. Oreja Ba, Goberna Ma, Verdú Mb, Navarro-Cano JAb, *"Constructed pine log piles facilitate plant establishment in mining drylands"* Department of Environment and Agronomy, Instituto Nacional de Investigación y Tecnología Agraria y Alimentaria (INIA), Ctra. de la Coruña, km 7,5 - 28040 Madrid

32. Jorge Castro, Craig D. Allen, Mercedes Molina-Morales, Sara Marañón-Jiménez, Ángela Sánchez-Miranda, Regino Zamora, *Salvage Logging Versus the Use of Burnt Wood as a Nurse Object to Promote Post-Fire Tree Seedling Establishment,* Restoration Ecology Volume 19, Issue 4 p. 537-544 as Refuges from fungal pathogens.

33. D. L. O'Hanlon-Manners, P. M. Kotanen, *LOGS AS REFUGES FROM FUNGAL PATHOGENS FOR SEEDS OF EASTERN HEMLOCK (TSUGA CANADENSIS)* Ecology, Volume 85, Issue 1 p. 284-289, January 2004

34. Self teaching resources for double-digging from Ecology Action: http://www.growbiointensive.org/Self_Teaching_2.html

35. Meta analysis, deep digging improved drought resistance and crop yields, especially where soil was compacted. Schneider et al, "The effect of deep tillage on crop yield – What do we really know? Soil and Tillage Research, 2017

36. One example of very positive yields for corn and lab lab. Iddi Mwanyoka, Dosteus Lopa, COMMUNITIES' PERCEPTION ON THE CONTRIBUTION OF SOIL AND WATER CONSERVATION MEASURES IN IMPROVING LAND PRODUCTIVITY IN THE DRY-LAND AREAS OF TANZANIA: THE CASE OF TERRACE, " FANYA JUU " and DOUBLE DIGGING IN SAME DISTRICT *International Journal of Agriculture and Environmental Research*

37. No benefit for two shallow rooted crops, "Small-scale, intensive cultivation methods: The effects of deep hand tillage on the productivity of bush beans and red beets" *American Journal of Alternative Agriculture,* Vol. 13, No. 1 (1998),

38. F.H King, *Farmers of Forty Centuries: Organic Farming in China, Korea, and Japan* 2004 edition, Dover.

39. Ruth Stout, Gardening Without Work, Norton Creek Press, 2011

40. Jean-Eves Humbert "Does delaying the first mowing date benefit biodiversity in meadowland?" *Environmental Evidence*, 2012

41. David Jacke and Eric Toensmier, *Edible Forest Gardens,* Chelsea Green, 2005. An encyclopedic two-volume set that was a great leap in the art of forest gardening. The "characteristics of a forest garden" presented here are largely adapted from this work with some additions from the academic literature.

42. Martin Crawford and Joanna Brown, *Creating a Forest Garden: Working with Nature to Grow Edible Crops,* Green Books, 2010. A classic from one of the major practitioners in the field.

43. Wayne Weiseman, Daniel Halsey, Bryce Ruddock, Integrated Forest Gardening, Chelsea Green, 2014.
44. Robert Hart, *Forest Gardening,*. Green Books, 1956.
45. D.J. McConnell, K.A.E. Dharmapala, S.R. Attanayake, *The Forest Farms of Kandy andOther Gardens of Complete Design,* Routledge, 2003
46. Allan Thomas and Sajan Kurien. *Home gardens of Kerala: Structural Configuration and Biodiversity*, Indian Journal of Research, 2012.
47. Beat, A.H., Chaubey, A.K. & Askary, T.H. Global distribution of entomopathogenic nematodes, *Steinernema* and *Heterorhabditis*. *Egypt J Biol Pest Control* **30,** 31 (2020).
48. R. E. Berry, J. Liu, G. Reed, Comparison of Endemic and Exotic Entomopathogenic Nematode Species for Control of Colorado Potato Beetle (Coleoptera: Chrysomelidae), Journal of Economic Entomology, Volume 90, Issue 6, 1 December 1997, Pages 1528–1533,
49. Strategies to Enhance Beneficials, SARE Outreach, 2005
50. I.J.PorterP.R.Merriman , Effects of solarization of soil on nematode and fungal pathogens at two sites in victoria, Plant Research Institute
51. Nethi Somasekhar, Parwinder S. Grewal, Elizabeth A. B. De Nardo, Benjamin R. Stinner, Non-target effects of entomopathogenic nematodes on the soil nematode community, Journal of Applied EcologyVolume 39, Issue 5
52. Shin Woong Kim, Walter R. Waldman, Tae-Young Kim, and Matthias C. Rillig Environmental Science & Technology 2020 54 (21),

Index

80/20 principle, 128, 177, 433
Adventure, 3, 25, 44, 74, 86, 122, 145, 184, 213, 256, 317, 366, 394, 446
Aesthetic Guilds, 395
Alan Chadwick, 155, 292
Alliums, 280
amaranths, 287
Anise hyssop, 286
Annual Garden Make-Over Guild, 422
Artichokes, 288
Asian pears, 252, 261, 278
Back to Eden, 198
Beginner's Annual Garden Make-Over Guild, 422
beginners' annual garden make-over bed, 123
Bellflowers, 290
Bill Mollison, 17, 70, 72, 77, 82, 84, 133, 141, 197, 211
Biodiversity, 117, 120, 231, 411, 453
BioIntensive, 130, 155, 182, 247, 292, 381
Black Panther Party, 56
Blackfoot Tipi, 37
Borage, 292
Brad Lancaster, 122
Bunias orientalis, 307
Camas, Camasia quamash, 293
Campanula. See Bellflowers
Captain Beefheart, 91
Cardoons, 288
Carl Jung, 97
catch and store, 213
Chadwick, Alan. See Alan Chadwick
Chives, 282
Chogyam Trungpa Rinpoche, 47
Christopher Alexander, 71, 73, 74, 76, 92, 225, 452
Climate Change, 117
color combinations, 345
Comfrey selections, Symphytum, 294
Common Soil Problems and What to do About Them, 179
Composting Toilets, 181
Cottage Core, 64
Cottage Garden, 4, 54, 232, 233, 235, 246
Cover Crop, 425
Crambe, 308
creative thinking, 36
David Holmgren, 70, 86, 452
Daylilies, Hemerocallis fulva, 296
Deep Mulch, 195
design, 69
Dogwood, 268
Donella Meadows', 176
double-digging, 136, 189, 194, 195, 452
Ecological Succession, 19
Edible Forest Gardens, 231
Edible Meadow Guild, 409
edible meadows, 23, 136, 141, 162, 179, 217, 283, 287, 293, 296, 419, 448
Egyptian Walking Onions, 281
Elderberry, 263, 278
elements of landscape design, 107
Elements of landscape design
Color, 107
Context, 107
Line, 107
Mark making, 107
Materiality, 107
Rythm, Repetition and ryme, 107
Scale, 107
Space, 107
Sybolism, 107
Theme, 107
Time, 107
Unifying features, 107
Views, 107
Enclosure, 329
Espaliered Fruit Trees, 277
F.H. King, 180
Farmers of Forty Centuries, 177, 180, 453
fertility systems, 156, 177, 184
Finley, Ron, 219
flowers and plant shapes, 111
forest gardens, 23, 123, 133, 137, 155, 162, 224, 229, 231, 251, 259, 272, 274, 296, 314, 375, 410, 448

FREE, 82
French Intensive Method, (FIM), 155
Frost channel, 149
Frost pockets, 149
Fruit Trees: How many to plant?, 260
Garden Aspects, 146
garlic spray, 135, 390
genii loci, 102, 103, 243
Gentrification, 245, 449
good enough, 129
goumi, 252, 353, 400, 436
Guild Design Basics, 397
Guild Matrix, 401
guilds, 22, 23, 64, 75, 123, 126, 136, 137, 141, 143, 161, 162, 163, 198, 199, 206, 217, 229, 231, 240, 241, 257, 283, 287, 290, 292, 298, 300, 302, 305, 307, 317, 330, 338, 347, 376, 395, 396, 397, 398, 404, 406, 413, 427, 430, 433, 434, 440, 442, 443, 445, 446, 448, 449
Haphazard Mulch Mandalas, 385
Hard and soft edges, 339
Hart, Robert, 231, 453
Hawthorns, 264
hedgerows, 23, 25, 120, 123, 136, 137, 141, 147, 155, 158, 162, 179, 182, 247, 248, 249, 250, 251, 253, 258, 259, 296, 306, 314, 315, 329, 338, 339, 358, 359, 360, 361, 376
Helianthus tuberosus, 313
Hemenway, Toby, 398, See Toby Hemenway
Hemerocallis fulva, 296
Herb Hedge, 342
herb spiral, 43, 86, 141, 211, 213, 214, 215
herbicide, 190
High Bio-diversity, 141
Holistic Natural Gardening Patterns, 130
Holistic Natural Gardening", 129
Holistic Natural Management, 3, 4, 127, 155, 177, 182
Holmgren, David, see David Holmgren
Holzer, Sepp, 205, 208, 406
How to Grow More Vegetables, 155, 156, 314
hugelkultur terraces, 123, 208
Hugelsculptures, 207
humus, 129
Integrated Forest Gardening, 231, 403, 453
invasive, 13
Jacke, David, 231, 453

Japanese Garden, 4, 242
Jardin de curé, 137, 220, 223, 232, 234, 237, 246, 290, 346
Jardin de Curé, 4, 234
John Frusciante,, 101
John Jeavons, 155
John Ruskin, 54
Kalamazoo, 258
Ken Wilbur, 97
keyhole, 156, 252, 379, 380, 382, 383, 384, 385, 422, 442
Keylines, 123
King, F.H., 180
Kyrgiez apple walnut forests, 224
Lawn Polyculture, 378
Leach Fields, 181
leverage points, 19, 27, 28, 33, 176, 180
Lillie House, 12, 22, 33, 75, 82, 122, 147, 151, 155, 172, 173, 186, 187, 214, 215, 220, 223, 234, 235, 240, 241, 246, 268, 277, 280, 314, 323, 331, 335, 358, 370, 430, 444
Log Erosion Barriers,, 167, 170
Lupines, Lupinus species, 300
magic, 33, 38, 47, 49, 102, 219, 239, 263, 265, 322, 373, 402, 405, 412
Mallows, Malva species, 301
Maslow Pyramid, 37
Meadows, Donella, 28, 31, 452
medlar, 253, 274
meta-patterns, 85
Microclimate Patterns, 147
Mike's Simplified Mulch Rules, 141
Mollison, Bill: see Bill Mollison
Mollison's Benchmarks, 80
Mosaic Landscape, 4, 247
Mosaic Landscapes, 361
mulberries, 260, 263
mulberry, 53, 201, 263, 273
Mulch Makers, 4, 195
Mulch Strip, 424
mulch-maker plants, 135, 311, 314
Mycohugelkultur, 207
No-Till, 139
Nurse log, 123, 126, 167, 169
Nurse Log Garden Edges, 168
Nurse Log Patterns, 166
Nurse Logs, 167

Okakura Kakuzō, 50
Pareto curve, 128
Path Hierarchy and Orders, 374
pattern languages, 71
Patterns for Beautiful Paths, 372
Patterns for Garden Color, 343
Patterns for Garden Layout, 338
Patterns for Natural Wildlife Habitat, 360
Patterns for Sacredness, 322
Patterns for simple ponds, 366
Permaculture Design, 69
Permaculture Design Certificate, 23, 128
Permaculture water-harvesting techniques, 126
Permaculture Zones, 136
place-making, 9
Planting into Resident Vegetation, 217
Plants for a Future, 257
Play, 3, 87
PLENTY, 84
Post Wild, 238
Pre-Raphaelite brotherhood, 54
punk, 60
Quinces, 273
Realistic Expectations, 80
Regenerative, 180, 220
Rikyu, 50
Sacred Geometry, 331
Scott Barry, 88
Sea kale, Crambe, 308
Sector analysis, 145
Serviceberry, 265
sheet-mulching, 136, 141, 189, 194, 197, 198, 199, 201, 217, 438
Sheet-Mulching, a guilde to, 197
Shungiku, 309
Skirret, 310
slash-mulch, 129, 135, 136, 158, 189, 294, 306, 411, 419
Solar Punk, 64
solid base, 26, 37, 41, 44, 85, 129, 449
Solid Base, 3, 4, 37, 42, 44, 85
Spielmaker, E. James, 207
Starting Gardens the Holistic Natural Way, 3, 189

sunchoke, 313, See helianthus
sunchokes, 179, 253, 313, 314, 376, 389, 400, 403, 411, 424, 432
Terminator Guilds, 433
the garden growing wild., 73
The Home Garden, 223
The Permaculture Design Process, 77
the petroleum, plastics, and poisons, 117
The Sacred Grove, 325
The Transformative Landscapes Recognition Criteria, 119
Themes, 85
Thermal belt, 149
Thermal mass, 149
Thermal walls, 149
three ethics, 70
Three Sisters Guild, 5, 417
Toby Hemenway, 8, 86, 127, 197, 199, 201, 367, 398, 406, 448, 452
Toensmier, Eric, 231, 453
Traditional Forest Garden, 223
transformation, 74
transformation., 23
Transformative Action, 23
transformative Adventures, 75
transformative landscapes, 8, 9, 227, 229, 243
Transforming Soil, 175
Tree Crops for Home Gardens, 278
Turkish Rocket, 307, 318, 423
uber-principle, 70
Urban heat island, 149
veggie garden circles, 162
Visual Mandalas, 335
Vivienne Westwood, 60
wabi-sabi, 50, 52, 53, 60, 232
Water Wise Design, 122
ways to start beds for holistic natural gardens, 136
Weiseman, Wayne, 231, 403, 453
Westwood, Vivienne. See Vivienne Westwood.
What to order: seeds, starts or plants?, 319
Wilbur, Ken, see Ken Wilbur
William Morris, 54

Beauty in Abundance

Usually, the opposite of a bad idea is not a good idea. More often, it's another bad idea.

*Yet we get stuck in these bad ideas, locked into conflict with others and within ourselves.
To find a better idea, we must transcend our false dichotomies, delusions, and traumas.*

Much of the work we must do to build better lives and a better world, appears to be a matter of this kind of transcendence. We must transcend the thinking and ideas which have brought us to this precipice.

This is our unique journey, the adventure of our time.

www.ingramcontent.com/pod-product-compliance
Lightning Source LLC
Chambersburg PA
CBHW081505080526
44589CB00017B/2656